Applications of Fractional-Order Systems to Automatic Control

Applications of Fractional-Order Systems to Automatic Control

Guest Editors

Germán Ardul Muñoz-Hernández
José Fermi Guerrero-Castellanos

MDPI

Basel • Beijing • Wuhan • Barcelona • Belgrade • Novi Sad • Cluj • Manchester

Guest Editors

Germán Ardul Muñoz-Hernández
Facultad de Ciencias de la Electrónica
Benemérita Universidad Autónoma de Puebla
Puebla
Mexico

José Fermi Guerrero-Castellanos
Facultad de Ciencias de la Electrónica
Benemérita Universidad Autónoma de Puebla
Puebla
Mexico

Editorial Office
MDPI AG
Grosspeteranlage 5
4052 Basel, Switzerland

This is a reprint of the Special Issue, published open access by the journal *Fractal and Fractional* (ISSN 2504-3110), freely accessible at: https://www.mdpi.com/journal/fractalfract/special_issues/JP5V366997.

For citation purposes, cite each article independently as indicated on the article page online and as indicated below:

Lastname, A.A.; Lastname, B.B. Article Title. *Journal Name* **Year**, *Volume Number*, Page Range.

ISBN 978-3-7258-5275-8 (Hbk)
ISBN 978-3-7258-5276-5 (PDF)
https://doi.org/10.3390/books978-3-7258-5276-5

Contents

About the Editors

Germán Ardul Muñoz-Hernández

Germán Ardul Muñoz-Hernández is an Industrial Engineer in Electronics from the Technological Institute of Veracruz. His qualifications include Master of Science in Electronics (National Institute of Astrophysics, Optics and Electronics: I.N.A.O.E.); thesis: "Evaluation of self-adjusting control algorithms with P.I.D. structure", and PhD in Control Systems (Bangor University, Wales, United Kingdom); thesis: "Application of Predictive Control in a Hydroelectric Plant by Pumping Station" (2005). He participated in a postdoctoral stay in Applied Physics at the European Center for Nuclear Research (CERN), Switzerland, and the Meritorious Autonomous University of Puebla (BUAP) from January to December 2007. He has worked at Fabrica de Papel Tuxtepec, S.A., in Computer Management and the Superintendence of Engineering, where he held the position of Superintendent. In his academic research, he has collaborated with a variety of different institutions. He has supervised 1 PhD thesis, 36 master's theses, and 46 theses and bachelor's degree works. He has both collaborated on and directed research and technological development projects. He has carried out research stays at the Institute of Electrical Research (1989); the University of South Carolina at Columbia (1995); the University of Michigan at Ann Arbor (1996); CERN Geneva, Switzerland (2007 and 2009); UPC Terrassa, Spain (2023); and INAOE, Mexico (2023). His area of research is instrumentation and control. He has published research in national and international conferences and in indexed journals, and is co-author of the books Analysis and Design of Electric Circuits: Theory and Practice (Alfaomega, 2014) and Modelling and Controlling Hydropower Plants (Springer, 2013). He is a Senior Member of the IEEE, and was President of the Puebla Section of the IEEE in the period 2016–2017. He is part of the National System of Researchers of Mexico.

José Fermi Guerrero-Castellanos

José Fermi Guerrero-Castellanos received his B.Sc. degree in Electronics from the "Benemérita Universidad Autónoma de Puebla (BUAP)", Mexico, in 2002 and his M.Sc. and Ph.D. in Automatic Control from the Grenoble Institute of Technology and Univ. Grenoble Alpes (former Joseph Fourier University), France, in 2004 and 2008, respectively. Between January and June 2008, he was a Postdoctoral Researcher at GIPSA-Lab Laboratory, Grenoble, France. After spending one year at the University Polytechnic of Puebla, Mexico, as an Assistant Professor, in 2009 he joined the Faculty of Electronics at BUAP as a full professor, where he established and directs the Control and Autonomous Systems Laboratory. His research interests include the modeling and control of mechatronics and unmanned vehicles, inertial navigation systems, event-based control, multi-agent systems, active-disturbance rejection control, Frugal Model Predictive Control, the application of fractional-order control, and renewable energy applications. He is a member of the IEEE and a member of the Conference Editorial Board of the IEEE-Control Systems Society. He was the general chair of the 2019 National Congress on Automatic Control in Mexico and since 2010, he has been a member of the National Researcher System of Mexico.

fractal and fractional

MDPI

Editorial

Closing Editorial for Applications of Fractional-Order Systems in Automatic Control

German A. Munoz-Hernandez * and Jose Fermi Guerrero-Castellanos

Electronic Sciences Faculty, Meritorious Autonomous University of Puebla (BUAP), Puebla de Zaragoza 72592, Mexico; fermi.guerrero@correo.buap.mx

* Correspondence: gmunoz@ieee.org; Tel.: +52-2222558932

Fractional-order systems have been applied in very diverse areas of science and engineering. The fractional-order calculus is a generalization of the integration and differentiation operators to non-integer order. The fractional-order provides additional flexibility and adjustments to the operation specifications. In automatic control, fractional systems allow forms of response that are impossible to achieve with classical control, e.g., improvement in noise measurement attenuation without unnecessarily deteriorating the disturbance rejection properties and, consequently, the time response of the closed-loop system.

The focus of this Special Issue is to show the advances in research on topics related to the theory, design, implementation, and application of fractional-order systems in automatic control. The works presented in this technical publication are those that have appeared in the *Fractal and Fractional* journal in a Special Issue of the same name, which included the thirteen individually published papers plus an Editorial signed by the editors of this reprint.

The papers of this reprint have been categorized into the following two main sections:

- Control Theory: Fractional-order Control Design and Stability Analysis;
- Control Engineering.

The later section, Control Engineering, has been divided into the following:

- ○ Applications of Fractional-Order Control for Automotive Control Systems;
- ○ Applications of Fractional-Order Control for Unmanned Vehicles;
- ○ Applications of Fractional-Order Control for Robotics;
- ○ Applications of Fractional-Order Control for Electric Power/Power Electronics.

The section on Control Theory: Fractional-Order Control Design and Stability Analysis consists of three papers. In the first work of this section entitled "Discrete-Time Design of Fractional Delay-Based Repetitive Controller with Sliding Mode Approach for Uncertain Linear Systems with Multiple Periodic Signals", the authors introduce a discrete-time design of a fractional internal model-based repetitive controller with a sliding mode approach presented for uncertain linear systems subject to repetitive trajectory and periodic disturbance. The proposed algorithm aims to enhance tracking accuracy, transient response, and robustness against parametric variations beyond what is offered by conventional repetitive controllers. Comparative simulation studies demonstrate the superior performance of the proposed controller. In the reference, namely, "Analysis of Error-Based Switched Fractional-Order Adaptive Systems: An Error Model Approach," the authors address a general analytical framework that allows proving the boundedness of the solutions and convergence of the estimation/tracking error with only particular analyses for specific schemes being accessible, presenting the analysis of four error models that can appear in the field of adaptive systems when these adaptive laws are chosen.

The boundedness of the solutions is proved for all cases, together with the convergence to zero of the estimation/tracking error. A numerical example is included to show the possible advantages of using switched adaptive laws in a model reference adaptive control application.

The work "Fractional-Order Control Method Based on Twin-Delayed Deep Deterministic Policy Gradient Algorithm" which is the third selected work of this section proposes a fractional-order control method based on the twin-delayed deep deterministic policy gradient (TD3) algorithm in reinforcement learning. A fractional-order disturbance observer is designed to estimate the disturbances, and the radial basis function network is selected to approximate system uncertainties. Then, a fractional-order sliding-mode controller is constructed to control the system, and the parameters of the controller are tuned using the TD3 algorithm, which can optimize the control effect. The results show that the fractional-order control method based on the TD3 algorithm can not only improve the closed-loop system performance under different operating conditions but also enhance the signal tracking capability.

The section on Control Engineering, Applications of Fractional-Order Control for Automotive Control Systems, consists of a manuscript "Vibration Suppression of the Vehicle Mechatronic ISD Suspension Using the Fractional-Order Biquadratic Electrical Network," with a positive real synthesis design method of vehicle mechatronic Inerter-Spring-Damper (ISD) suspension based on the fractional-order biquadratic transfer function is proposed. In this work, the positive real condition of the fractional-order biquadratic transfer function is given. Then, a quarter dynamic model of the vehicle mechatronic ISD suspension is established, and the parameters of the fractional-order biquadratic transfer function and vehicle suspension are obtained by a genetic algorithm. The main objective of the work is to break the constriction of the integer-order transfer function in vehicle ISD suspension design.

The section on Control Engineering, Applications of Fractional-Order Control for Unmanned Vehicles consists of two papers. In the paper "Fractional-Order Controller for the Course Tracking of Underactuated Surface Vessels Based on Dynamic Neural Fuzzy Model," the authors proposes a control algorithm based on the dynamic neural fuzzy model (DNFM). The DNFM simultaneously adjusts the structure and parameters during learning and fully approximates the inverse dynamics of ships. Aiming at the uncertainty problem caused by the time-varying modeling parameters associated with ship speed in the course tracking control of underactuated surface vessels (USVs). Simulation experiments were conducted, respectively, for course tracking, course tracking under wind and wave interferences, and comparison with five different controllers. This proposed controller can overcome the influence of the uncertainty of modeling parameters, tracking the desired course quickly and effectively. The second paper of this section is titled "Hover Flight Improvement of a Quadrotor Unmanned Aerial Vehicle Using PID Controllers with an Integral Effect Based on the Riemann–Liouville Fractional-Order Operator: A Deterministic Approach," which proposes a set of PID controllers with an integral effect based on the Riemann–Liouville fractional-order approach to improve the hovering flight of a quadrotor UAV. This research introduces a set of fractional-order PID controllers for UAV hover stability, which offer better adaptability to non-linear dynamics and robustness than traditional PID controllers. The numerical simulation results show the effectiveness of the proposed Fractional Integral Action PID Controller in the control of UAV hovering flight, while comparative analyses against a classical controller emphasize the benefits of the fractional-order approach in terms of control accuracy.

The section, Control Engineering, Applications of Fractional-Order Control for Robotics has three papers. In the first work of this section one finds the work "Robust Fractional-Order PI/PD Controllers for a Cascade Control Structure of Servo Systems," where a cascade control structure is suggested to control servo systems that normally include a servo motor in coupling with two kinds of mechanism elements, a translational or rotational movement. These kinds of systems have high demands for performance in terms of the fastest response and no overshoot/oscillation to a ramp function input. The fractional-order proportional integral (FOPI) and proportional derivative (FOPD) controllers are addressed to deal with those control problems due to their flexibility in tuning rules and robustness. Simulation studies are considered for two kinds of plants that prove the effectiveness of the proposed method, with good tracking of the ramp function input under the effects of the disturbances. The next work, namely, "A Fractional-Order ADRC Architecture for a PMSM Position Servo System with Improved Disturbance Rejection," proposes an active disturbance rejection control (ADRC) architecture for a permanent magnet synchronous motor (PMSM) position servo system. The presented method achieved enhanced tracking and disturbance rejection performance with a limited observer bandwidth. The model-aided extended state observer (MESO)-based ADRC was designed for the current, speed, and position loops of the PMSM position servo system. The results demonstrate not only that the proposed method achieved superior tracking performance but also that the position error of the proposed strategy decreases to 2.25% when the constant disturbance was input, significantly improving the disturbance rejection performance. The third work of the section Control Engineering, Applications of fractional-order control for Robotic is "Fractional Control of a Class of Underdamped Fractional Systems with Time Delay—Application to a Teleoperated Robot with a Flexible Link," where the authors addresses the robust control of processes modeled by a fractional-order time delay transfer function. A new method for tuning fractional-order PI and PD controllers is developed. The stability is assessed based on the frequency domain tuning of the regulators to control such delayed fractional-order underdamped processes. In order to analyze the closed-loop stability and robustness, the new concept of Robust High-Frequency Condition is introduced. The analysis based on that demonstrates that each controller has a different region of feasible frequency specifications, and, in all cases, they depend on their fractional integral or derivative actions. Finally, an application example, the position control of a teleoperated manipulator with a flexible link, is presented.

Finally, the section in Control Engineering, Applications of Fractional-Order Control for Electric Power/Power Electronics, is composed of four papers. The first work of this section is "Fractional-Order Phase Lead Compensation Multirate Repetitive Control for Grid-Tied Inverters". In this work, a fractional-order phase lead proportional-integral multi-resonant multi-rate repetitive control (FPL-PIMR-MRC) is proposed for grid-tied inverters. The proposed method can provide a suitable fractional phase lead step to achieve a wide stability region, minor tracking errors, and low hardware costs. The IIR fractional-order lead filter design, stability analysis, and step-by-step parameter tuning of the FPL-PIMR-MRC system are derived in detail. Finally, the simulation performed confirms the feasibility and effectiveness of the proposed scheme. The next work is titled "Optimal Fractional-Order Controller for the Voltage Stability of a DC Microgrid Feeding an Electric Vehicle Charging Station" suggests a fractional-order proportional integral (FOPI) controller to improve the performance and energy management of a standalone electric vehicle charging stations (EVCSs) microgrid. The microgrid is supplied mainly by photovoltaic (PV) energy and utilizes a battery as an energy storage system (ESS). The FOPI's settings are best created utilizing the grey wolf optimization (GWO) method to attain the highest performance possible. The results demonstrate that the suggested FOPI controller significantly improves the transient responsiveness of the EVCSs performance compared to the standard PI controller. The results support the suggested FOPI control's robustness to parameter mismatches. The third work, namely, "Automatic Voltage Regulator Betterment Based on a New Fuzzy FOPI+FOPD Tuned by TLBO," presents a novel Fuzzy Logic Controller (FLC) framework aimed at enhancing the performance and stability of Automatic Voltage Regulators (AVRs). The proposed system combines fuzzy control theory with the Fractional-Order Proportional Integral Derivative (FOPID) technique and employs cascading control theory to significantly improve reliability and robustness. The methodology is validated through comparative analyses with controllers reported in prior studies, highlighting substantial improvements in performance metrics. Key findings demonstrate significant reductions in peak overshoot, peak undershoot, and settling time, emphasizing the proposed controller's effectiveness. The last work presented in the section is "Applying a Gain Scheduled Fractional Order Proportional Integral and Derivative Controller to a Quadratic Buck Converter," which is also the last paper of the editon. This work present a fractional-order proportional integral and derivative controller with adaptation characteristics in the control parameters depending on the required output, gain scheduling fractional-order PID (GS-FO-PID). In this work a fractional-order PID is applied to the voltage control of a DC–DC buck quadratic converter (QBC). The performance of the GS-FO-PID is compared with the one from a classic PID. The GS-FO-PID presents better performance when the reference voltage is changed and also with load changes.

Thus, this Special Issue, which is the subject of our Editorial, brings together some novel ideas on the application of fractional calculus in automatic control systems; it is a significant and relevant volume for our field of study and will be appreciated as a useful reference within the specialized literature.

Conflicts of Interest: The authors declare no conflicts of interest.

List of Contributions:

1. Kurniawan, E.; Burrohman, A.M.; Purwowibowo, P.; Wijonarko, S.; Maftukhah, T.; Prakosa, J.A.; Rustandi, D.; Pratiwi, E.B.; Az-Zukhruf, A. Discrete-Time Design of Fractional Delay-Based Repetitive Controller with Sliding Mode Approach for Uncertain Linear Systems with Multiple Periodic Signals. *Fractal Fract.* **2025**, *9*, 41.
2. Aguila-Camacho, N.; Gallegos, J.A.; Chen, Y.; Travieso-Torres, J.C. Analysis of Error-Based Switched Fractional-Order Adaptive Systems: An Error Model Approach. *Fractal Fract.* **2024**, *8*, 706.
3. Jiao, G.; An, Z.; Shao, S.; Sun, D. Fractional-Order Control Method Based on Twin-Delayed Deep Deterministic Policy Gradient Algorithm. *Fractal Fract.* **2024**, *8*, 99.
4. Shen, Y.; Li, Z.; Tian, X.; Ji, K.; Yang, X. Vibration Suppression of the Vehicle Mechatronic ISD Suspension Using the Fractional-Order Biquadratic Electrical Network. *Fractal Fract.* **2025**, *9*, 106.
5. Li, G.; Li, Y.; Li, X.; Liu, M.; Zhang, X.; Jin, H. Fractional-Order Controller for the Course Tracking of Underactuated Surface Vessels Based on Dynamic Neural Fuzzy Model. *Fractal Fract.* **2024**, *8*, 720.
6. Delgado-Reyes, G.; Valdez-Mart, J.S.; Guevara-Lopez, V.-M.J.S.P.; Hernandez-Perez, M.A. Hover Flight Improvement of a Quadrotor Unmanned Aerial Vehicle Using PID Controllers with an Integral Effect Based on the Riemann–Liouville Fractional-Order Operator: A Deterministic Approach. *Fractal Fract.* **2024**, *8*, 634.

7. Chuong, V.L.; Nam, N.H.; Giang, L.H.; Vu, T.N.L. Robust Fractional-Order PI/PD Controllers for a Cascade Control Structure of Servo Systems. *Fractal Fract.* **2024**, *8*, 244.
8. Wang, S.; Gan, H.; Luo, Y.; Luo, X.; Chen, Y. A Fractional-Order ADRC Architecture for a PMSM Position Servo System with Improved Disturbance Rejection. *Fractal Fract.* **2024**, *8*, 54.
9. Gharab, S.; Batlle, V.F. Fractional Control of a Class of Underdamped Fractional Systems with Time Delay—Application to a Teleoperated Robot with a Flexible Link. *Fractal Fract.* **2023**, *7*, 646.
10. Liang, F.; Lee, H.-J.; Zhang, H. Fractional-Order Phase Lead Compensation Multirate Repetitive Control for Grid-Tied Inverters. *Fractal Fract.* **2023**, *7*, 848.
11. Zaid, S.A.; Bakeer, A.; Albalawi, H.; Alatwi, A.M.; AbdelMeguid, H.; Kassem, A.M. Optimal Fractional-Order Controller for the Voltage Stability of a DC Microgrid Feeding an Electric Vehicle Charging Station. *Fractal Fract.* **2023**, *7*, 677.
12. Shouran, M.; Alenezi, M. Automatic Voltage Regulator Betterment Based on a New Fuzzy FOPI+FOPD Tuned by TLBO. *Fractal Fract.* **2025**, *9*, 21.
13. Hernandez, G.A.M.; Guerrero-Castellanos, J.F.; Acosta-Rodriguez, R.A. Applying a Gain Scheduled Fractional Order Proportional Integral and Derivative Controller to a Quadratic Buck Converter. *Fractal Fract.* **2025**, *9*, 160.

fractal and fractional

MDPI

Article

Discrete-Time Design of Fractional Delay-Based Repetitive Controller with Sliding Mode Approach for Uncertain Linear Systems with Multiple Periodic Signals

Edi Kurniawan *, Azka M. Burrohman, Purwowibowo Purwowibowo, Sensus Wijonarko, Tatik Maftukhah, Jalu A. Prakosa, Dadang Rustandi, Enggar B. Pratiwi and Amaliyah Az-Zukhruf

Research Center for Photonics, National Research and Innovation Agency, Tangerang Selatan 15314, Indonesia; a.m.burohman@ieee.org (A.M.B.); purw001@brin.go.id (P.P.); sens001@brin.go.id (S.W.); tati006@brin.go.id (T.M.); jalu002@brin.go.id (J.A.P.); dada003@brin.go.id (D.R.); enggar.banifa@ui.ac.id (E.B.P.); skpbo.fotonik@brin.go.id (A.A.-Z.)

* Correspondence: edik004@brin.go.id

Abstract: In this paper, a discrete-time design of a fractional internal model-based repetitive controller with a sliding mode approach is presented for uncertain linear systems subject to repetitive trajectory and periodic disturbance. The proposed algorithm, named a fractional delay-based repetitive sliding mode controller (FD-RSMC), aims to enhance tracking accuracy, transient response, and robustness against parametric variations beyond what is offered by conventional repetitive controllers. First, a fractional delay-based repetitive controller (FD-RC) that allows the periodic delay steps to be noninteger is presented to improve the trajectory tracking accuracy and good disturbance compensation of multiple periodic signals. Second, a sliding mode control (SMC) with a discrete-time reaching law is systematically incorporated into FD-RC to improve transient response, especially during the learning period of FD-RC, and also to provide system robustness against model uncertainties. Finally, the stability proof of the closed-loop system with the proposed controller is assessed based on a delayed-sliding mode-reaching condition. Finally, comparative simulation studies are presented to demonstrate the superior performance of the proposed controller.

Keywords: repetitive control; internal model; fractional delay; sliding mode control; uncertain linear systems; repetitive trajectory; periodic disturbance

1. Introduction

Tracking and rejecting periodic signals are common challenges in many control applications. Repetitive control (RC) is an effective learning control strategy for tracking and/or compensating a periodic signal. RC outperforms classical control schemes, such as proportional-integral-derivative (PID), because RC can learn repetitive signal values and then use them as control input [1]. Recently, RC has become increasingly popular and has been applied to a variety of applications, including optical-stabilized systems [2], piezo-actuated nanoscanners [3], mechanical ventilation systems [4], substrate carrier systems [5], and robotic leg prostheses [6]. Additionally, RC has gained popularity in active vibration damping control, where it is effectively used to suppress periodic disturbances, as demonstrated in [7–9].

Inou et al. [10] first introduced a continuous-time RC model based on an internal model principle of Wonham and Francis [11]. According to the internal model principle, the periodic signal model must be incorporated into the closed-loop system for error-free tracking of the periodic signal. The continuous-time RC model in [10] was represented by an internal model $I_c(s) = e^{-sT_r}/(1-e^{-sT_r})$, where T_r is a period of trajectory $r(t)$. The continuous-time internal model $I_c(s)$ is marginally stable and exhibits an infinite dimensional structure as the model has infinite poles located at an imaginary axis: $\pm jk\omega$, where $k = 1, 2, \cdots, \infty$. Due to marginally stable poles and infinite dimension structure, the continuous-time model $I_c(s)$ is sensitive to high-frequency disturbances and model uncertainties. Additionally, the internal model can only be used for a properly stable plant with a relative degree of zero. A strict proper plant, on the other hand, is modeled in many engineering situations [12]. In addition, the internal model is limited to a properly stable plant with a relative degree of zero, while many engineering applications are modeled as strictly proper plants. In order to avoid an infinite-dimensional structure, a discrete-time internal model-based RC was introduced in [13] and formulated as $I_d(z) = z^{-N_r}/(1-z^{-N_r})$, where $N_r = T_r/T \in \mathbb{Z}$ is an integer number of samples per trajectory period and T is a sampling period. The discrete-time RC possesses a finite-dimensional structure due to the fact that its poles are finite and evenly spaced and located at a unit circle. For repetitive signals with the frequency h/T_r (where $h = 1, 2, \cdots, N_r/2$), this discrete model generates a null tracking error. This means that the tracking is limited to the Nyquist components of the periodic signal, encompassing only the fundamental frequency and its harmonics h/T_r.

However, the tracking performance and stability of a discrete-time RC system depend on the two following assumptions: (1) the trajectory period T_r is precisely known, which gives an integer delay length, N_r; (2) an accurate model of the plant is also known. The first common assumption in the design of discrete-time RC is that the trajectory period (T_r) is known and fixed. In practice, however, the trajectory and sampling periods are subject to variation, which can change the delay length (N_r) from an integer to a fractional value. Rounding the delay length to the nearest integer value is the simplest method for implementing fractional delay length. However, the actual reference and the internal model will experience a frequency mismatch due to this solution. As a result, the tracking accuracy of the system is compromised. The second assumption implies that the stability of the RC system relies on the known plant model. This is due to the fact that RC is a model-based learning control scheme requiring an accurate model of the plant, especially one needed to design a compensator. A compensator is part of RC besides an internal model, which is used to compensate plant model dynamics and stabilize the closed-loop RC system. Hence, the tracking and robust performance of a standalone RC system is generally susceptible to the trajectory's period variation and plant model uncertainties.

Several studies have focused on improving the performance of RCs when subjected to period variation [14–19]. The adaptive-based RC approach is demonstrated in the works of [14–16] since the proposed scheme operates by adapting the sampling period T in order to preserve the integer value N_r. The adaptive-based RC solution gives rise to system complexity and implementation due to the requirement to adjust the sampling period. Moreover, the adjustment in the sampling period will also change the dynamics of the discrete-time plant model. Other than adaptive-based RC, higher-order-based RC found in [17–19] was also presented to enhance the tracking performance of the systems under period variation. Instead of using one-period delay, z^{-N_r}, such as with general RC, higher-order-based RC employs weighting multiple delays ($w_1z^{-N_r}, w_2z^{-2N_r}, \cdots, w_mz^{-pN_r}$) in constructing the internal model. Here, $w_1, w_2, ...w_m$ are the weights, and m is the highest order of RC. The higher-order-based RC aims to extend the high gains of the internal model to the neighboring region around the nominal targeted frequency $(1/T_r)$. Thus, higher-order-based RC can improve the trajectory-tracking performance at the neighboring frequencies, e.g., $(1/T_r) \pm \Delta_f$, where Δ_f represents a minor frequency shift. However, due to multiple delays $z^{-N_r}, z^{-2N_r}, \cdots, z^{-mN_r}$, the higher-order-based solution requires large memories and converges slower compared to the general RC.

In contrast to adaptive-based RC, which seeks to maintain the integer value N_r, and higher-order-based RC, which extends the internal model's high gains to the neighboring region of the targeted frequency, we implement fractional internal model-based RC in this study. The fractional internal model-based RC is later referred to as fractional delay-RC (FD-RC), emphasizing that in situations involving period variation, the delay length, N_r, constructing the internal model does not need to be an integer. Here, the fractional time delay is also rooted in fractional calculus, which has been extensively studied and widely used in controller design and dynamics modeling in various applications, as evidenced in [20–25]. In [20], the Riemann-Liouville definition of fractional derivatives for the purpose of recreating the spatial structure of the fractional Schrödinger equation for electrical screening potentials was investigated. An explicit solution formula for fractional delay difference systems and a discrete delayed Mittag-Leffler matrix function were derived in [21]. Practical applications of fractional calculus for synthesizing a fractional-order PID controller to control bifurcation in fractional-order small-world networks were demonstrated in [22]. Additionally, the work [23] designed a reduced-order discrete-time fractional-order PID controller. Furthermore, the authors of [24] developed model-based fractional order controllers for time-delay processes in servo systems, while the authors of [25] dealt with the design of complex fractional order speed controllers for induction motors. These studies demonstrate the wide range of applications and potential benefits of fractional calculus in enhancing controller performance and dynamics modeling. In contrast to [20–25], we concern ourselves with the fractional delay design in discrete time, which is required to construct an accurate internal model of the RC system.

In order to address the second limitation of general RC, an integrated sliding mode controller (SMC) is utilized to ensure robustness in the presence of model uncertainties. SMC is a widely recognized control scheme that operates on the principle of switching control and attempts to steer a sliding function in the direction of a designated sliding surface. A discrete-time SMC has been widely exploited in much of the literature [26–32], and here, the discrete-time SMC with a linear switching function will be incorporated into the FD-RC. Numerous research studies have been conducted on the utilization of sliding mode control with a supplementary control scheme, such as repetitive control, as reported in [33–38]. In [33], a repetitive sliding mode controller incorporating an exponential-based bi-power reaching law for three-phase voltage source inverters was introduced. An integral sliding mode control scheme based on repetitive control was developed in [34] for uncertain repetitive processes with the presence of matched uncertainties, external disturbances, and norm-bounded nonlinearities. A repetitive sliding mode controller with a modified RC was designed in [35] for uncertain linear systems. An adaptive smooth second-order sliding mode repetitive control method for a class of nonlinear systems with unknown disturbances and uncertainties was proposed [36]. In [37], a discrete-time low-order sliding mode repetitive controller was presented for a class of uncertain linear systems affected by band-limited periodic disturbances and parametric uncertainties. Lu et al. [38] presented an enhanced sliding-mode repetitive learning control scheme using the wavelet transform, which speeds up the learning process.

In contrast to previous studies (references [33–38]), the main aim of this study is to address the issue of tracking and/or rejecting disturbances within a particular class of uncertain linear systems that exhibit multiple periodic signals. These systems may exhibit variability in both the reference signal and disturbance signal periods, which results in non-integer discrete-time delay lengths. Therefore, in the context of this study, our aim is to develop a novel control approach known as the fractional delay-based repetitive sliding mode controller (FD-RSMC). The objective of this controller is also to overcome the primary constraints of conventional RC systems by combining fractional delay-based repetitive control (FD-RC) with sliding mode control (SMC), as previously described. Here, FD-RC is applied to keep the delay length fixed to the period of the actual reference or disturbance. Thus, the period disparity between the controller and the actual reference due to period variation can be avoided, as the RC system's tracking accuracy depends on this condition. The fractional delay term is realizable and straightforward to use because it can be implemented as an independent, stable, and causal IIR filter. The coefficients of the corresponding filter can be obtained by utilizing the Thiran formula [39,40], as illustrated in the works in [41,42]. In contrast to the internal model examined in this study, which deals with the reference or disturbance model, the research described in [41,42] focused on the fractional order design of the RC compensator with the objective of developing a closed-loop RC system stabilizing controller. Next, SMC is systemically incorporated into FD-RC to improve the robustness of the system against parametric uncertainties. Moreover, SMC is also capable of enhancing the transient response of the system, especially during the learning process of the RC part. In this study, a systematic integration between FD-RC and SMC is described, and a stability analysis is performed. Simulation and comparison studies over an integer delay-based repetitive sliding mode controller (ID-RSMC) and standalone RC are provided to show the effectiveness of the proposed design. In order to emphasize the significance of our work, the main contributions are highlighted as follows:

(1) A novel and systemic integration is established between FD-RC and SMC with the objective of overcoming the constraints of general RC. The proposed approach subsequently addresses the two primary limitations of conventional RC, namely its vulnerability to fluctuations in tracking or disturbance periods and the uncertainties associated with plant models.

(2) Stability analysis of the FD-RSMC system is carried out based on the reaching condition, and it is proved that the evolution of the sliding function decreases and moves toward the sliding surface. Furthermore, every control parameter utilized in the formulation of the proposed control law is proven to be stable and realizable.

(3) The implementation of the FD-RC component in the proposed FD-RSMC is feasible due to the substitution of a stable and causal IIR filter for the fractional portion of the RC delay.

The rest of this paper is structured as follows: Section 2 describes problem formulation and preliminary, which comprises the class of the system, control objective, assumptions used in the design, and the overview of single and two-period general RCs. Section 3 illustrates the main results of the FD-RC and FD-RSMC designs and the stability analysis of the FD-RSMC closed-loop system. Section 4 discusses the simulation and comparison studies, and Section 5 presents the conclusions of the paper.

2. Problem Formulation and Preliminary

2.1. Problem Formulation

We consider a class of uncertain linear systems with internal disturbance in the following form:

$$\{A(z) + \delta_A(z)\}y_k = z^{-d}\{B(z) + \delta_B(z)\}(u_k + f_k) \tag{1}$$

where u_k, y_k, and f_k correspond to control input, system output, and input disturbance at time $t = kT$, respectively, k is a time step, T is the sampling period, z^{-d} is an input delay with d step, $\mathrm{A}(z)$ and $\mathrm{B}(z)$ are denumerator and numerator polynomials, respectively, and $\delta_{\mathrm{A}}(z)$ and $\delta_{\mathrm{B}}(z)$ denote the parametric uncertainties of $\mathrm{A}(z)$ and $\mathrm{B}(z)$, respectively.

The polynomials $\mathrm{A}(z)$ and $\mathrm{B}(z)$ are given as

$$\mathrm{A}(z)= 1 + \mathrm{a}_1 z^{-1} + \cdots + \mathrm{a}_{\mathrm{m}} z^{-\mathrm{m}}, \tag{2}$$

$$\mathrm{B}(z)= \mathrm{b}_0 + \mathrm{b}_1 z^{-1} + \cdots + \mathrm{b}_{\mathrm{n}} z^{-\mathrm{n}}, \tag{3}$$

where $\mathrm{n} < \mathrm{m}$, m, and n are the degrees of $\mathrm{A}(z)$ and $\mathrm{B}(z)$, and $(\mathrm{m} - \mathrm{n})$ is a relative degree of the system (1). Here, $\{\mathrm{a}_1, \mathrm{a}_2, \cdots, \mathrm{a}_{\mathrm{m}}\}$, and $\{\mathrm{b}_0, \mathrm{b}_1, \cdots, \mathrm{b}_{\mathrm{n}}\}$ are the nominal parameters of $\mathrm{A}(z)$ and $\mathrm{B}(z)$, which are subject to parametric uncertainties $\delta_{\mathrm{A}}(z)$, $\delta_{\mathrm{B}}(z)$, respectively. Note that $\delta_{\mathrm{A}}(z)$ and $\delta_{\mathrm{B}}(z)$ are polynomials that represent parameter changes in $\mathrm{A}(z)$ and $\mathrm{B}(z)$, respectively, and can also be viewed as linear functions. The expressions for $\delta_{\mathrm{A}}(z)$ and $\delta_{\mathrm{B}}(z)$ are as follows:

$$\delta_{\mathrm{A}}(z)= \Delta\mathrm{a}_1 z^{-1} + \cdots + \Delta\mathrm{a}_{\mathrm{m}} z^{-\mathrm{m}}, \tag{4}$$

$$\delta_{\mathrm{B}}(z)= \Delta\mathrm{b}_0 + \Delta\mathrm{b}_1 z^{-1} + \cdots + \Delta\mathrm{b}_{\mathrm{n}} z^{-\mathrm{n}}. \tag{5}$$

Let us consider the tracking error e_k, defined as

$$e_k := r_k - y_k, \tag{6}$$

where r_k corresponds to the desired trajectory or a reference signal. By rearranging the system (1), we have the representation of system output, y_k, and disturbances v_k, p_k, and $\mathcal{L}_k$ as

$$y_k= \frac{z^{-d}\mathrm{B}(z)}{\mathrm{A}(z)} u_k + \frac{z^{-d}\mathrm{B}(z)}{\mathrm{A}(z)} f_k - \frac{\delta_{\mathrm{A}}(z)}{\mathrm{A}(z)} y_k + \frac{z^{-d}\delta_{\mathrm{B}}(z)}{\mathrm{A}(z)} u_k + \frac{z^{-d}\delta_{\mathrm{B}}(z)}{\mathrm{A}(z)} f_k, \tag{7}$$

$$v_k= \frac{z^{-d}\mathrm{B}(z)}{\mathrm{A}(z)} f_k, \tag{8}$$

$$p_k= -\frac{\delta_{\mathrm{A}}(z)}{\mathrm{A}(z)} y_k + \frac{z^{-d}\delta_{\mathrm{B}}(z)}{\mathrm{A}(z)} u_k + \frac{z^{-d}\delta_{\mathrm{B}}(z)}{\mathrm{A}(z)} f_k, \tag{9}$$

$$\mathcal{L}_k= v_k + p_k, \tag{10}$$

where $(v_k, f_k, p_k, \mathcal{L}_k) \in \mathbb{R}$, v_k is a filtered periodic disturbance f_k, p_k is a parametric disturbance, and $\mathcal{L}_k$ is a lump disturbance affecting the system. Then, the system output y_k (7) and tracking error (6) can be rewritten as

$$y_k= \frac{z^{-d}\mathrm{B}(z)}{\mathrm{A}(z)} u_k + \mathcal{L}_k. \tag{11}$$

$$e_k= r_k - \frac{z^{-d}\mathrm{B}(z)}{\mathrm{A}(z)} u_k - \mathcal{L}_k. \tag{12}$$

The objective of this research is to develop a control law u_k for the system (1), which is subject to repetitive trajectory, r_k, parametric uncertainties, $(\delta_{\mathrm{A}}, \delta_{\mathrm{B}})$, and periodic disturbance, f_k, such that the system output y_k precisely follows the trajectory r_k and faster system response and robustness are achieved. In addition, the design also permits a fractional time delay in the control law, which is later proven to be capable of improving the tracking accuracy. The aforementioned goals also imply that the lump disturbance $\mathcal{L}_k$ appearing in (12) is maximally compensated, resulting in the tracking error converging to a nearly zero value, $e_k \approx 0$. The following assumptions are made prior to designing the proposed controller to ensure that the research objective is met.

Assumption 1. *The desired trajectory r_k and the disturbance f_k are both periodic with fundamental periods T_r and T_f, respectively, where T_r and T_f are known. It is assumed that the trajectory frequency $\frac{1}{T_r}$ and the disturbance frequency $\frac{1}{T_f}$ are uncorrelated, implying that $(\frac{1}{T_r}) \neq h(\frac{1}{T_f})$ or $(\frac{1}{T_f}) \neq h(\frac{1}{T_r})$ for any harmonic number h.*

Assumption 2. *The input delay z^{-d}, as well as the nominal polynomials $A(z)$ and $B(z)$, are assumed to be known. Furthermore, $A(z)$ and $B(z)$ are stable polynomials, meaning that all their roots lie inside the unit circle. The perturbed polynomials $\{A(z) + \delta_A(z)\}$ and $\{B(z) + \delta_B(z)\}$ remain stable, indicating that the uncertain system (1) is stable and minimum-phase. If the open-loop system is either unstable or nonlinear, a different approach would be necessary for formulating the control law.*

2.2. Single and Dual Internal Model-Based RCs

RC is one of the learning control methods that are capable of tracking repetitive trajectory, r_k, or/and rejecting the periodic disturbance f_k with zero tracking error. This is due to the inclusion of the internal model representing the model of r_k or/and f_k into the controller. Let a desired trajectory r_k and disturbance f_k be both periodic with basis periods of T_r and T_f, respectively. Note that the trajectory frequency $\frac{1}{T_r}$ and disturbance frequency $\frac{1}{T_f}$ are uncorrelated, implying that $\frac{1}{T_r} \neq h\frac{1}{T_f}$ or $\frac{1}{T_f} \neq h\frac{1}{T_r}$ for harmonic number h. The numbers of samples per trajectory and disturbance periods are denoted as N_r and N_f and calculated as $N_r = T_r/T$ and $N_f = T_f/T$, where T is a sampling period. A generic form of RC for a single purpose, e.g., tracking r_k is given by the following control update:

$$u_k = u_{k-N_r} + C(z)e_{k-N_r}, \tag{13}$$

which can be transformed into an input-output transfer function as

$$\frac{u(z)}{e(z)} = \underbrace{\left[\frac{z^{-N_r}}{1 - z^{-N_r}}\right]}_{\text{single internal model}} C(z). \tag{14}$$

Here, u_{k-N_r} and e_{k-N_r} in (13) are one-cycle delays of input u_k and error e_k, respectively, $C(z)$ is a compensator generally designed as an inverse plant model, and z^{-N_r} is a delay term with the length of N_r. The term $z^{-N_r}/(1 - z^{-N_r})$ in (14) is denoted as a single-period internal model, as the model only targets one basis frequency and its harmonics.

For both tracking r_k and rejecting f_k, the following two-periods RC law is required:

$$u_k = [\mathrm{D}(z)]u_k + C(z)[\mathrm{D}(z)]e_k, \tag{15}$$

which is equivalent to the transfer function as follows:

$$\frac{u(z)}{e(z)} = \underbrace{\left[\frac{\mathrm{D}(z)}{1-\mathrm{D}(z)}\right]}_{\text{dual internal model}} \mathcal{C}(z). \tag{16}$$

In contrast to single-period RC, which employs a one-cycle delay denoted as z^{-N_r}, two-period RC is constructed using a delay term denoted as $\mathrm{D}(z)$. In this context, $\mathrm{D}(z)$ is represented as multiple time delays and is derived to ensure that the denominator of the internal model matches the following expression:

$$1-\mathrm{D}(z) = \prod_{j=1}^{L} 1-z^{-N_j}, \tag{17}$$

where L denotes the number of targeted basis frequencies. As our focus is on two fundamental frequencies, specifically the trajectory frequency $1/T_r$ and the disturbance frequency $1/T_f$, $D(z)$ is consequently modeled as

$$\begin{aligned} 1-\mathrm{D}(z) &= \prod_{j=1}^{2} 1-z^{-N_j}, \\ \mathrm{D}(z) &= 1-\left(1-z^{-N_r}\right)\left(1-z^{-N_f}\right), \\ \mathrm{D}(z) &= z^{-N_r}+z^{-N_f}-z^{-N_r-N_f}. \end{aligned} \tag{18}$$

A complete diagram of two-period RC embedding $\mathrm{D}(z)$ is shown in Figure 1. The other difference between single-period and two-period RCs can be seen from the magnitude-frequency characteristics of their internal models. For instance, given that $T_r = 1$ s ($f_r = 1$ Hz) and $T_f = 0.8$ s ($f_f = 1.25$ Hz), the magnitude responses of single- and two-period internal models are plotted in Figure 2. It can be seen from Figure 2 that two-period RC exhibits extra peaks showing the targeted disturbance frequencies $f_f = h \times 1.25$ Hz ($h = 1, 2, \cdots, N_f/2$). Therefore, two-period RC here can be applied to simultaneously track the trajectory r_k and reject the disturbance f_k.

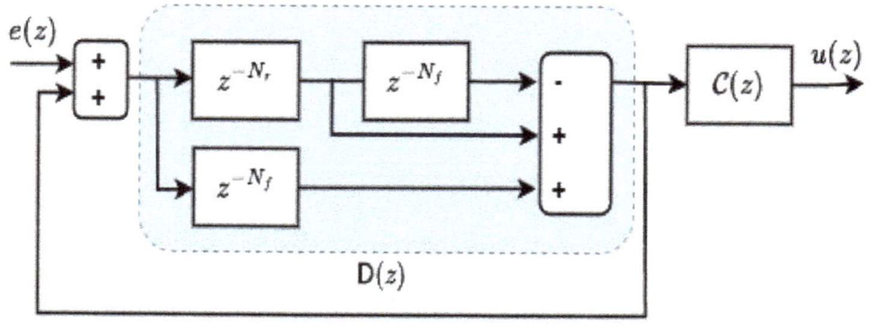

Figure 1. Block diagram of two-period RC embedding the delay term $\mathrm{D}(z)$.

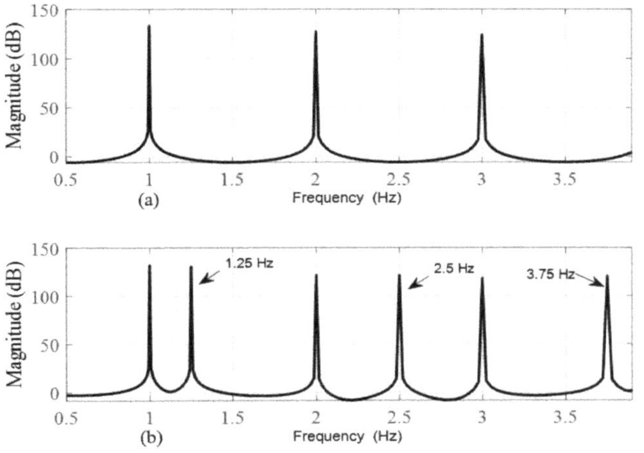

Figure 2. Magnitude-frequency characteristics: (**a**) single-period RC, (**b**) two-period RC.

In order to ensure the asymptotic stability of system (1) with the single internal model-based RC (14), the following conditions must be satisfied [43,44]:

- The plant model $\mathcal{G}(z)$ is a stable transfer function. Based on (1), $\mathcal{G}(z)$ is given by

$$\mathcal{G}(z) = z^{-d}\frac{\mathrm{A}(z)+\delta_{\mathrm{A}}(z)}{\mathrm{B}(z)+\delta_{\mathrm{B}}(z)}. \tag{19}$$

- The compensator $\mathcal{C}(z)$ in (14) is such that

$$\| [1-\mathcal{C}(z)\mathcal{G}(z)]z^{-N_r} \|_{\infty} < 1. \tag{20}$$

Similarly, for the dual internal model-based RC (16), the asymptotic stability of system (1) is guaranteed by

- The plant model $\mathcal{G}(z)$ is a stable transfer function.
- The compensator $\mathcal{C}(z)$ in (16) is such that

$$\| [1-\mathcal{C}(z)\mathcal{G}(z)]\mathrm{D}(z) \|_{\infty} < 1. \tag{21}$$

Consider the nominal model of (19), which is given by $\mathcal{G}_o(z) = z^{-d}\frac{\mathrm{A}(z)}{\mathrm{B}(z)}$ and is stable and a minimum-phase system. Assuming this nominal model is known, the compensator $\mathcal{C}(z)$ can be selected as the inverse of the nominal model. That is

$$\mathcal{C}(z) = z^{d}\frac{\mathrm{B}(z)}{\mathrm{A}(z)}. \tag{22}$$

However, the presence of uncertainties $\delta_{\mathrm{A}}(z)$ and $\delta_{\mathrm{B}}(z)$ may cause the condition (21) to be violated, potentially leading to system instability. This issue highlights a potential drawback of standalone RC when dealing with model uncertainties.

Next, we define a repetition model $\mathrm{E}(z)$, which is a denominator of the internal model of the two-period RC. The repetition model $\mathrm{E}(z)$ is given by

$$\mathrm{E}(z) = 1 - \mathrm{D}(z), \tag{23}$$

By assuming that r_k and f_k are perfectly periodic with periods T_r and T_f, respectively, then the following properties apply:

$$\mathrm{E}(z)r_k = 0 \rightarrow r_k - [\mathrm{D}(z)]r_k = 0, \tag{24}$$

$$\mathrm{E}(z)f_k = 0 \rightarrow f_k - [\mathrm{D}(z)]f_k = 0. \tag{25}$$

The repetition model $\mathrm{E}(z)$ is an important part of the dual internal model-based RC, which will later be incorporated into SMC to obtain the proposed controller.

3. Controller Design

3.1. Fractional Delay-Based Repetitive Controller (FD-RC)

In the majority of RC designs, the delay lengths N_r and N_f in (18) must be integers, making the delay terms z^{-N_r} and z^{-N_f} realizable. In the case where N_r and N_f are non-integer, then N_r and N_f are rounded to the closest integer values. For instance, consider the trajectory period to be 1.03 s and the sampling period to be 0.004 s. Then, we obtain N_r as 257.5. In order to implement the fractional time-delay $z^{-257.5}$, the straightforward solution is obtained by rounding the delay length to the closest integer value, that is z^{-257}. However, the rounded delay does not represent the actual trajectory period (i.e., $0.004 \times 257 = 1.028$ s $\neq 1.03$ s), which later will introduce certain steady-state errors. In this work, the delay lengths N_r and N_f are not required to be integers, so they can both be fractional. Suppose that both delay lengths N_r and N_f are fractional numbers that can be decomposed into

$$N_r = N_{ri} + \mathrm{n}_r, \tag{26}$$

$$N_f = N_{fi} + \mathrm{n}_f, \tag{27}$$

where $N_{ri} < N_r$ and $N_{fi} < N_f$, $\{N_{ri}, N_{fi}\}$ are both integers, and $\{\mathrm{n}_r, \mathrm{n}_f\}$ are both fractional numbers. Consequently, the delay terms z^{-N_r} and z^{-N_f} can be separated into

$$z^{-N_r} = z^{-N_{ri}}\left[z^{-\mathrm{n}_r}\right], \tag{28}$$

$$z^{-N_f} = z^{-N_{fi}}\left[z^{-\mathrm{n}_f}\right], \tag{29}$$

As illustrated in [41,42], the fractional delays $\{z^{-\mathrm{n}_r}, z^{-\mathrm{n}_f}\}$ can be approximated by infinite impulse response (IIR) filters $\{\mathrm{F}_r(z), \mathrm{F}_f(z)\}$, the coefficients of which are calculated based on the Thiran formula as follows [39,40]:

$$z^{-\mathrm{n}_r} \approx \mathrm{F}_r(z) = \frac{\mathrm{c}_p z^p + \mathrm{c}_{p-1} z^{p-1} + \cdots + \mathrm{c}_1 z + 1}{z^p + \mathrm{c}_1 z^{p-1} + \cdots + \mathrm{c}_p}, \tag{30}$$

where $p = \mathrm{ceil}(\mathrm{n}_r)$, and $\{\mathrm{c}_1, \mathrm{c}_2, \cdots, \mathrm{c}_p\}$ are calculated according to

$$\mathrm{c}_i = (-1)^i \binom{p}{i} \prod_{j=0}^{p} \frac{\mathrm{n}_r - p + j}{\mathrm{n}_r - p + j + i}, \ \forall\, i : 1, 2, \cdots, p, \ \text{where } \binom{p}{i} = \frac{p!}{i!(p-i)!}. \tag{31}$$

Then, similar procedures are applied to obtain $F_f(z)$. Finally, $D(z)$ in (18) can be rewritten as

$$D_R(z) = F_r(z)z^{-N_{ri}} + F_f(z)z^{-N_{fi}} - F_r(z)F_f(z)z^{-N_{ri}-N_{fi}}, \tag{32}$$

where $D_R(z)$ represents the fractional delay model of $D(z)$ (18), and the realization of $D_R(z)$ is illustrated in Figure 3.

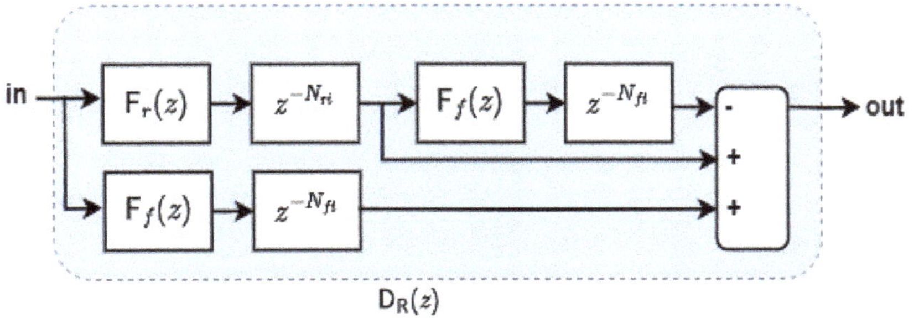

Figure 3. Realization of delay term $D_R(z)$.

Note that from a practical perspective, the fractional delay-based RC is feasible and straightforward to implement in real-time applications. The controller is constructed using the fractional delay terms, which are divided into integer and fractional components, both of which are realizable. The fractional part of the delay can be implemented by replacing the delay with a stable and causal IIR filter, with coefficients obtained using the Thiran formula. Meanwhile, the integer portion is also realizable, as it is essentially a transport delay.

3.2. Fractional Delay-Based Repetitive Sliding Mode Controller (FD-RSMC)

In this subsection, we describe how to integrate FD-RC with a nonlinear control scheme such as a sliding mode controller (SMC). Incorporating SMC into the controller design requires the establishment of a sliding surface and reaching law functions. The following linear sliding surface is considered

$$s_k = H(z)e_k, \tag{33}$$

where $H(z)$ is a polynomial with zeros inside the unit circle, e_k is the tracking error, and s_k is a sliding surface. Based on the studies in [26,45], the discrete-time delayed reaching law is used as follows:

$$s_{k+d} - s_k = -k_1 s_k - k_2 \text{sat}\left(\tfrac{s_k}{k_3}\right), \tag{34}$$

where k_1, k_2, and k_3 are positive constants and $0 < (1 - k_1) < 1$. The expression (34) is identical to (35) and (36) as follows:

$$s_{k+d} = (1 - k_1)s_k - k_2 \text{sat}\left(\frac{s_k}{k_3}\right), \tag{35}$$

$$s_k = (1 - k_1)s_{k-d} - k_2 \text{sat}\left(\frac{s_{k-d}}{k_3}\right). \tag{36}$$

The saturation function $\text{sat}\left(\frac{s_{k-d}}{k_3}\right)$ used in (36) is selected over the original switching function, which employs a sign function to reduce the chattering phenomenon. This saturation function, commonly known as a boundary layer, smooths the sliding mode control signal. Here, $\mathrm{sat}\left(\frac{s_{k-d}}{k_3}\right)$ is characterized as

$$\mathrm{sat}\left(\frac{s_{k-d}}{k_3}\right) = \begin{cases} 1, & s_{k-d} > k_3, \\ \frac{1}{k_3} s_{k-d}, & -k_3 < s_k < k_3, \\ -1, & s_{k-d} < -k_3, \end{cases} \tag{37}$$

and is illustrated in Figure 4.

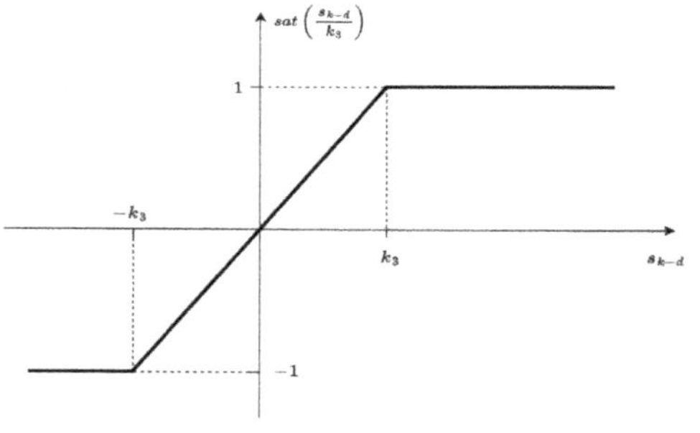

Figure 4. Plot of the saturation function (37).

Let us rewrite (12) to have the following error dynamics:

$$A(z)e_k = -z^{-d}B(z)u_k + A(z)r_k - \mathcal{W}_k, \tag{38}$$

where $\mathcal{W}_k = A(z)\mathcal{L}_k$. By multiplying the repetitive model (23) by the sliding surface (33), we have

$$s_k - [D(z)]s_k = H(z)e_k - H(z)[D(z)]e_k, \tag{39}$$

where $[D(z)]s_k$ is equivalent to

$$[D(z)]s_k = s_{k-N_r} + s_{k-N_f} - s_{k-N_r-N_f}. \tag{40}$$

For simplicity, we will use the notation $[D(z)]s_k$ in the derivation instead of those on the right-hand side of (40). By adopting minimum control variance in [46], the polynomial $H(z)$ is made equivalent to

$$H(z) = A(z)R(z) + z^{-d}S(z), \tag{41}$$

where $R(z)$ and $S(z)$ are solutions to polynomial matching (41). Then, (39) can be expressed as

$$s_k - [D(z)]s_k = A(z)R(z)e_k - A(z)R(z)[D(z)]e_k + S(z)e_{k-d} - S(z)[D(z)]e_{k-d} \tag{42}$$

Consider the nominal plant with no uncertainties and disturbances, i.e., $\mathcal{L}_{Ak} = 0$, which yields the error dynamics as follows:

$$A(z)e_k = -z^{-d}B(z)u_k + A(z)r_k. \tag{43}$$

Substituting (43) to (42) yields:

$$\begin{aligned} s_k - [D(z)]s_k =& -B(z)R(z)u_{k-d} + A(z)R(z)r_k + B(z)R(z)[D(z)]u_{k-d} \\ & -A(z)R(z)[D(z)]r_k + S(z)(e_{k-d} - [D(z)]e_{k-d}). \end{aligned} \tag{44}$$

By defining $P(z) = B(z)R(z)$ and $Q(z) = A(z)R(z)$, (44) can be simplified and rearranged to

$$\begin{aligned} P(z)u_{k-d} =& P(z)[D(z)]u_{k-d} + Q(z)(r_k - [D(z)]r_k) + S(z)(e_{k-d} - [D(z)]e_{k-d}) \\ & + [D(z)]s_k - s_k. \end{aligned} \tag{45}$$

By multiplying both sides of (45) by z^d, we obtain

$$\begin{aligned} P(z)u_k =& P(z)[D(z)]u_k + Q(z)(r_{k+d} - [D(z)]r_{k+d}) + S(z)(e_k - [D(z)]e_k) \\ & + [D(z)]s_{k+d} - s_{k+d}. \end{aligned} \tag{46}$$

Based on property (24), the term $Q(z)(r_{k+d} - [D(z)]r_{k+d})$ in (46) can be neglected. That is

$$\begin{aligned} Q(z)(r_{k+d} - [D(z)]r_{k+d}) =& Q(z)(1 - [D(z)])r_{k+d} \\ =& Q(z)\{E(z)r_{k+d}\} = 0. \end{aligned} \tag{47}$$

Therefore, (46) can be expressed as

$$P(z)u_k = P(z)[D(z)]u_k + S(z)(e_k - [D(z)]e_k) + [D(z)]s_{k+d} - s_{k+d}. \tag{48}$$

By dividing each side of (48) by $P(z)$ and substituting s_{k+d} (35) into (48), we obtain the following control law expression:

$$\begin{aligned} u_k =& [D(z)]u_k + \frac{S(z)}{P(z)}(e_k - [D(z)]e_k) + P^{-1}(z)[D(z)]s_{k+d} \\ & - P^{-1}(z)\left\{(1-k_1)s_k - k_2 \mathrm{sat}\left(\frac{s_k}{k_3}\right)\right\}. \end{aligned} \tag{49}$$

Considering the fact that the delay terms $\left\{z^{-N_r}, z^{-N_f}\right\}$ in $D(z)$ are fractional, then the control law (49) is implemented as follows:

$$\begin{aligned} u_k =& [D_R(z)]u_k + \frac{S(z)}{P(z)}(e_k - [D_R(z)]e_k) + P^{-1}(z)\left[z^d D_R(z)\right]s_k \\ & - P^{-1}(z)\left\{(1-k_1)s_k - k_2 \mathrm{sat}\left(\frac{s_k}{k_3}\right)\right\}, \end{aligned} \tag{50}$$

where $D_R(z)$ is given in (32). Then, the control law (50) can be broken down into two parts, i.e., a delayed control input u_d and an SMC input u_s, as shown below:

$$\begin{aligned} u_k =& \underbrace{[D_R(z)]u_k - \frac{S(z)}{P(z)}[D_R(z)]e_k + P^{-1}(z)\left[z^d D_R(z)\right]s_k}_{\text{a delayed control input, } u_d} \\ & \underbrace{+\frac{S(z)}{P(z)}e_k - P^{-1}(z)\left\{(1-k_1)s_k - k_2 \mathrm{sat}\left(\frac{s_k}{k_3}\right)\right\}}_{\text{an SMC input, } u_s}. \end{aligned} \tag{51}$$

Finally, a block diagram of the system (1) regulated by the proposed control law (51) is depicted in Figure 5.

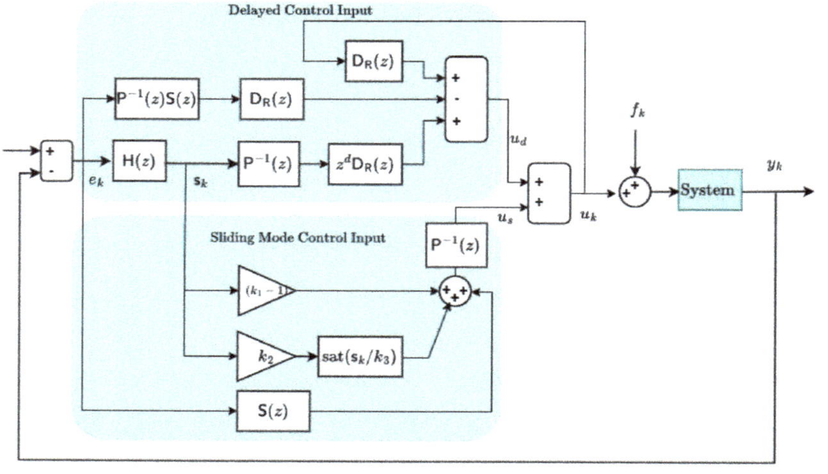

Figure 5. Block diagram of the proposed FD-RSMC scheme.

Remark 1. *Note that the transfer functions $D_R(z)$, $z^dD_R(z)$, $P^{-1}(z)$, and $P^{-1}(z)S(z)$ are involved in the construction of the proposed control law (51). Based on (32), $D_R(z)$ is basically a chain of the filtered delays applied to the input signals u_k, e_k, and s_k. The filters $F_r(z)$ and $F_f(z)$ composing $D_R(z)$ are formulated based on Thiran formula and designed to be stable and causal; then, $D_R(z)$ is also causal and bounded-input bounded-output (BIBO) stable. The term $z^dD_R(z)$ remains stable and causal since $D_R(z)$ is stable and has delay lengths of N_r and N_f, which are $N_r >> d$ and $N_f >> d$.*

Remark 2. *The inverse model $P^{-1}(z)$ is stable given that the polynomial $B(z)$ has stable zeros, implying that system (1) is a class of minimum-phase plant. For the nonminimum phase plant, one can formalize the polynomial $P(z)$ based on the work* [47] *in order to obtain a stable inverse model $P^{-1}(z)$. Since $P^{-1}(z)$ is stable and $S(z)$ plays as a numerator, then the term $P^{-1}(z)S(z)$ is also stable and causal.*

Hence, based on Remarks 1 and 2, all the terms constructing the proposed controller (51) are causal and stable, indicating that the proposed controller is realizable for real-time applications and generates bounded output when given the bounded input. In addition, all the signals required to compute the control law are measurable.

3.3. Stability Analysis of FD-RSMC System

In order to evaluate the system's stability, we first derive the sliding mode dynamics by substituting the polynomial $H(z)$ (41) and the control law (49) into the sliding surface (33). Let us rewrite the sliding surface (33) by substituting (41) into (33) to obtain

$$s_k = A(z)R(z)e_k + z^{-d}S(z)e_k. \tag{52}$$

By replacing $A(z)e_k$ in (52) with error dynamics (38), we obtain

$$s_k = -z^{-d}B(z)R(z)u_k + A(z)R(z)r_k - R(z)\mathcal{W}_k + z^{-d}S(z)e_k. \tag{53}$$

Shifting forward both sides of (53) by d steps and considering that $\mathrm{P}(z) = \mathrm{B}(z)\mathrm{R}(z)$ and $\mathrm{Q}(z) = \mathrm{A}(z)\mathrm{R}(z)$, (53) can be expressed as

$$s_{k+d} = -\mathrm{P}(z)u_k + \mathrm{Q}(z)r_{k+d} - \mathrm{R}(z)\mathcal{W}_{k+d} + \mathrm{S}(z)e_k. \tag{54}$$

Substituting the control law (49) into the sliding dynamics (54) gives

$$\begin{aligned} s_{k+d} =& -\mathrm{P}(z)[\mathrm{D}(z)]u_k - \mathrm{S}(z)(e_k - [\mathrm{D}(z)]e_k) - [\mathrm{D}(z)]s_{k+d} + (1-k_1)s_k \\ & -k_2\mathrm{sat}\left(\frac{s_k}{k_3}\right) + \mathrm{Q}(z)r_{k+d} - \mathrm{R}(z)\mathcal{W}_{k+d} + \mathrm{S}(z)e_k. \end{aligned} \tag{55}$$

Let us also multiply (54) by $[\mathrm{D}(z)]$ to yield

$$\begin{aligned} [\mathrm{D}(z)]s_{k+d} =& -\mathrm{P}(z)[\mathrm{D}(z)]u_k + \mathrm{Q}(z)[\mathrm{D}(z)]r_{k+d} - \mathrm{R}(z)[\mathrm{D}(z)]\mathcal{W}_{k+d} \\ & +\mathrm{S}(z)[\mathrm{D}(z)]e_k. \end{aligned} \tag{56}$$

By subtracting (56) from (55), we have

$$s_{k+d} = (1-k_1)s_k - k_2\mathrm{sat}\left(\tfrac{s_k}{k_3}\right) - \mathrm{R}(z)(\mathcal{W}_{k+d} - [\mathrm{D}(z)]\mathcal{W}_{k+d}). \tag{57}$$

Finally, the sliding dynamics (57) can be rewritten as

$$s_{k+d} = (1-k_1)s_k - k_2\mathrm{sat}\left(\tfrac{s_k}{k_3}\right) + \omega_{k+d}, \tag{58}$$

where

$$\omega_{k+d} = -\mathrm{R}(z)(\mathcal{W}_{k+d} - [\mathrm{D}(z)]\mathcal{W}_{k+d}). \tag{59}$$

By shifting backward (58) by d steps, we obtain

$$s_k = (1-k_1)s_{k-d} - k_2\mathrm{sat}\left(\tfrac{s_{k-d}}{k_3}\right) + \omega_k, \tag{60}$$

After obtaining the final sliding dynamics (60), the following reaching conditions [26,27] are used to evaluate the stability of the closed-loop system with control law (49):

$$[s_k - s_{k-d}]\mathrm{sgn}(s_{k-d}) < 0 \tag{61}$$

$$[s_k + s_{k-d}]\mathrm{sgn}(s_{k-d}) > 0, \tag{62}$$

where $\mathrm{sgn}(s_k)$ represents a signum of signal s_k. Note that the reaching conditions (61) and (62) are identical to

$$|s_k| < |s_{k-d}|. \tag{63}$$

Remark 3. *The inclusion of the sliding mode approach with the reaching law (36) results in a nonlinear controller, making frequency domain analysis no longer applicable. Consequently, stability is assessed in the time domain by ensuring the evolution of the sliding surface s_k as specified in Equation (63). The following theorem demonstrates that the reaching condition $|s_k|<|s_{k-d}|$ is satisfied when the sliding function s_{k-d} is outside the region, i.e., $|s_{k-d}|>k_3$. Additionally, it shows that the reaching condition is still met when the sliding surface s_{k-d} has entered the region, i.e., $|s_{k-d}|\leq k_3$.*

Then, Theorem 1 is established for the system (1) governed by the control law (49).

Theorem 1. *Define ω_k in (60) to be $|\omega_k| < \mu$, and*

$$\mu < k_2,\ 0 < (1-k_1) < 1,\ k_2 < k_3(1-k_1), \tag{64}$$

where μ, k_1, k_2, k_3 are some positive constants. Then, the sliding dynamics s_k meets the following conditions:

$$C1 : |s_k| < |s_{k-d}| \quad \textit{for } |s_{k-d}| > k_3, \tag{65}$$

$$C2 : |s_k| < k_3 \quad \textit{for } |s_{k-d}| = k_3, \tag{66}$$

$$C3 : |s_k| < k_3 \quad \textit{for } |s_{k-d}| < k_3, \tag{67}$$

Proof. The proof is completed for the three conditions listed below.

C1: $|s_k| < |s_{k-d}|$ for $|s_{k-d}| > k_3$

The state $|s_{k-d}| > k_3$ is identical to $s_{k-d} > k_3$ and $s_{k-d} < -k_3$. Then, we need to demonstrate that when $s_{k-d} > k_3$ or $s_{k-d} < -k_3$, the condition $|s_k| < |s_{k-d}|$ is met. The condition $|s_k| < |s_{k-d}|$ is equivalent to reaching conditions (61) and (62). Because $s_{k-d} > k_3$, $\text{sgn}(s_{k-d})$ is equal to 1. Then, (61) and (62) can be expressed, respectively, as

$$s_k - s_{k-d} < 0 \tag{68}$$

$$s_k + s_{k-d} > 0. \tag{69}$$

On the basis of (37) and (64), we are able to rewrite (60) as

$$\begin{aligned} s_k - s_{k-d} &= -k_1 s_{k-d} - k_2 \text{sat}\left(\frac{s_{k-d}}{k_3}\right) + \omega_k \\ &= -k_1 s_{k-d} - k_2 + \omega_k \\ &< -k_2 + \omega_k < -k_2 + \mu < 0. \end{aligned} \tag{70}$$

Based on (70), condition (68) is satisfied. Next, we prove condition (69) by deriving (60) as

$$\begin{aligned} s_k + s_{k-d} &= (1-k_1) s_{k-d} - k_2 \text{sat}\left(\frac{s_{k-d}}{k_3}\right) + \omega_k + s_{k-d} \\ &= (2-k_1) s_{k-d} - k_2 + \omega_k \\ &> (2-k_1) k_3 - k_2 + \omega_k \\ &\geq k_3 + (1-k_1) k_3 - k_2 + \omega_k \\ &> k_3 + \omega_k > k_2 + \omega_k > \mu + \omega_k > 0. \end{aligned} \tag{71}$$

It follows from (71) that condition (69) is met. Next, we examine the evolution of a sliding surface when $s_{k-d} < -k_3$. Since $s_{k-d} < -k_3$, then (61) and (62) are given by

$$s_k - s_{k-d} > 0 \tag{72}$$

$$s_k + s_{k-d} < 0. \tag{73}$$

Following (70) and (71), we have $[s_k - s_{k-d}]$ and $[s_k + s_{k-d}]$ for $s_{k-d} < -k_3$ as

$$\begin{aligned} s_k - s_{k-d} &= -k_1 s_{k-d} - k_2 \text{sat}\left(\frac{s_{k-d}}{k_3}\right) + \omega_k \\ &= -k_1 s_{k-d} + k_2 + \omega_k \\ &> k_1 k_3 + k_2 + \omega_k \\ &> k_1 k_3 + (\mu + \omega_k) > k_1 k_3 > 0, \end{aligned} \tag{74}$$

$$\begin{aligned} s_k + s_{k-d} &= (2 - k_1) s_{k-d} - k_2 \text{sat}\left(\frac{s_{k-d}}{k_3}\right) + \omega_k \\ &= (2 - k_1) s_{k-d} + k_2 + \omega_k \\ &< -(2 - k_1) k_3 + k_2 + \omega_k \\ &= -k_3 - (1 - k_1) k_3 + k_2 + \omega_k \\ &< -k_3 - k_2 + k_2 + \omega_k \\ &< -k_3 + \omega_k < \mu - k_3 < 0. \end{aligned} \tag{75}$$

According to (74) and (75), the conditions (72) and (73) for $s_{k-d} < -k_3$ are fulfilled. This completes the proof for C1.

C2: $|s_k| < k_3$ for $|s_{k-d}| = k_3$

The case $|s_k| < k_3$ is also identical to

$$s_k < k_3 \tag{76}$$

$$s_k > -k_3. \tag{77}$$

In addition, the state $|s_{k-d}| = k_3$ implies that $s_{k-d} = k_3$ and $s_{k-d} = -k_3$. By evaluating s_k respectively when $s_{k-d} = k_3$ and $s_{k-d} = -k_3$, we obtain

$$\begin{aligned} s_k &= (1 - k_1) s_{k-d} - k_2 \text{sat}\left(\frac{s_{k-d}}{k_3}\right) + \omega_k \ \rightarrow s_k = k3 \\ &= (1 - k_1) k_3 - k_2 + \omega_k \\ &< (1 - k_1) k_3 - k_2 + \mu \\ &< (1 - k_1) k_3 < k_3, \end{aligned} \tag{78}$$

$$\begin{aligned} s_k &= -(1 - k_1) k_3 + k_2 + \omega_k \ \rightarrow s_k = -k3 \\ &> -(1 - k_1) k_3 > -k_3. \end{aligned} \tag{79}$$

It follows from (78) and (79) that (76) and (77) are satisfied. This concludes the proof for C2.

C3: $|s_k| < k_3$ for $|s_{k-d}| < k_3$

Here, $|s_{k-d}| < k_3$ represents the situation where s_{k-d} is within the bands, i.e., $0 \leq s_{k-d} < k_3$ and $-k_3 < s_{k-d} < 0$.

By examining s_k when s_{k-d} is inside $0 \leq s_{k-d} < k_3$ and $-k_3 < s_{k-d} < 0$, we obtain

$$\begin{aligned} s_k &= (1-k_1)s_{k-d} - k_2\frac{s_{k-d}}{k_3} + \omega_k \ \rightarrow 0 \leq s_{k-d} < k_3 \\ &= \left(1-k_1-\frac{k_2}{k_3}\right)s_{k-d} + \omega_k \\ &< \left(1-k_1-\frac{k_2}{k_3}\right)k_3 + \omega_k \\ &= (1-k_1)k_3 - k_2 + \omega_k \\ &< (1-k_1)k_3 < k_3, \end{aligned} \tag{80}$$

$$\begin{aligned} s_k &= \left(1-k_1-\frac{k_2}{k_3}\right)s_{k-d} + \omega_k \ \rightarrow -k_3 < s_{k-d} < 0 \\ &> -\left(1-k_1-\frac{k_2}{k_3}\right)k_3 + \omega_k \\ &= -(1-k_1)k_3 + k_2 + \omega_k \\ &> -(1-k_1)k_3 > -k_3. \end{aligned} \tag{81}$$

Based on (80) and (81), the proof for C3 is concluded. Thus, Theorem 1 and its proof is completely established. □

Remark 4. *By obtaining the complete proof of Theorem 1, we have demonstrated that the reaching condition (63) is ensured. This implies that the control law u_k in (51) guarantees the convergence of the sliding surface s_k. The tracking error e_k is also vanishing because the sliding surface is a linear function of the tracking error. This means that accurate reference tracking and good disturbance compensation are achievable, even though the system is subject to model uncertainties. Here, the fast transient response is feasible due to the presence of SMC input, as shown in (51), which handles the tracking and rejection tasks when the delayed control input is still in the learning cycles.*

Finally, the design of the proposed FD-RSMC is presented in the following step-by-step algorithm (see Table 1):

Table 1. Design algorithm of the proposed FD-RSMC.

Step 1: Obtain the nominal polynomials A(z), B(z), and input delay step d from the system (1).
Step 2: Choose the stable polynomial H(z) for the sliding surface (33).
Step 3: Solve Equation (41) to obtain R(z) and S(z), and compute $\mathrm{P}^{-1}(z)$ as $\mathrm{P}^{-1}(z) = 1/[\mathrm{B}(z)\mathrm{R}(z)]$
Step 4: Pick the positive constants k_1, k_2, and k_3 such that $0 < (1-k_1) < 1$, $k_2 < k_3(1-k_1)$ as in (64).
Step 5: Obtain the trajectory period T_r and disturbance period T_f, and calculate N_r and N_f.
Step 6: Compute $z^{-N_{ri}}$ and $z^{-\mathrm{n}_r}$ according to (28), and approximate $z^{-\mathrm{n}_r}$ by using IIR filter $\mathrm{F}_r(z)$, the coefficients of which are obtained from (30) and (31).
Step 7: Compute $z^{-N_{fi}}$ and $z^{-\mathrm{n}_f}$ according to (29), and approximate $z^{-\mathrm{n}_f}$ by using IIR filter $\mathrm{F}_f(z)$, the coefficients of which are obtained from (30) and (31).
Step 8: Construct the delay term $\mathrm{D_R}(z)$ according to (32).
Step 9: Synthesize the control law u_k according to (51).

4. Results and Discussion

4.1. Controller Configurations

A continuous-time model of a quanser servo in [15] sampled at $T = 0.03$ s is utilized in the simulations. The model is subject to repetitive trajectory r_k, periodic input disturbance f_k, and parametric uncertainties $\delta_{\mathrm{A}}(z), \delta_{\mathrm{B}}(z)$, with the nominal model expressed as follows:

$$(1 - 1.132z^{-1} + 0.4783z^{-2})y_k = z^{-1}(0.2042 + 0.1424z^{-1})(u_k + f_k). \quad (82)$$

It follows from (82) that $\mathrm{A}(z)$, $\mathrm{B}(z)$, and d are given by $\mathrm{A}(z) = 1 - 1.132z^{-1} + 0.4783z^{-2}$, $\mathrm{B}(z) = 0.2042 + 0.1424z^{-1}$, and $d = 1$. Let us choose the stable polynomial for the sliding surface (33) as $\mathrm{H}(z) = 1 - 0.6z^{-1} + 0.09z^{-2}$, which has two zeros located at 0.3. Then, by solving (41), we can obtain

$$\mathrm{R}(z) = 1,\ \mathrm{S}(z) = 0.532 - 0.388z^{-1},\ \mathrm{P}(z) = \mathrm{R}(z)\mathrm{B}(z) = 0.2042 + 0.1424z^{-1}. \quad (83)$$

In the simulations, the desired trajectory and input disturbance are respectively modeled as (84) and (85), and this is illustrated in Figure 6. The trajectory and disturbance models are given as

$$r_k = 0.5\ \sin(2\pi kT) + 0.4\ \sin(4\pi kT) + 0.1\ \sin(6\pi kT), \quad (84)$$

$$f_k = 0.1\ \sin(2\pi(1/0.95)kT) + 0.1\ \sin(4\pi(1/0.95)kT) + 0.1\ \sin(6\pi(1/0.95)kT). \quad (85)$$

According to (84) and (85), both trajectory and disturbance consist of the first three harmonics with distinct fundamental frequencies, with the fundamental frequencies of trajectory and disturbance being 1 Hz and 1.053 Hz, respectively. Hence, the trajectory period T_r and disturbance T_f are obtained as $T_r = 1$ s and $T_f = 0.95$ s, respectively. Then, N_r and N_f can be calculated as

$$N_r = 1/0.03 = 33.333;\ N_f = 0.95/0.03 = 31.667, \quad (86)$$

From (86), it is clear that the actual delay lengths N_r and N_f are fractional, implying that z^{-N_r} and z^{-N_f} cannot be implemented directly. Therefore, we apply the fractional delay-based repetitive control design strategy presented in Section 3.2. Then, Tables 2 and 3 are provided to summarize the approximations of z^{-N_r} and z^{-N_f} for various IIR filter orders. From Tables 2 and 3, we can understand that the typical solution utilizes rounding the fractional N_r and N_f to the nearest integer values, resulting in the integer delays z^{-33} and z^{-32}, which are simple to resolve. However, this solution introduces certain steady-state errors due to the period being mismatched, and this error will be reflected in the simulation's output. Here, Table 2 also demonstrates that the filter $\mathrm{F}_r(z)$ approximated by the Thiran Formula (30) has causal form for $p = 1, 2, 3$. Moreover, all of the filters for various p have poles located within the unit circle, i.e., $\mathrm{F}_r(z)$ with $p = 1, 2, 3$ has poles at $\{-0.5\}$, $\{-0.617, 0.045\}$, and $\{-0.677, 0.038 \pm i0.067\}$. Hence, the filters used to approximate the fractional delay component in z^{-N_r} can be concluded to be stable and causal. Likewise, Table 3 reveals that the filter $\mathrm{F}_f(z)$ used to replace the fractional delay component in z^{-N_f} is stable and causal for various filter orders. The pole positions of $\mathrm{F}_f(z)$ for $p = 1, 2, 3$ are at $\{-0.199\}$, $\{-0.321, 0.072\}$, and $\{-0.398, 0.062 \pm 0.079\}$, respectively.

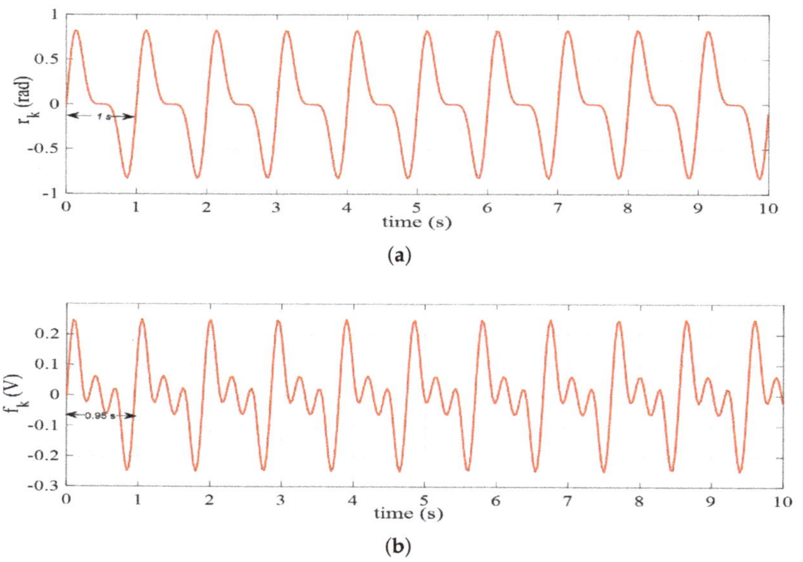

Figure 6. (**a**) Desired repetitive trajectory r_k and (**b**) periodic input disturbance f_k.

Table 2. Approximations of the fractional delay z^{-N_r}.

N_r	**Round(** N_r**)**	N_{ri}	n_r	**IIR Order (p)** $p = \mathrm{ceil}(\mathrm{n}_r)$	$\mathrm{F}_r(z)$
33.333	33	33	0.333	1	$\frac{0.500z+1}{z+0.500}$
		32	1.333	2	$\frac{-0.028z^2+0.572z+1}{z^2+0.572z-0.028}$
		31	2.333	3	$\frac{0.004z^3-0.046z^2+0.6z+1}{z^3+0.6z^2-0.046z+0.004}$

Table 3. Approximations of the fractional delay z^{-N_f}.

N_f	**Round(** N_f**)**	N_{fi}	n_f	**IIR Order (p)** p**=ceil(n**$_r$**)**	$\mathrm{F}_f(z)$
31.667	32	31	0.667	1	$\frac{0.199z+1}{z+0.199}$
		30	1.667	2	$\frac{-0.023z^2+0.249z+1}{z^2+0.249z-0.023}$
		29	2.667	3	$\frac{0.004z^3-0.039z^2+0.272z+1}{z^3+0.272z^2-0.039z+0.004}$

4.2. *Tracking Performance and Comparison Studies*

Now, we aim to examine the tracking performance of FD-RSMC with $p = 1$ for both $\mathrm{F}_r(z)$ and $\mathrm{F}_r(z)$. The values of control parameters synthesizing the proposed FD-RSMC are presented in Table 4. Please note that the parameter k_3 shown in Table 4 corresponds to the boundary layer of the reaching law (36). While decreasing the boundary layer parameter k_3 to a very small value can accelerate the transient response, it also raises the risk of chattering. Therefore, we have carefully adjusted k_3 and set it to 0.1. The trajectory and output responses of the system (82) regulated by FD-RSMC are given in Figure 7. As depicted in Figure 7, the system begins to respond during the first repetition, even though the output is less than the desired value. Then, the output of the system exceeds the intended value during the second repetition. Lastly, the output precisely follows the intended trajectory starting at the third repetition.

Table 4. Values of controller parameters.

Parameter	Value
$D_R(z)$	$F_r(z)z^{-33} + F_f(z)z^{-31} - F_r(z)F_f(z)z^{-64}$
$F_r(z), F_f(z)$	see Tables 2 and 3 (IIR order = 1)
$P(z)$	$0.2042 + 0.1424z^{-1}$
$S(z)$	$0.532 - 0.388z^{-1}$
k_1, k_2, k_3	$0.05, 0.03, 0.1$

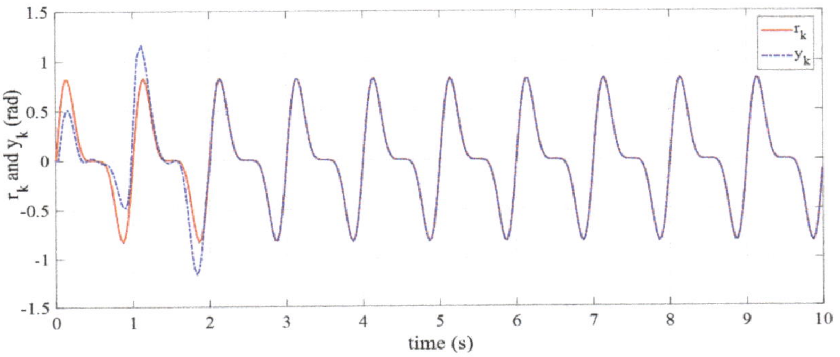

Figure 7. Desired trajectory r_k and output y_k with FD-RSMC.

The performance of the system employing the FD-RSMC can be further evaluated using the tracking error e_k and sliding surface s_k plotted in Figure 8. Figure 8a depicts the evolution of both tracking error e_k and sliding surface s_k during the transient and steady-state periods. Meanwhile, e_k and s_k in the steady state only are plotted in Figure 8b. It is evident from Figure 8a that e_k and s_k exhibit similar trends during the transient period, with the exception that e_k has a larger peak than s_k. This is understandable given that s_k is essentially an attenuated version of e_k as a result of the definition given in (33), where $s_k = H(z)e_k$. Here, polynomial $H(z)$ behaves as a linear sliding function that reduces the tracking error amplitude. Now, the error and sliding signals during the steady state are depicted in Figure 8b, spanning from $t = 3$ s to 10 s. The error plot indicates that the error values during the steady state are extremely small, i.e., within ± 0.0015. Similar to the behavior observed during the transient response, the sliding values are also smaller than the error values.

Next, we investigate the system's tracking performance using FD-RSMC with different order approximations ($p = 2, 3$). Figure 9 displays the tracking errors of FD-RSMC for three distinct approximations ($p = 1, 2, 3$), with Figure 9a showing the errors during the transient and steady-state periods and Figure 9b exhibiting the errors during the steady state only. From Figure 9a, the differences between the error values of FD-RSMC for three distinct approximations are not noticeable. However, Figure 9b reveals error discrepancies, indicating that steady-state errors using FD-RSMC with $p = 2, 3$ are less than steady-state errors using FD-RSMC with $p = 1$. Moreover, the steady-state errors of FD-RSMC with $p = 2, 3$ are very small and nearly overlapping, and ranged within ± 0.001.

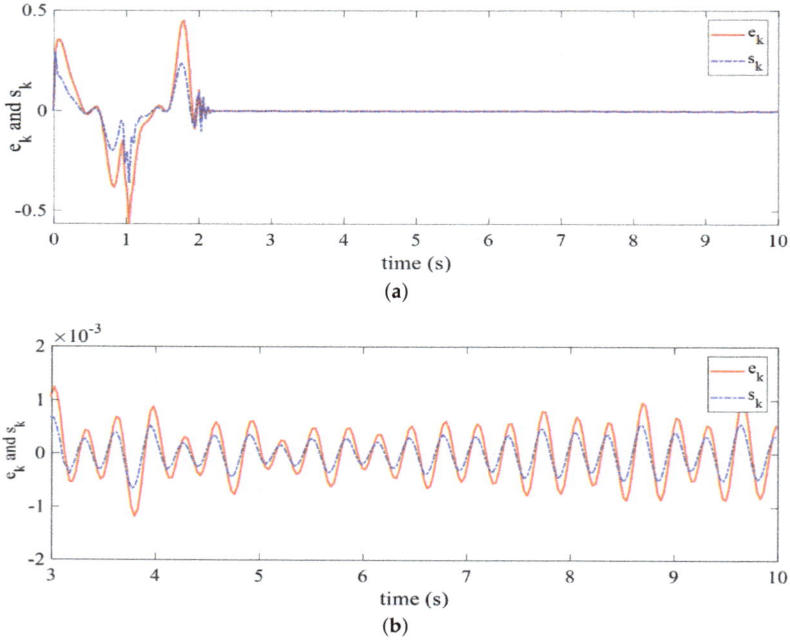

Figure 8. (**a**) Evolution of the tracking error e_k and sliding surface s_k with FD-RSMC (**b**) e_k and s_k during the steady-state period.

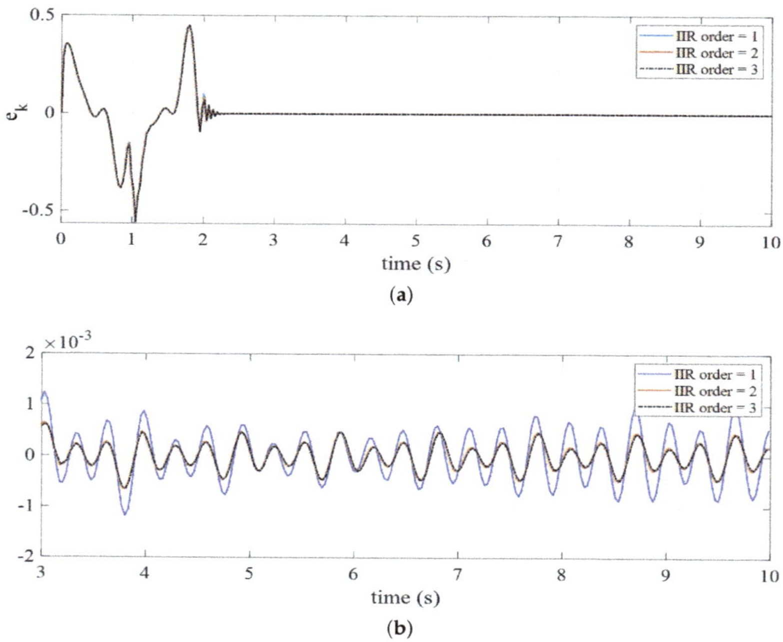

Figure 9. (**a**) Tracking errors e_k of FD-RSMC for different order approximations (**b**) e_k during the steady-state period.

Furthermore, we summarize the utilized delay term $D_R(z)$ for various approximations in Table 5. In addition, Table 5 presents a root mean square error (erms) and the root mean square of steady-state error (ess-rms). Here, erms and ess-rms are expressed as

$$\text{erms}=\sqrt{\frac{1}{\tau_e}\sum_{k=1}^{\tau_e} e_k^2}, \tag{87}$$

$$\text{ess}-\text{rms}=\sqrt{\frac{1}{(\tau_e-\tau_{ss})}\sum_{k=\tau_{ss}}^{\tau_e} e_k^2}, \tag{88}$$

where $\tau_e = \text{round}(10/0.03) = 333$ and $\tau_{ss} = 3/0.03 = 100$. Table 5 shows that the erms and ess-rms of the system using FD-RSMC with $p = 2,3$ are even smaller than those of the system using FD-RSMC with $p = 1$. These results indicate that the tracking accuracy of FD-RSMC with higher-order approximations, particularly during the steady state, is superior to that of FD-RSMC with first-order approximations.

Table 5. Delay term $D_R(z)$ and rms error for different approximation.

Method	IIR Order	$D_R(z)$	erms $\times 10^{-1}$	ess-rms $\times 10^{-4}$
FD-RSMC	1	see Table 4	1.017	6.83
	2	$F_r(z)z^{-32} + F_f(z)z^{-30} - F_r(z)F_f(z)z^{-62}$	1.016	4.37
	3	$F_r(z)z^{-31} + F_f(z)z^{-29} - F_r(z)F_f(z)z^{-60}$	1.016	4.26

A first comparison is made to integer delay-based RSMC (ID-RSMC), which corresponds to control strategy (50) with the delay term $D_R(z)$ replaced by

$$D_I(z) = z^{-33} + z^{-32} - z^{-65}. \tag{89}$$

Figure 10 depicts the tracking output of the ID-RSMC system, which exhibits a similar transient response to the output of the FD-RSMC system shown in Figure 7. In the meantime, Figure 11 displays the tracking error e_k and sliding surface s_k of ID-RSMC. From Figure 11b, it is clear that the ID-RSMC yields a larger steady-state error than the FD-RSMC, with max$|\text{ess}|$ having a value of 0.03. These results indicate that FD-RSMC provides more accurate tracking than ID-RSMC. Additionally, the magnitude-frequency characteristics of the internal models $I_{in}(z)$ and $I_R(z)$ can be used to explain the superior tracking performance of FD-RSMC over ID-RSMC. Based on the internal model formula in (16), $I_{in}(z)$ and $I_R(z)$ can be expressed as $I_{in}(z) = D_I(z)/(1 - D_I(z))$ and $I_R(z) = D_R(z)/(1 - D_R(z))$, respectively. The magnitude-frequency characteristics of $I_{in}(z)$ and $I_R(z)$ are plotted in Figure 12.

As shown in Figure 12, the peak frequencies of $I_R(z)$ are kept at 1 Hz and 1.053 Hz, which correspond to the fundamental frequencies of trajectory r_k and disturbance f_k modeled in (84) and (85). In addition, their harmonics remained at h $\times$ 1 Hz and h $\times$ 1.053 Hz, where h is the harmonic number ($h = 1,2,3,\cdots$). On the contrary, the peak of the basis frequencies of $I_{in}(z)$ are shifted to 1.01 Hz and 1.042 Hz, indicating that their harmonics are increasingly displaced. The frequency changes produced by the integer-based internal model are relatively small: $\Delta f_r = 0.01$ Hz and $\Delta f_f = -0.011$ Hz. However, as indicated by previous results, these small shifts impact the tracking accuracy of the controlled system, particularly during the steady state.

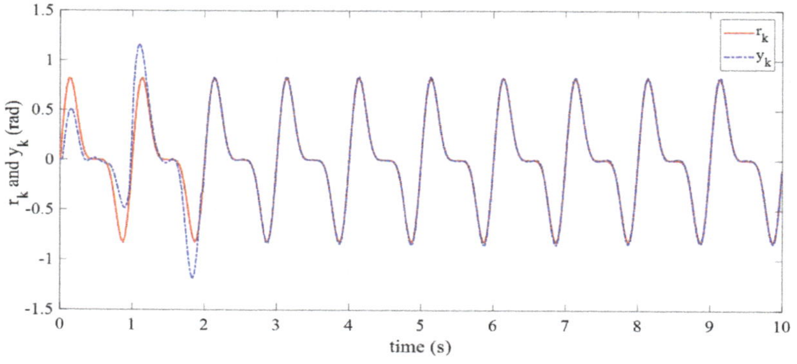

Figure 10. Desired trajectory r_k and output y_k with ID-RSMC.

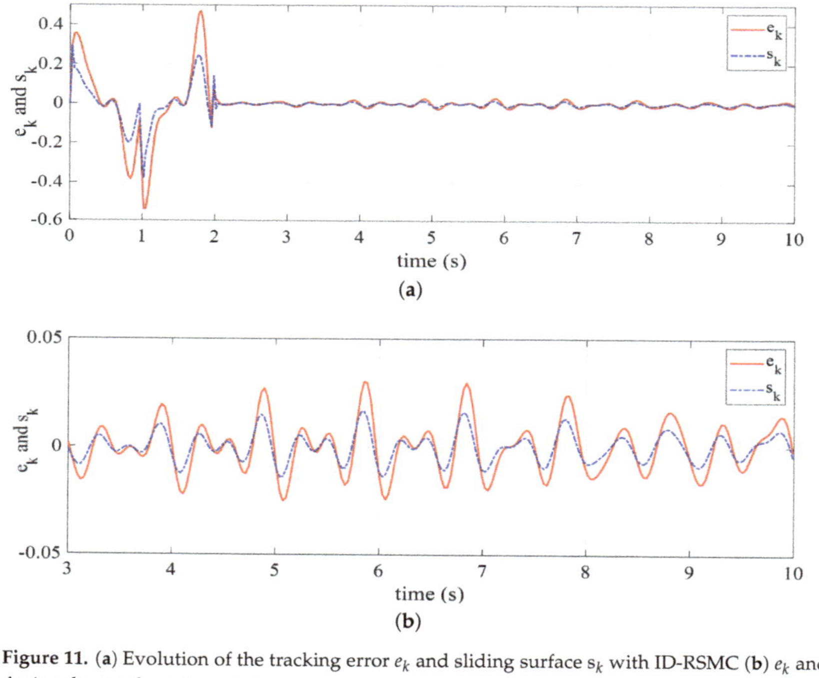

Figure 11. (**a**) Evolution of the tracking error e_k and sliding surface s_k with ID-RSMC (**b**) e_k and s_k during the steady-state period.

Other comparisons are made to ID-RC and FD-RC to evaluate the tracking performance of the system when the system (82) is governed solely by the RC scheme. In Table 6, the control laws and parameters of ID-RC and FD-RC are enumerated. The tracking outputs and errors of the system using ID-RC and FD-RC are plotted in Figures 13 and 14. Figure 13 demonstrates that the output y_k indicates a small amplitude during the first repetition, and then during the second repetition, the output y_k rises to a larger peak value with an amplitude that is twice that of the trajectory r_k, i.e., $\max|y_k| = 1.512$. The transient behaviors can also be seen from Figure 14. We can observe that the error convergences are achieved after 2 s, which are almost similar to the cases of FD-RSMC and ID-RSMC. However, the error responses during the transient period, i.e, first two repetitions, exhibit a larger error value reaching a maximum absolute error, $\max|e_k|$, of 1.8. These findings demonstrate that FD-RSMC offers superior transient responses compared to both ID-RC and FD-RC. It can be understood since during the transient period, ID-RC and FD-RC operates in the learning phase due to the presence of delay terms $D_I(z)$ and $D_R(z)$. Meanwhile, FD-RSMC utilizes the reaching law of SMC to generate a control signal during the learning period. Therefore, the FD-RSMC yields a better transient response than the ID-RC and FD-RC. From Figure 14, it can be clearly seen that the steady-state error of FD-RC is smaller than that of ID-RC, indicating the benefit of using the fractional-based delay.

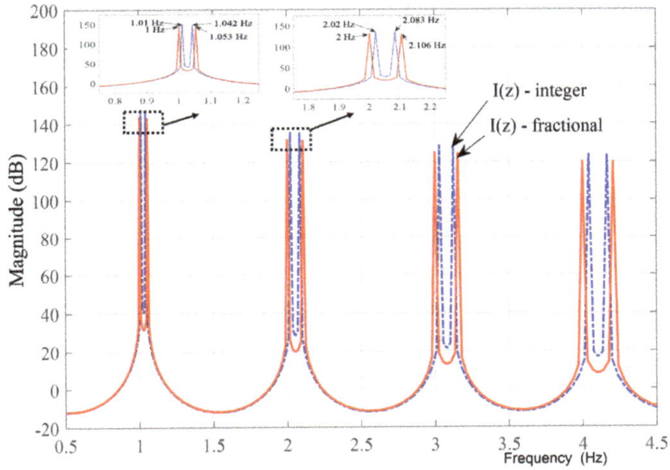

Figure 12. Magnitude-frequency characteristics of integer- and fractional-based internal models.

Table 6. Control laws and parameters of ID-RC and FD-RC for comparisons.

Method	Control Law (u_k)	Parameters
ID-RC	$u_k = [D_I(z)]u_k + \mathcal{C}(z)[D_I(z)]e_k$	$D_I(z)$: see (89), $\mathcal{C}(z) = \frac{1-1.132z^{-1}+0.4783z^{-2}}{0.2042z^{-1}+0.1424z^{-2}}$
FD-RC	$u_k = [D_R(z)]u_k + \mathcal{C}(z)[D_R(z)]e_k$	$D_R(z)$: see Table 5, $\mathcal{C}(z)$: similar as above

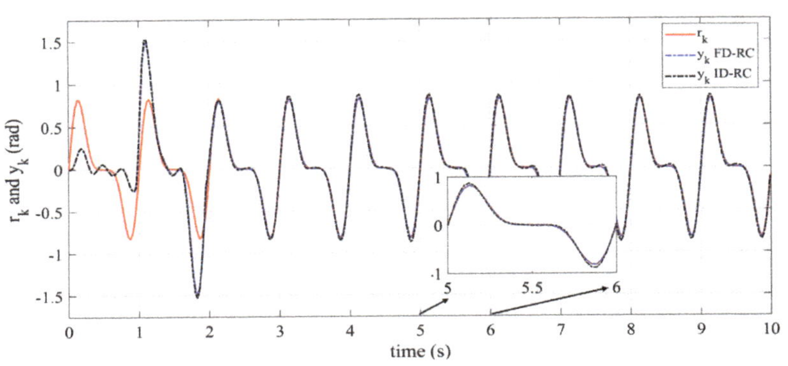

Figure 13. Desired trajectory r_k and output y_k with FD-RC and ID-RC.

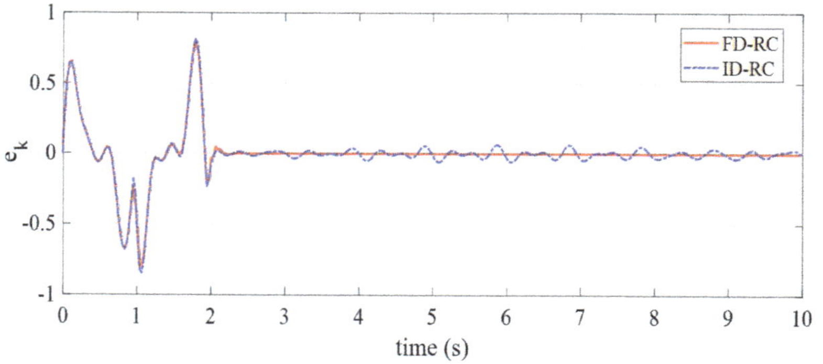

Figure 14. Evolution of the tracking error e_k with FD-RC and ID-RC.

4.3. Robustness Performance and Comparison Studies

Lastly, we investigate the robustness of FD-RSMC, ID-RSMC, FD-RC, and ID-RC against system parametric uncertainties. For this purpose, the nominal parameters of the system in (82) are substantially varied by 10%, 20%, and 30%. Table 7 enumerates the parameters of the perturbed systems with 10%, 20%, and 30% variations, along with the zero-pole positions. It can be seen from Table 7 that the parametric uncertainties alters the system's poles. However, the locations remain inside the unit circle, indicating that the perturbed models remain open-loop stable. The error graphs in Figure 15 depict the robustness behaviors of the FD-RSMC and ID-RSMC. As shown in Figure 15a,c,e, the error convergence of FD-RSMC becomes slower as the parameter variation increases. However, the stability of the FD-RSMC system is maintained and a high tracking accuracy during the steady state is also preserved.

Table 7. Parameters variations and zero-pole locations of the perturbed system.

Variation	a_1	a_2	b_0	b_1	$p_{1,2}(A(z)) \rightarrow \mid p_{1,2} \mid$	$z_1(B(z))$
0%	−1.132	0.478	0.204	0.142	$0.566 \pm j0.397 \rightarrow 0.691$	−0.697
10%	−0.929	0.392	0.168	0.117	$0.464 \pm j0.420 \rightarrow 0.626$	−0.697
20%	−0.632	0.267	0.114	0.079	$0.316 \pm j0.409 \rightarrow 0.517$	−0.698
30%	−0.442	0.187	0.079	0.056	$0.221 \pm j0.371 \rightarrow 0.432$	−0.697

The tracking errors of the ID-RSMC system under parameter variations are presented in Figure 15b,d,f, demonstrating a similar trend to that of the FD-RSMC case, particularly in the stability of a closed-loop system. However, in contrast to the FD-RSMC case, the steady-state error of the ID-RSMC system is amplified as the variation grows. Therefore, increasing parameter variation in the ID-RSMC system considerably degrades tracking accuracy, especially in the steady-state period.

In the meantime, Figure 16 illustrates the robustness performance of FD-RC and ID-RC systems. As shown in Figure 16a,c,e, FD-RC is capable of maintaining system stability with a 20% increase in parameter variation. When the variation reaches 30%, however, the error response slowly diverges, indicating that the system output becomes unstable. The worst robust performance is given by the ID-RC system as indicated in Figure 16b,d,f. The ID-RC system remains stable at a 10% variation; however, the steady-state error significantly degrades in comparison to the nominal case. Then, the system output becomes unstable when the variation reaches 20%, as indicated by the error response rapidly diverging to relatively large values. The magnitude of the error divergence increases when the variation reaches 30%.

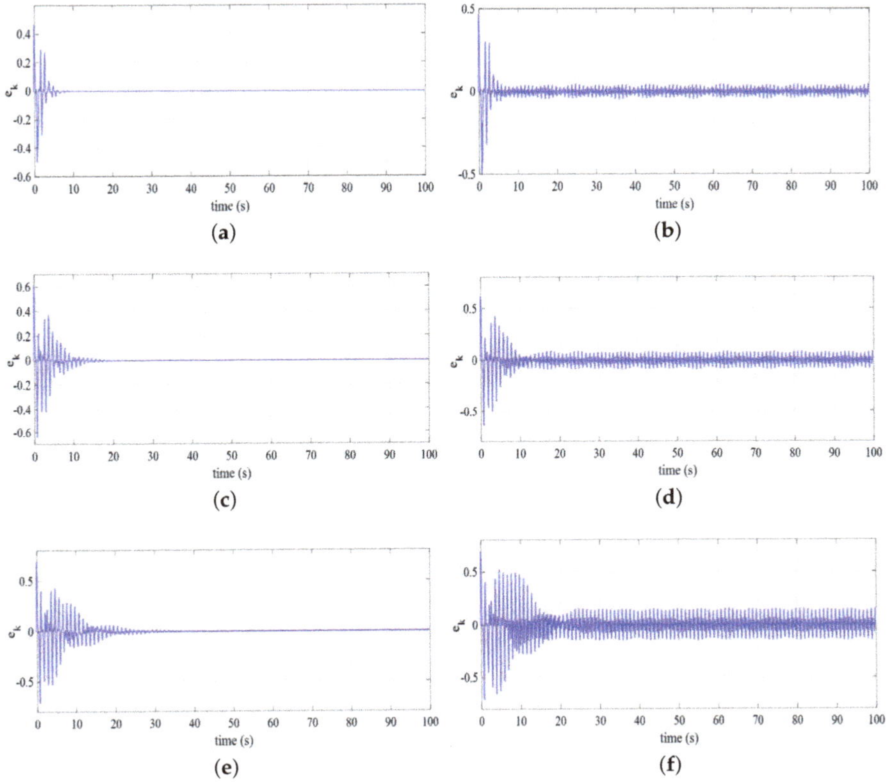

Figure 15. Tracking error e_k of the system under parameter variations: (**a**) 10% FD-RSMC; (**b**) 10% ID-RSMC; (**c**) 20% FD-RSMC; (**d**) 20% ID-RSMC; (**e**) 30% FD-RSMC; (**f**) 30% ID-RSMC.

The robust performance provided by each controller is then evaluated in terms of root mean square error (erms), maximum of absolute error (max$|e|$), and stability, as enumerated in Tables 8 and 9.

Table 8. Erms, max $|e|$, and stability of FD-RSMC and ID-RSMC systems under model variations.

Variation	FD-RSMC(p=2)			ID-RSMC		
	erms	**max $\|e\|$**	**Stable**	**erms**	**max $\|e\|$**	**Stable**
10%	0.036	0.497	Yes	0.041	0.497	Yes
20%	0.053	0.638	Yes	0.068	0.638	Yes
30%	0.073	0.709	Yes	0.113	0.709	Yes

Table 9. Erms, max $|e|$, and stability of FD-RC and ID-RC systems under model variations.

Variation	FD-RC			ID-RC		
	erms	**max** $|e|$	**Stable**	**erms**	**max** $|e|$	**Stable**
10%	0.054	0.706	Yes	0.069	0.706	Yes
20%	0.082	0.757	Yes	9.702	56.103	No
30%	0.142	0.785	No	63.96	556.43	No

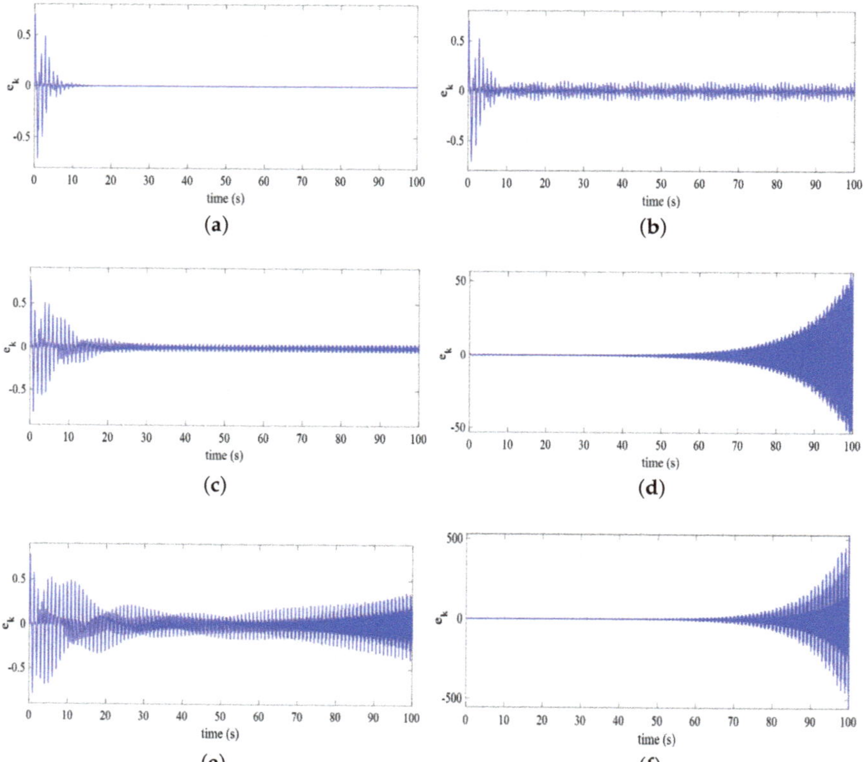

Figure 16. Tracking error e_k of the system under parameter variations: (**a**) 10% FD-RC; (**b**) 10% ID-RC; (**c**) 20% FD-RC; (**d**) 20% ID-RC; (**e**) 30% FD-RC; (**f**) 30% ID-RC.

The findings presented in Tables 8 and 9 demonstrate that the FD-RSMC outperforms the other three controllers in terms of tracking precision and robustness performance. It is evident that the ID-RSMC also maintains closed-loop stability despite the presence of parametric variations. However, the proposed FD-RSMC yields erms values that are significantly smaller in magnitude compared to ID-RSMC. Furthermore, it is observed that the system under FD-RC and ID-RC exhibits large and diverge error values, suggesting that the controlled system lacks robustness and is susceptible to parameter variations. The stability of the closed-loop system cannot be maintained when the parameters for ID-RC and FD-RC are varied by 20% and 30%, respectively. As clearly indicated in Figure 16d,f, instability occurs in the system controlled with ID-RC when the model parameters are varied by 20% and 30%. In contrast, stability is maintained for both FD-RSMC and ID-

RSMC. This outcome is understandable, as the compensator $\mathcal{C}(z)$ in (22) is designed based on the nominal polynomials $\mathrm{A}(z)$ and $\mathrm{B}(z)$, leading to a violation of the stability conditions, particularly in relation to condition (21). Let $\mathcal{G}_{20}(z)$ and $\mathcal{G}_{30}(z)$ represent the plant model (82) with 20% and 30% parameter variations, respectively, as also outlined in Table 7. According to the H_∞ calculations in (21), $\| [1-\mathcal{C}(z)\mathcal{G}_{20}(z)]\mathrm{D}_I(z) \|_\infty = 2.193$ and $\| [1-\mathcal{C}(z)\mathcal{G}_{30}(z)]\mathrm{D}_I(z) \|_\infty = 2.491$, both of which exceed 1, indicate that asymptotic stability is not preserved when using the standalone ID-RC. However, in the cases of both FD-RSMC and ID-RSMC, the parameters $\mathrm{S}(z)$ and $\mathrm{P}(z)$ used in synthesizing the controller are also based on the nominal polynomials $\mathrm{A}(z)$ and $\mathrm{B}(z)$. Nevertheless, the inclusion of the SMC component successfully maintains stability even when there is a mismatch between the model used in the controller design and the actual system model. Based on the findings, we can conclude that the transient response, tracking accuracy, stability, and robustness of the controlled system are enhanced when fractional delay-based repetitive control and sliding mode control are properly combined.

5. Conclusions

In this article, an FD-RSMC has been proposed for uncertain linear systems with periodic reference and disturbance. It has been demonstrated that the proposed algorithm consists of a fractional delay repetitive control scheme and a nonlinear control scheme, namely the sliding mode control technique. First, fractional delay repetitive control is employed to deal with trajectory tracking and disturbance rejection, as well as improving the tracking accuracy during the steady state. The fractional delay component can be realized with a stable IIR filter. Sliding mode control is incorporated to enhance the transient response of the RC-controlled system and to provide robustness in the face of parametric uncertainties. Simulation and comparison studies have verified the excellent performance of the proposed FD-RSMC. Extending the FD-RSMC design to handle unstable systems or nonlinear systems with various types of disturbances, supported by rigorous analysis and results, will be the focus of our future research.

Author Contributions: Methodology, E.K.; Software, J.A.P., E.B.P. and A.A.-Z.; Validation, A.M.B.; Resources, P.P.; Data curation, D.R.; Writing—original draft, E.K.; Writing—review & editing, A.M.B. and J.A.P.; Visualization, D.R., E.B.P. and A.A.-Z.; Supervision, S.W.; Project administration, T.M.; Funding acquisition, P.P., S.W. and T.M. All authors have read and agreed to the published version of the manuscript.

Funding: The authors received financial support for publication from the UPLAND Project, with the support of IsDB and IFAD, that focus to increase smallholders' agriculture productivity, incomes, livelihoods, and resilience in the targeted area. Meanwhile, this research was supported by the Research Center for Photonics and the Deputy for Research and Innovation Infrastructure (DIRI) under the National Research and Innovation Agency (BRIN), Indonesia.

Data Availability Statement: The original contributions presented in the study are included in the article, further inquiries can be directed to the corresponding author.

Acknowledgments: The authors would like to express sincere thanks, particularly to Rahmanto, the Director of Agriculture Irrigation at the Ministry of Agriculture, Republic of Indonesia.

Conflicts of Interest: The authors declare no conflicts of interest.

Abbreviations

The following abbreviations are used in this manuscript:

RC	Repetitive control
SMC	Sliding mode control
IIR	Infinite impulse response
FD-RC	Fractional delay-based RC
ID-RC	Integer delay-based RC
FD-RSMC	Fractional delay-based repetitive sliding mode control
ID-RSMC	Integer delay-based repetitive sliding mode control

References

1. Hillerstrom, G.; Walgama, K. Repetitive control theory and applications-a survey. In Proceedings of the 13th IFAC World Congress, San Francisco, CA, USA, 30 June–5 July 1996; Volume 29, pp. 1446–1451.
2. Feng, N.; Ruan, Y.; Xu, T.; Tang, T. Enhanced Observer-Based Repetitive Control for Tip-Tilt Disturbance Rejections Beyond Control Bandwidth in Optical-Stabilized Systems. *IEEE Photonics J.* **2023**, *15*, 1–7.
3. Li, L.; Huang, W.-W.; Wang, X.; Zhu, L. Dual-Notch-Based Repetitive Control for Tracking Lissajous Scan Trajectories with Piezo-Actuated Nanoscanners. *IEEE Trans. Instrum. Meas.* **2022**, *71*, 1–12.
4. Reinders, J.; Giaccagli, M.; Hunnekens, B.; Astolfi, D.; Oomen, T.; van de Wouw, N. Repetitive Control for Lur'e-Type Systems: Application to Mechanical Ventilation. *IEEE Trans. Control Syst. Technol.* **2023**, 1819–1829.
5. Mooren, N.; Witvoet, G.; Oomen, T. Gaussian Process Repetitive Control with Application to an Industrial Substrate Carrier System with Spatial Disturbances. *IEEE Trans. Control Syst. Technol.* **2023**, *31*, 344–358.
6. Li, Q.; Zhang, T.; Li, G.; Li, Z.; Xia, H.; Su, C.-Y. Neural-Dynamics Optimization and Repetitive Learning Control for Robotic Leg Prostheses. *IEEE/ASME Trans. Mechatron.* **2022**, *27*, 811–822.
7. Lin, C.-Y.; Chang, C.-M. Hybrid proportional derivative/repetitive control for active vibration control of smart piezoelectric structures. *J. Vib. Control* **2013**, *19*, 992–1003. [CrossRef]
8. Jin, W.; Li, Y.; Sun, G.; Bu, L.H. Repetitive Control Based on Active Damping with Reduced Computation Delay for LCL-Type Grid-Connected Inverters. *Energies* **2017**, *10*, 586. [CrossRef]
9. Daley, S.; Hätönen, J.; Owens, D.H. Active Vibration Isolation in A Smart Spring Mount Using a Repetitive Control Approach. *IFAC Proc.* **2005**, *38*, 55–60. [CrossRef]
10. Inoue, T.; Nakano, M.; Iwai, S. High accuracy control of a proton synchrotron magnet power supply. *IFAC Proc. Vol.* **1981**, *14*, 3137–3142.
11. Francis, B.; Wonham, W. The internal model principle for linear multivariable regulators. *Appl. Math. Optim.* **1975**, *2*, 170–194.
12. Chuei, R.; Cao, Z. Extreme learning machine-based super-twisting repetitive control for aperiodic disturbance, parameter uncertainty, friction, and backlash compensations of a brushless DC servo motor. *Neural Comput. Appl.* **2020**, *32*, 14483–14495.
13. Nakano, M.; Hara, S. *Microprocessor-Based Repetitive Control. Microprocessor-Based Control Systems*; Springer: Dordrecht, The Netherlands, 1986; pp. 279–296.
14. Olm, J.M.; Ramos, G.A.; Costa-Castelló, R. Adaptive compensation strategy for the tracking/rejection of signals with time-varying frequency in digital repetitive control systems. *J. Process Control* **2010**, *20*, 551–558. [CrossRef]
15. Kurniawan, E.; Cao, Z.; Man, Z. Design of Robust Repetitive Control with Time-Varying Sampling Periods. *IEEE Trans. Ind. Electron.* **2014**, *61*, 2834–2841. [CrossRef]
16. Kurniawan, E.; Cao, Z.; Man, Z. Digital design of adaptive repetitive control of linear systems with time-varying periodic disturbances. *IET Control Theory Appl.* **2014**, *8*, 1995–2003. [CrossRef]
17. Steinbuch, M.; Weiland, S.; Singh, T. Design of noise and period-time robust high-order repetitive control, with application to optical storage. *Automatica* **2007**, *43*, 2086–2095.
18. Jamil, M.; Waris, A.; Gilani, S.O.; Khawaja, B.A.; Khan, M.N.; Raza, A. Design of Robust Higher-Order Repetitive Controller Using Phase Lead Compensator. *IEEE Access* **2020**, *8*, 30603–30614. [CrossRef]
19. Prakosa, J.A.; Purwowibowo, P.; Kurniawan, E.; Wijonarko, S.; Maftukhah, T.; Sari, F.; Pratiwi, E.B.; Rustandi, D. Discrete-Time Design of Dual Internal Model-Based Repetitive Control Systems. *Appl. Sci.* **2022**, *12*, 11746. [CrossRef]
20. Al-Raeei, M. Applying fractional quantum mechanics to systems with electrical screening effects. *Chaos Solitons Fractals* **2021**, *150*, 111209.
21. Du, F.; Jia, B. Finite time stability of fractional delay difference systems: A discrete delayed Mittag-Leffler matrix function approach. *Chaos Solitons Fractals* **2020**, *141*, 110430.
22. Xiao, M.; Tao, B.; Zheng, X.W.; Jiang, G. Fractional-Order PID Controller Synthesis for Bifurcation of Fractional-Order Small-World Networks. *IEEE Trans. Syst. Man Cybern. Syst.* **2021**, *51*, 4334–4346. [CrossRef]

23. Stanisławski, R.; Rydel, M.; Li, Z. A New Reduced-Order Implementation of Discrete-Time Fractional-Order PID Controller. *IEEE Access* **2022**, *10*, 17417–17429. [CrossRef]
24. Arya, P.P.; Chakrabarty, S. A Robust Internal Model-Based Fractional Order Controller for Fractional Order Plus Time Delay Processes. *IEEE Control Syst. Lett.* **2020**, *4*, 862–867. [CrossRef]
25. Adigintla, S.; Aware, V.M.; Bingi, K.; Das, S. Novel Complex Fractional Order Speed Controller for IM Drive under Varying Operating Conditions with Enhanced Robustness. *IEEE Trans. Ind. Electron.* **2024**, *71*, 8438–8447. [CrossRef]
26. Gao, W.; Wang, Y.; Homaifa, A. Discrete-time variable structure control systems. *IEEE Trans. Ind. Electron.* **1995**, *42*, 117–122.
27. Bartoszewicz, A. Discrete-time quasi-sliding-mode control strategies. *IEEE Trans. Ind. Electron.* **1998**, *45*, 633–637.
28. Qu, S.; Xia, X.; Zhang, J. Dynamical behaviors of an euler discretized sliding mode control systems. *IEEE Trans. Autom. Control* **2014**, *59*, 2525–2529.
29. Ma, H.; Wu, J.; Xiong, Z. A novel exponential reaching law of discrete-time sliding-mode control. *IEEE Trans. Ind. Electron.* **2017**, *64*, 3840–3850.
30. Ma, H.; Li, Y.; Xiong, Z. Discrete-time sliding-mode control with enhanced power reaching law. *IEEE Trans. Ind. Electron.* **2019**, *66*, 4629–4638.
31. Chen, X.; Li, Y.; Ma, H.; Tang, H.; Xie, Y. A novel variable exponential discrete time sliding mode reaching law. *IEEE Trans. Circuits Syst. II Express Briefs* **2021**, *68*, 2518–2522.
32. Ren, Q.; Kao, Y.; Zhu, Y.; Xia, H.; Wang, C. A new discrete reaching condition and generalized discrete reaching law with different convergence rates. *Automatica* **2021**, *132*.
33. Zheng, L.; Jiang, F.; Song, J.; Gao, Y.; Tian, M. A Discrete-Time Repetitive Sliding Mode Control for Voltage Source Inverters. *IEEE J. Emerg. Sel. Top. Power Electron.* **2018**, *6*, 1553–1566.
34. Toujeni, N.; Mnasri, C.M.; Gasmi, M. Repetitive control based on integral sliding mode control of matched uncertain systems. *Int. J. Adv. Comput. Sci. Appl.* **2018**, *9*, 446–455.
35. Kurniawan, E.; Wang, H.; Sirenden, B.H.; Prakosa, J.A.; Adinanta, H.; Suryadi, S. Discrete-time modified repetitive sliding mode control for uncertain linear systems. *Int. J. Adapt. Control Signal Process.* **2021**, *35*, 2245–2258.
36. Jia, C.; Longman, R.W. An adaptive smooth second-order sliding mode repetitive control method with application to a fast periodic stamping system. *Syst. Control Lett.* **2021**, *151*, 104912.
37. Kurniawan, E.; Pratiwi, E.B.; Harno, H.G.; Prakosa, J.A.; Purwowibowo, P.; Adinanta, H.; Widiyatmoko, B.; Ula, R.K.; Hanto, D. Low-order sliding mode repetitive control for uncertain linear systems perturbed by band-limited periodic disturbances. *Proc. Inst. Mech. Eng. Part I J. Syst. Control Eng.* **2023**, *237*, 1558–1572.
38. Lu, Y.S.; Wu, B.X.; Lien, S.F. An improved sliding-mode repetitive learning control scheme using wavelet transform. *Asian J. Control* **2012**, *14*, 991–1001.
39. Laakso, T.; Valimaki, V. Splitting the Unit Delay. *IEEE Signal Process. Mag.* **1996**, *13*, 30–60.
40. Thiran, J.P. Recursive digital filters with maximally flat group delay. *IEEE Trans. Circuit Theory* **1971**, *18*, 659–664.
41. Chen, S.; Zhao, Q.; Ye, Y.; Qu, B. Using IIR Filter in Fractional Order Phase Lead Compensation PIMR-RC for Grid-Tied Inverters. *IEEE Trans. Ind. Electron.* **2023**, *70*, 9399–9409.
42. Kurniawan, E.; Prakosa, J.A.; Wang, H.; Wijonarko, S.; Maftukhah, T.; Purwowibowo, P.; Septanto, H.; Pratiwi, E.B.; Rustandi, D. Design of Fractional Order Odd-Harmonics Repetitive Controller for Discrete-Time Linear Systems with Experimental Validations. *Sensors* **2022**, *22*, 8873. [CrossRef]
43. Tomizuka, M.; Tsao, T.-C.; Chew, K.-K. Analysis and Synthesis of Discrete-Time Repetitive Controllers. *J. Dyn. Syst. Meas. Control Trans. ASME* **1989**, *111*, 353–358.
44. Longman, R.W. On the theory and design of linear repetitive control systems. *Eur. J. Control* **2010**, *16*, 447–496.
45. Sun, M.; Wang, Y.; Wang, D. Variable-structure repetitive control: A discrete-time strategy. *IEEE Trans. Ind. Electron.* **2005**, *52*, 610–616. [CrossRef]
46. Furuta, K. VSS type self-tuning control. *IEEE Trans. Ind. Electron.* **1993**, *40*, 37–44. [CrossRef]
47. Kucera, V. Minimum variance control: A homage to peterka. *Int. J. Adapt. Control Singnal Process.* **1999**, *13*, 433–449.

fractal and fractional

MDPI

Article

Analysis of Error-Based Switched Fractional-Order Adaptive Systems: An Error Model Approach

Norelys Aguila-Camacho [1,*], Javier Alejandro Gallegos [2], Yangquan Chen [3] and Juan Carlos Travieso-Torres [4]

1 Department of Electricity, Universidad Tecnológica Metropolitana, Av. José Pedro Alessandri 1242, Santiago 7800002, Chile
2 Department of Electrical Engineering, Pontificia Universidad Católica de Chile, Av. Vicuña Mackenna 4860, Santiago 7820436, Chile; javier.gallegos@uc.cl
3 School of Engineering, University of California Merced, 5200 Lake Rd, Merced, CA 95343, USA; ychen53@ucmerced.edu
4 Department of Industrial Technologies, Universidad de Santiago de Chile, El Belloto 3735, Santiago 9170125, Chile; juancarlos.travieso@usach.cl
* Correspondence: norelys.aguila@utem.cl

Abstract: Switched adaptive laws for parameter estimation have been proposed in recent years to improve the balance between control energy and system performance in adaptive schemes, which is often a big issue when using traditional integer-order or fractional-order adaptive laws in adaptive identification and control. These switched adaptive laws are represented as fractional-order differential equations whose order can switch between a number within the range $(0, 1)$ and 1. However, a general analytical framework that allows proving the boundedness of the solutions and convergence of the estimation/tracking error is not yet available, with only particular analyses for specific schemes being accessible. This paper address this issue, presenting the analysis of four error models that can appear in the field of adaptive systems when these adaptive laws are chosen. The boundedness of the solutions is proved for all cases, together with the convergence to zero of the estimation/tracking error. Additionally, sufficient conditions for parameter convergence are presented, showing that the excitation condition required for parameter convergence in the vector case is also sufficient for parameter estimation in the matrix case. A numerical example is included to show the possible advantages of using switched adaptive laws in a Model Reference Adaptive Control application. Results show that controller parameters can be found for the switched controller, enabling us to obtain an overall improvement of 7.75% with respect to the non-switched integer-order controller and 34.6% with respect to the non-switched fractional-order controller.

Keywords: fractional calculus; switched fractional-order adaptive system; switched fractional-order error models

1. Introduction

The main goal in developing adaptive methods is to address problems such as estimation or tracking in the presence of parametric uncertainties [1]. Many different adaptive schemes have been proposed and improved over the years, such as Model Reference Adaptive Control (MRAC) [2], Self-Tuning Regulators [3], Adaptive Pole Placement Control [4], Adaptive Passivity-Based Control [5], and Adaptive Observers [2], among others. Applications span process control, automotive systems, positioning systems, propulsion systems, and a great deal of effort in flight control (see [6] and the references therein).

Technically speaking, the term *adaptive system* has been adopted to refer to the wide set of feedback systems involving estimation and control, regardless of the specific adaptive scheme used. For analysis purposes, adaptive systems are commonly represented in the form of differential and algebraic equations, referred to as *error models*, which contain two different types of errors. The first error (e) represents a tracking or estimation error, and the second error (ϕ) represents a parameter estimation error.

The parameter error ϕ is often stated as the difference between a designed parameter θ and a vector θ^* of the form $\phi = \theta - \theta^*$ [7]. This vector θ^* is any vector such that if $\theta = \theta^*$, then the underlying problem (e.g., identification, estimation, control, optimization, etc.) is solved. Thereby, it is referred to as an ideal vector, which could be not unique. Due to uncertainties, θ^* is unknown and the adaptive problem is to tune θ to solve the underlying problem. This tuning can be represented by a differential equation or inclusion, called adaptive law, which uses the available signal e, providing a measure of the disagreement in the solution of the underlying problem. Since θ^* is usually constant, the time derivative of the parameter error ϕ coincides with the time derivative of the parameter θ, leading to a mathematical representation of the dynamics of ϕ. Some examples of measurable functions e are the tracking error, which is the control (or output) error in MRAC schemes, defined as the difference between the plant output and the output of the reference model ($e = y_p - y_m$) [8]; also the output error in APBC controllers, defined as the difference between a desired reference output trajectory and the plant output ($e = y^* - y$) [5], among others. Some examples of estimation error are the output error in adaptive observers, defined as the difference between the observer's output and the plant output ($e = \hat{y} - y_p$) [2], the estimation error in an identification scheme [2], etc.

For continuous-time and standard-derivative dynamics, a general enough description of an error model can be given as follows [2]:

$$\dot{e} = f(e, \phi, \omega, t) \tag{1}$$

$$\dot{\phi} = g(e, \omega, t), \tag{2}$$

where ω contains measurable input–output data.

In the early 2000s, fractional order operators (FOs) [9] started to be introduced to model real systems, due to the link between these operators and certain physical phenomena such as damping coefficients in a variety of viscoelastic materials, diffusion, and electrical impedance of biological tissues [10]. From their definition, FOs inherently incorporate memory, which is useful when modeling systems whose current state depends on a history of past states. This includes, for instance, materials with viscoelastic properties and processes with hysteresis. Some other real processes can be better modeled using FOs, such as biological processes such as drug diffusion and cardiac rhythms, whose nature is irregular and memory-driven. The modeling of complex systems is also a very large field of research where FOs have been proved to be very useful [11]. The application of FOs to modeling real systems has been widely reported in the literature. Just to mention some of them, we can find fractional-order models for food science [12], signal/speech processing [13,14], biology [15–17], disease/infection rates [18–21], economics [22,23], batteries [24–26], twin rotor system using fractional-order Hammerstein models [27], energy production and consumption [28], etc. FOs have also been introduced in the control field, leading to fractional-order controllers with better performance than classic integer-order controllers. For instance, we can mention the fractional-order sliding mode control [29,30], iterative optimized fractional-order PID control [31], fractional-order iterative learning control [32], etc. The reader can check many survey papers detailing fractional-order controllers and the reported results in different areas, such as applications of fractional-order controllers to unmanned aerial and ground vehicles [33], to temperature control [34], to speed control of brushless direct current motor drives [35], to the control of Multiple-Input Multiple-Output (MIMO) processes [36], etc. Adaptive control has not been an exception, with a large number of works reporting the use of FOs in the adaptive controller/observer design, and/or to model the system to be controlled or identified (see, for instance, [37–39]). Within that, the analytical analysis of the resulting fractional-order adaptive schemes became relevant, leading to the adoption of the error model concept in a generalized way, e.g., fractional-order error models (FOEMs). The general representation of a FOEM is the same as in (1) and (2), but using fractional-orders derivatives for e, ϕ, instead of the classic integer-order derivatives.

The analysis of FOEMs remains challenging, mainly because of the unique properties of fractional operators and the limited availability of effective analytical tools. Still, significant advances have been made over the years, such as the analysis of the four more common FOEMs (e.g., FOEM1 to FOEM4) in [40], for cases where the fractional order is the same for differential equations of e and ϕ. Modifications of these FOEMs were introduced in [41], addressing the analysis of FOEM1 to FOEM4 in the presence of bounded disturbances and parameter variations. Later in [42], mixed-order cases were analyzed for FOEM2 and FOEM3, where the fractional order for differential equations of e and ϕ does not need to be the same, as long as the fractional order in the ϕ equation is lower or equal than the fractional order in the e equation. Although these advances have been made, there are still many open questions to the analysis of FOEMs, such as how to choose the fractional orders for the adaptive laws (which leads to the fractional order in the ϕ equation), how to prove the convergence to zero of the estimation or control error e for some cases, how to deal with FOEMs where all the components of differential equations for e,ϕ can have different fractional orders, etc.

A particular open problem has arisen recently due to the introduction of switched adaptive laws in adaptive schemes, aimed at improving system performance and controlling energy use and leading to a differential equation for ϕ (e.g., (2)) whose order can switch among a fractional-order α and the integer-order 1, according to some switching rule. Indeed, the problem of analyzing the boundedness and error convergence in these switched equations has only been addressed for particular adaptive schemes. For instance, in [43], an identification problem has been addressed, as well as a Switched Fractional-Order Model Reference Adaptive Control (SFOMRAC) scheme to control Linear Time Invariant (LTI) Single-Input Single-Output (SISO) plant. In this work, the switching rule in the adaptive law is time-based. This means that, at some time instant defined by the designer, the order changes from fractional to integer. Within this switched adaptive law, the authors reported that in the identification problem, the estimated parameters got closer and faster to the ideal unknown values and with lesser overshoot than in the non-switched cases. In the SFOMRAC scheme, switched adaptive laws kept system behavior as desired and simultaneously spent a lower amount of control energy. Recently, a couple of works have been published where the order of the adaptive law uses an error-based switching rule, that is, the order switches between a fractional order and the integer order according to the magnitude of the control error e. These two works correspond to a high-gain adaptive control scheme [44] and a SFOMRAC for Multiple-Input Multiple-Output (MIMO) systems [8]. In the former, the authors found that the switching strategy permits obtaining lower values for controller gains, resulting in improved robustness, while simultaneously guaranteeing fast convergence of the state to the origin, leading to an improved balance between system performance and control energy efficiency. In the latter, switching error levels were introduced into the MRAC framework, aimed at obtaining an improved balance between control energy and system performance, compared to the non-switched equivalent strategies. Although these works have provided analytical guarantees on the performance of the resulting switched schemes, the potential use of switched adaptive laws in different problems and applications leads to the need of a general framework to establish the boundedness of signals and convergence of errors.

This is precisely the main goal of this paper, where the error model approach is adopted to analyze adaptive schemes using switched fractional-order adaptive laws. The main contributions are stated in the following section.

- In the context of the error model approach and including the four error equations for e that appear more often in the study of control and estimation adaptive systems [2], this paper proposes differential equations for the parameter error ϕ that can switch between a fractional order and integer order, using an error-based switching mechanism.

This leads to the analysis of four switched fractional-order error models (SFOEMs), specifically SFOEM1 to SFOEM4. The only previous work addressing something within this scope is [8], where a SFOMRAC is proposed and analyzed (without proving parametric convergence), whose structure coincides with the SFOEM2 in this paper, constituting a particular case of this paper.

- A complete analytical proof of stability and convergence is provided for each of the presented SFOEMs, allowing its future application to any switched adaptive scheme that can be put in their form, which is the fundamental advantage of the error model approach.
- In contrast to all the revised literature, such as [2,3,40,45], where the error models are analyzed for the case where ϕ is a vector, the analysis in this paper is made considering ϕ as a matrix (multi-variable case) for three out of four error models. We found that the same excitation condition for the vector case is sufficient to estimate the matrix parameters. Overall, this is due to the excitation for each column vector of the parameter matrix being shared.

This paper is organized as follows. Section 2 contains some concepts, definitions, and analytical tools to be used in the analysis throughout this paper. Sections 3–6 present the Switched Fractional-Order Error Models 1 to 4 with their corresponding analytical analyses. Finally, Section 7 presents a numerical simulation to illustrate some of the results, while Section 8 presents the main conclusions of this work.

2. Basic Concepts

This section presents some concepts of fractional calculus and analytical tools that are used within this paper. In what follows, $\mathbb{R}_{>(\geq)0}$ denotes the set of positive (non-negative) real numbers, and $\mathbb{R}^n$ represents the Euclidean space of dimension n. For $x \in \mathbb{R}^n$ and $A \in \mathbb{R}^{n\times n}$, a positive-definite matrix $|x|_A := x^\top Ax$ defines a norm. In the case of matrices, $|B|_A := B^\top AB$ for $B \in \mathbb{R}^{n\times n}$. PE($n$) denotes the set of persistently exciting signals of order n [2].

2.1. Definitions from Fractional Calculus

Fractional calculus studies integrals and derivatives of orders that can be any real or complex numbers [9]. The Riemann–Liouville fractional integral is one of the main concepts of fractional calculus, and it is presented in Definition 1.

Definition 1 (Riemann–Liouville fractional integral [46]).

$$I_a^\alpha f(t) := \frac{1}{\Gamma(\alpha)}\int_a^t (t-\tau)^{\alpha-1} f(\tau)d\tau, \tag{3}$$

where $\alpha > 0$ and $\Gamma(\alpha) = \int_0^\infty t^{\alpha-1}e^{-t}dt$ is the Gamma function [9].

Several alternative definitions exist for fractional derivatives. Definition 2 represents the Caputo fractional derivative, which is commonly applied in engineering problems and is the one adopted in this paper.

Definition 2 (Caputo fractional derivative [46]).

$$D_a^\alpha f(t) := I_a^{m-\alpha} D^m f(t), \tag{4}$$

where $\alpha > 0$ and $m = \lceil \alpha \rceil$.

2.2. Analytical Tools

Property 1. *For any continuous function $f : [0, \infty) \to \mathbb{R}$, any real numbers $a, \alpha, \beta > 0$. Moreover, for any real number $t > 0$, the following holds [46]:*

$$I_a^{\alpha} I_a^{\beta} f(t) = I_a^{\alpha+\beta} f(t), \tag{5}$$

and

$$I_a^{\alpha} D_a^{\alpha} f(t) = f(t) - f(a). \tag{6}$$

The following result establishes the existence and uniqueness in the set of continuous functions for the solutions of a kind of fractional equations.

Theorem 1 ([47])**.** *Consider the following system of fractional-order integral equations:*

$$y(t) = p(t) + I^{\alpha}[f(., y(.))](t), \tag{7}$$

where α is a vector with components α_i, and y, f, and p are vectors of components y_i, f_i, and p_i, respectively, for $i = 1, \ldots, n$. If $p : [0, T] \to \mathbb{R}^n$ is a continuous function and $f_i(.,.)$ are continuous functions in their first variables and Lipschitz-continuous functions in their second variables for $i = 1, \ldots, n$, then

- i *There exists a unique continuous solution $y \in C[0, T]$ to System* (7)*.*
- ii *$y \in C[0, T]$ is a solution to System* (7) *for*

$$p_i(t) := \sum_{k=0}^{\lceil\alpha\rceil-1} \frac{t_k}{k!} y_{i_0}^{(k)} \tag{8}$$

if and only if each of its components y_i is a solution to $D^{\alpha_i} y_i = f_i(t, y)$ with an initial condition $y_i^{(k)}(0) = y_{i_0}^{(k)}$ for $k = 1, \ldots, \lceil\alpha\rceil - 1$ and $i = 1, \ldots, n$.

Finally, a useful property for the derivative of composite functions is presented.

Theorem 2 ([48])**.** *For a given $x \in \mathbb{R}^d$, let $u \in \{x\} + I_{0+}^{\alpha} C([0, T]), \mathbb{R}^d$, and $V : \mathbb{R}^d \to \mathbb{R}$ satisfies the following conditions:*

- *Function V is convex on $\mathbb{R}^d$ and $V(0) = 0$.*
- *Function V is differentiable in $\mathbb{R}^d$.*

Then, the following inequality holds for all $t \in [0, T]$:

$$D_{0+}^{\alpha} V(u(t)) \leq \langle \nabla V(u(t)), D_{0+}^{\alpha} u(t) \rangle. \tag{9}$$

Theorem 1 will be used to prove the continuity of the solutions of the switched fractional-order error models addressed in Sections 3–6. This will allow us to use Theorem 2 in the proof of Theorem 3 (which is also valid for the proof of Theorem 6) and Theorem 5 (which includes Theorem 4 as a special case. Property 1 is also used in these proofs.

3. Analysis of Switched Fractional-Order Error Model 1

Many problems in adaptive estimation and control may be expressed as

$$y(t) = \theta^{*T} \omega(t), \tag{10}$$

where $\theta^* \in \mathbb{R}^{m \times n}$ represents an unknown parameter matrix, $\omega \in \mathbb{R}^m$ is a measurable C^1-function called a regressor, and $y \in \mathbb{R}^n$ is a measurable signal, called *plant output* hereafter, which can be determined at each time instant [6]. For instance, when designing adaptive observers for linear plants, the plant output can be represented as an algebraic combination of filtered inputs and outputs [2]. In addition, the plant model used in some combined/composite approaches to adaptive control can be written in this form in (10), as reported in [45,49].

In these problems, an estimator for plant output is usually formulated in the form $\hat{y}(t) = \theta^T(t)\omega(t)$, with $\theta \in \mathbb{R}^{m\times n}$ being the estimated parameter matrix. If the estimation error is defined as $e = \hat{y} - y$ and the parameter error as $\phi = \theta - \theta^*$, we can easily derive that the relationship between e and ϕ can be stated as

$$e(t) = \phi^T(t)\omega(t). \tag{11}$$

The most common approach for adjusting the estimate θ at each time t is to use an adaptive law that is obtained using the gradient-descent approach to minimize e^Te [2,50], leading to an equation for the parameter error ϕ as

$$\dot{\phi}(t) = -\gamma\omega(t)e^T(t), \tag{12}$$

with $\gamma \in \mathbb{R}^+$ being the adaptive gain. Equations (11) and (12) constitute the classic Error Model 1, which is completely studied in ([2], Chapter 7) for the case where $e \in \mathbb{R}$. The case where $e \in \mathbb{R}^n$ is briefly analyzed in ([2], Chapter 10). Fractional-Order Error Model 1 (FOEM1) uses a fractional-order derivative instead of the integer one in Equation (12), whose analysis for the scalar case $e \in \mathbb{R}$ can be found in [40].

In this paper, a switched derivation order approach is used to adjust the estimated parameter θ, resulting in a different error dynamics, called Switched Fractional-Order Error Model 1 (SFOEM1) hereafter, which has not been previously analyzed. The definition of SFOEM1 corresponds to

$$e(t) = \phi^T(t)\omega(t), \quad e(0) = e_0 \tag{13}$$

$$D_{a(t)}^{\alpha(t)}\phi(t) = -\gamma(t)\omega(t)e^T(t), \quad \phi(0) = \phi_0 \tag{14}$$

$$\gamma(t) = \frac{\gamma_i}{1 + tr[\theta(t_i)\theta^T(t_i)]}, \quad \gamma(0) = \gamma_0. \tag{15}$$

This yields

$$D_{a(t)}^{\alpha(t)}\phi(t) = -\gamma(t)\omega(t)\omega^T(t)\phi(t). \tag{16}$$

The adaptive gain γ, aimed to manipulate the speed of (14) and to normalize, takes a constant value $\gamma_i/(1 + tr[\theta(t_i)\theta^T(t_i)])$ in each switching interval $t \in [t_i, t_{i+1})$, with $\{\gamma_i\}_{i\in\mathbb{N}}$ being a designer-chosen sequence of real numbers, such that $0 < \gamma_l < \gamma_i$ for any i.

The fractional-order $\alpha(t)$ is varied using the following switched strategy for any $t > 0$:

$$\alpha(t) = \begin{cases} \alpha_0 & if \quad \|e(t)\| > \epsilon \quad \text{and} \quad h(t) \\ 1 & if \qquad \text{otherwise} \end{cases} \tag{17}$$

where $\epsilon > 0$ and $\alpha_0 \in (0,1)$ are designer choices. Function $h(\cdot)$ is a logic function (i.e., taking true/false values) defined for any $t \in [t_i, t_{i+1})$ as

$$h(t) = \begin{cases} \text{False} \quad if \quad (\ i > C, \quad \|e(t)\| < \epsilon + \delta \quad \text{and} \quad \mathcal{T}(t) \le i \cdot \delta\ ) \\ \text{True} \quad if \qquad \text{otherwise} \end{cases} \tag{18}$$

where $C \ge 1$ is a designer constant, and $\mathcal{T}$ is the largest interval of time in which the fractional mode is active at time t, i.e.,

$$\mathcal{T}(t) := \max_j\{|t_{j+1} - t_j| : t_{j+1} \le t \ \ \& \ \ \alpha(t_j) = \alpha_0\}. \tag{19}$$

The definition of the function h encodes, on the one hand, a hysteresis mechanism to avoid Zeno solutions. This is accomplished by fixing a sufficiently small $\delta > 0$, so the switching from the integer mode to the fractional mode occurs only when $\|e(t_i)\| = \epsilon + \delta$, while the switching from the fractional mode to the integer mode will occur only when $\|e(t_j)\| = \epsilon$. On the other hand, the function h also encodes a mechanism to promote that, in disturbed or transient stages ($\|e\| > \epsilon$), the adaptation with α_0 should occur. Conversely, it promotes the situation where staying close to the aim ($\|e\| < \epsilon$) calls for a lesser need of switching to α_0. To this aim, Condition $\mathcal{T}(t) \leq i \cdot \delta$ means that the transitions to the fractional mode remain active as long as the fractional mode is needed, as measured in terms of time quantification $\mathcal{T}$ and relative to a measure of the overall time given by $i \cdot \delta$, where we use the same δ for simplicity; however, it can be scaled differently. The use of a sufficiently large constant C in (18) ensures that both mechanisms previously explained are triggered after a finite number C of switches, allowing the transient to be as unaffected as possible.

The differential Equation (14) is understood in the resetting mode; that is, every time a switch occurs, the initial time a of the fractional operator is set equal to the switching time t_i. This defines the initial time function $a = a(t)$ in (14). Moreover, every time a switch occurs, $\theta(t_i) = \theta(t_i^-)$, which implies that $\phi(t_i) = \phi(t_i^-)$. Thus, discontinuities are excluded.

Boundedness of the Signals and Convergence of the Estimation Error in SFOEM1

The following result states the boundedness of the signals and convergence of the estimation error e in SFOEM1.

Theorem 3. *Consider SFOEM1* (13) *and* (14) *specified by* (15) *and* (17)–(19)*. Then, θ remains bounded, and $e \in \mathcal{L}^2$. If ω is additionally assumed to be a bounded function, then the error e is bounded. If $\dot{\omega}$ is also assumed to be bounded, then the error e converges asymptotically to zero, i.e.,* $\lim_{t\to\infty} e(t) = 0$*. Moreover, if $\omega \in \mathrm{PE}(m)$, then* $\lim_{t\to\infty} \phi(t) = 0$*, i.e.,* $\lim_{t\to\infty} \theta(t) = \theta^*$.

Proof of Theorem 3. This proof relies on proving several claims as in the following.

i. *There exists $T_i \in \mathbb{R}_{\geq 0} \cup \{+\infty\}$, such that the existence and uniqueness of continuous solutions holds on $[t_i, t_i + T_i)$, where t_i is any switching time.*

Proof of Claim i. Since $\omega \in \mathcal{C}^1$ and γ is constant between switches times, the right-hand side of (16) is locally Lipschitz-continuous as a function of ϕ. Then, we can find a small enough T_i, such that the right-hand side is Lipschitz-continuous with respect to ϕ and continuous with respect to t when restricted to $\mathcal{C}[t_i, t_i + T]$. This enable us to apply Theorem 1 on the space $\mathcal{C}[t_i, t_i + T]$, by writing (16) in the form (7). It also allows us to conclude that there is a unique continuous solution ϕ on $[t_i, t_i + T_i)$. Since ω is also continuous, by (13), the estimation error e is also a unique continuous function on $[t_i, t_i + T_i)$. □

ii. *The fractional mode, i.e., $\alpha = \alpha_0 < 1$, can remain activated only on time intervals of finite lengths.*

Proof of Claim ii. Via Part (i), we can consider that a solution to (16) exists, and that it is unique and continuous at least in a small neighborhood $[t_i, t_i + T_i)$ after the switching time. Let us consider that the fractional mode is active in the interval $[t_i, t_i + T_i)$ with $T_i \in \mathbb{R}_{\geq 0} \cup \{+\infty\}$. Note that, since (16) operates in the resetting mode, the fractional derivative must be started in each t_i. For notation convenience, we set $I_i^\alpha = I_{t_i}^\alpha$ and $D_i^\alpha = D_{t_i}^\alpha$.

Since the solution of (16) is continuous in $[t_i, t_i + T_i)$ and γ remains constant in each switching interval, we can apply Theorem 2 to the function $tr[\phi^T \phi]$ and Property 6. Then, by using (16), it follows that

$$\frac{1}{\gamma} D_i^{\alpha_0} tr\left[\phi^T \phi\right] \leq \frac{1}{\gamma} 2tr\left[\phi^T D_i^{\alpha_0} \phi\right] = -2tr\left[ee^T\right]. \tag{20}$$

Applying Property 6 on (20), we obtain the following for any $t \in [t_i, t_i + T_i)$:

$$\frac{1}{\gamma}tr\left[\phi^T(t)\phi(t)\right] - \frac{1}{\gamma}tr\left[\phi^T(t_i)\phi(t_i)\right] \leq -2I_i^{\alpha_0}tr[ee^T](t). \quad (21)$$

From (21), since the right-hand side is always less than or equal to zero, it can be concluded that $tr\left[\phi^T(t)\phi(t)\right] \leq tr\left[\phi^T(t_i)\phi(t_i)\right], \forall t > t_i$. Note that this will also hold for intervals in the integer-order mode, because an expression similar to (21) can be established, using the integer-order integral on the right-hand side. By the resetting mode and the no jump condition in switching times on θ, for any time interval, it holds that $tr\left[\phi^T(t)\phi(t)\right] \leq tr\left[\phi^T(0)\phi(0)\right]$. Since $\phi(0)$ is bounded, it follows that $\phi(t)$ is bounded and $I_i^{\alpha_0}tr[ee^T](t)$ is bounded by a number C, depending only on the initial condition $\phi(0)$.

Now, we need to prove that there exists a time instant $\bar{t} < \infty$, $\bar{t} \geq t_i$ such that $\|e(\bar{t})\| = \epsilon$ so that the fractional mode is deactivated in finite time. If that were not the case and $\|e(t)\| > \epsilon$ for all $t \geq t_i$, then $I_i^{\alpha_0}tr[ee^T](t) \to \infty$, contradicting the fact stated in the previous paragraph, which states that $I_i^{\alpha_0}tr[ee^T](t) < C < \infty$ for any t. This proves Claim (ii).

□

iii. *There is no finite escape time in each mode of operation.*

Proof of Claim iii. To prove this claim, we will differentiate between the intervals in fractional mode and the intervals in integer mode.

When fractional mode is active, then the following candidate Lyapunov function can be proposed:

$$V = \frac{1}{\gamma}tr\left[\phi^T\phi\right]. \quad (22)$$

By applying the α_0 fractional-order derivative to (22), as well as using Theorem 2 and (16), it follows that

$$D_i^{\alpha_0}V = \frac{1}{\gamma}D_i^{\alpha_0}tr\left[\phi^T\phi\right] \leq \frac{1}{\gamma}2tr\left[\phi^T D_i^{\alpha_0}\phi\right] = -2tr\left[ee^T\right]. \quad (23)$$

From (23), it can be seen that $D_i^{\alpha_0}V \leq 0$ for all $t \in [t_i, t_i + 1]$. Therefore, using ([48], Theorem 3), it can be concluded that ϕ remains bounded. Thus, there is no finite-time escape in fractional mode.

The proof in integer-order mode is similar, using the same candidate Lyapunov function (22) and taking its first-order derivative, leading to $\dot{V} = -2tr\left[ee^T\right] \leq 0$. This proves the claim. □

iv. *There exists a finite number of switches, after which the mode becomes integer. In particular, there is no Zeno solution.*

Proof of Claim iv. It follows from Part (ii) that if the number of switches is finite, then the final mode is necessarily integer. Also, if the number of switches is finite, then there is no Zeno solution. Thus, we still need to prove that the number of switches is finite.

To this aim, it is enough to prove that $\mathcal{T}$ (19) is bounded by a constant that does not depend on i, because due to the choice of the hysteresis function h (18), this would imply that there exists a large enough i such that $\mathcal{T}(t) \leq i\delta$ for any $t > 0$, making h false after finite switches. Consequently, this would deactivate the fractional mode.

To prove that $\mathcal{T}$ is bounded, let us use (21), which holds for every time interval when the fractional-order mode is active. Since in fractional mode $\|e\| > \epsilon$, then we can state the following from (21):

$$tr\left[\phi^T(t)\phi(t)\right] - tr\left[\phi^T(t_i)\phi(t_i)\right] \leq -2\gamma I_i^{\alpha_0}\epsilon^2 = \frac{-2\gamma\epsilon^2}{\alpha_0\Gamma(\alpha_0)}(t - t_i)^{\alpha_0}. \quad (24)$$

Since $\mathcal{T}$ is the time the fractional mode remains active—with it already being proved in (ii) that for every time interval, it holds that $tr\left[\phi^T(t)\phi(t)\right] \leq tr\left[\phi^T(0)\phi(0)\right]$—it can be stated from (24) that

$$\mathcal{T}^{\alpha_0} \leq \frac{\alpha_0\Gamma(\alpha_0)}{2\gamma_l\epsilon^2} tr\left[\phi^T(0)\phi(0)\right]. \tag{25}$$

Inequality (25) establishes an upper bound for $\mathcal{T}$ that does not depend on the number of switches i, and the claim follows. □

v. *Statement of Theorem 3 holds.*

Proof of Claim v. From Claims (iii) and (iv), and since ϕ remains bounded by a bound independent of i, it is enough to prove the statement for the integer mode. As in the proof of item (ii), with V as in (22), one obtains $\dot{V} = -2tr\left[ee^T\right] \leq 0$ for the integer mode. This implies $\phi \in \mathcal{L}_\infty$ and, by integrating this inequality, $e \in \mathcal{L}_2$ since $tr\left[ee^T\right] = |e|^2$. If $\omega \in \mathcal{L}_\infty$, then $e \in \mathcal{L}_\infty$; hence $\dot{\phi} \in \mathcal{L}_\infty$. Therefore, if $\dot{\omega} \in \mathcal{L}_\infty$, then $\dot{e} \in \mathcal{L}_\infty$. Applying the Barbalat Lemma, we get $\lim_{t\to\infty} e(t) = 0$. On the other hand, according to (16), the vector obtained from the i-th column of ϕ, denoted as $\phi^{(i)}$, is satisfactory for $i = 1, \ldots, n$

$$\dot{\phi}^{(i)} = \gamma\omega\omega^T\phi^{(i)}.$$

So, according to ([2], Theorem 2.16) and ([2], Comment 2.10), if $\omega \in \mathrm{PE}(m)$, then $\lim_{t\to\infty} \phi(t) = 0$. □

This completes the proof of Theorem 3. □

4. Analysis of Switched Fractional-Order Error Model 2

Switched Fractional-Order Error Model 2 (SFOEM2) arises when the system to be controlled or identified has all the elements in its state vector accessible, and the adaptive laws use a switched approach. In contrast to SFOEM1, where the relationship between $\phi^T\omega$ and e is algebraic, it is dynamic in SFOEM2.

The structure of the SFOEM2 corresponds to

$$\dot{e}(t) = A_m e(t) + B_m\phi^T(t)\omega(t), \quad e(0) = e_0. \tag{26}$$

$$D_{a(t)}^{\alpha(t)}\phi(t) = -\gamma(t)B_m P e(t)\omega^T, \quad \phi(0) = \phi_0 \tag{27}$$

$$\gamma(t) = \frac{\gamma_i}{1 + tr[\theta(t_i)\theta^T(t_i)]}, \quad \gamma(0) = \gamma_0 \tag{28}$$

$$\alpha(t) = \begin{cases} \alpha_0 & if \quad \|e(t)\| > \epsilon \quad \text{and} \quad h(t) \\ 1 & if \qquad \text{othewise} \end{cases} \tag{29}$$

$$h(t) = \begin{cases} \text{False} & if \quad (\ i > C, \quad \|e(t)\| < \epsilon + \delta \quad \text{and} \quad \mathcal{T}(t) \leq i \cdot \delta\ ) \\ \text{True} & if \qquad \text{otherwise} \end{cases} \tag{30}$$

$$\mathcal{T}(t) := \max_j\{|t_{j+1} - t_j| : t_{j+1} \leq t \quad \& \quad \alpha(t_j) = \alpha_0\}. \tag{31}$$

In Equations (26) and (27), $e \in \mathbb{R}^n$ is the tracking/estimation error, assumed to be accessible. $A_m \in \mathbb{R}^{n\times n}$ is a known Hurwitz matrix, e.g., a square matrix whose eigenvalues have negative real parts. $B_m \in \mathbb{R}^{n\times q}$ is a constant matrix, either positive- or negative-definite, the sign of which is assumed known and, without a loss of generality, taken to be positive. $\phi \in \mathbb{R}^{m\times q}$ is the parameter error, given by $\phi(t) = \theta(t) - \theta^*$, with $\theta^* \in \mathbb{R}^{m\times q}$ being the ideal (true) parameter, which is assumed to be unknown. In addition, $\theta \in \mathbb{R}^{m\times q}$ is the adjustable parameter that estimates θ^*. $\omega \in \mathbb{R}^m$ is a measurable $\mathcal{C}^1$ function. $P \in \mathbb{R}^{n\times n}$ is a symmetric positive-definite matrix, such that $A_m^T P + PA_m = -Q$, with $Q \in \mathbb{R}^{n\times n}$ being an arbitrary positive-definite matrix. The existence of such P is ensured by the Hurwitz property of A_m. Structure and parameters of (28)–(31) are the same as explained in SFOEM1 and are thus not repeated here.

As in the case of SFOEM1, the differential Equation (27) is understood in the resetting mode; that is, every time a switch occurs, the initial time a of the fractional operator is set equal to the switching time t_i. Also, every time a switch occurs, $\phi(t_i) = \phi(t_i^-)$ to avoid discontinuities.

Boundedness of the Signals and Convergence of the Estimation Error in SFOEM2

In [8], a particular adaptive control problem was addressed, namely a Switched Fractional-Order Model Reference Adaptive Controller for Multiple-Input Multiple-Output systems, where the plant and the reference model were described by integer-order differential equations and the order of the adaptive laws switched between a fractional order and the integer order. The closed loop description of the adaptive system in [8] is identical to the structure of SFOEM2 (26)–(31), but the claim below is more general than ([8], Theorem 3), particularly in providing conditions for parametric convergence.

Theorem 4. *Consider the Switched Fractional-Order Error Model 2 defined by* (26) *and* (27)*, with* (28)–(31)*. Then, θ and e remain bounded, and $e \in \mathcal{L}^2$. If ω is also assumed to be a bounded function, then the error e asymptotically converges to zero, i.e., $\lim_{t\to\infty} e(t) = 0$. Moreover, if $\omega \in \mathrm{PE}(m)$, then $\lim_{t\to\infty} \phi(t) = 0$, i.e., $\lim_{t\to\infty} \theta(t) = \theta^*$.*

Proof of Theorem 4. The reader is referred to the proof of Theorem 5, as SFOEM2 is a particular case of SFOEM3 by taking $C_m = I$ in the latter. □

5. Analysis of Switched Fractional-Order Error Model 3

Switched Fractional-Order Error Model 3 (SFOEM3) has a very similar structure to SFOEM2. However, there is a fundamental difference between them, because while the whole estimation/tracking error e is accessible in SFOEM2, only an algebraic combination of the components of e, namely e_o, is accessible in SFOEM3. This error model typically applies when only the plant's output, rather than its full state vector, is accessible for control or identification. Consequently, SFOEM3 can be used in a broader range of problems compared to SFOEM2.

The structure of SFOEM3 has the following form:

$$\begin{array}{rcl} \dot{e}(t) &=& A_m e(t) + B_m \phi^T(t)\omega(t), \quad e(0) = e_0. \\ e_o(t) &=& C_m\, e(t) \end{array} \tag{32}$$

$$D_{a(t)}^{\alpha(t)} \phi(t) = -\gamma(t)\omega(t)e_o^T(t), \quad \phi(0) = \phi_0 \tag{33}$$

$$\gamma(t) = \frac{\gamma_i}{1 + tr[\theta(t_i)\theta^T(t_i)]}, \quad \gamma(0) = \gamma_0 \tag{34}$$

$$\alpha(t) = \begin{cases} \alpha_0 \;\; if \;\; \|e_o(t)\| > \epsilon \quad \text{and} \quad h(t) \\ 1 \quad if \qquad\qquad \text{othewise} \end{cases} \tag{35}$$

$$h(t) = \begin{cases} \text{False} \;\; if \;\; (\; i > C, \;\; \|e_o(t)\| < \epsilon + \delta \;\; \text{and} \;\; \mathcal{T}(t) \le i \cdot \delta \;) \\ \text{True} \;\; if \qquad\qquad \text{otherwise} \end{cases} \tag{36}$$

$$\mathcal{T}(t) := \max_j \{|t_{j+1} - t_j| : t_{j+1} \le t \;\; \& \;\; \alpha(t_j) = \alpha_0\}. \tag{37}$$

In Equations (32) and (33), $e \in \mathbb{R}^n$ is the tracking/estimation error, which is not accessible in this case, while $e_o \in \mathbb{R}^q$ is the output error, which is accessible. $A_m \in \mathbb{R}^{n\times n}$ is a known Hurwitz matrix, $C_m \in \mathbb{R}^{q\times n}$ is a constant matrix, and $B_m \in \mathbb{R}^{n\times q}$ is a constant matrix, either positive- or negative-definite, the sign of which is assumed to be known and, without a loss of generality, taken to be positive. $\phi \in \mathbb{R}^{m\times q}$ is the parameter error, e.g., $\phi(t) = \theta(t) - \theta^*$, with $\theta^* \in \mathbb{R}^{m\times q}$ being the unknown parameter and $\theta \in \mathbb{R}^{m\times q}$ being the adjustable parameter. $\omega \in \mathbb{R}^m$ is the measurable $\mathcal{C}^1$ regressor function. On the other hand, $\gamma \in \mathbb{R}^+$ is an adaptive gain. The structure and parameters of (34)–(37) are the same as explained in SFOEM1 and are therefore not repeated here.

As stated in [2], it can be expected that having no access to e will imply that more stringent conditions must be imposed on the transfer function between $\phi^T(t)\omega(t)$ and $e_o(t)$. In that sense, it is assumed that positive-definite matrices $P = P^T \in \mathbb{R}^{n\times n}$ and $Q = Q^T \in \mathbb{R}^{n\times n}$ [2] exist, such that

$$\begin{aligned} A_m^T P + PA_m &= -Q \\ PB_m &= C_m^T, \end{aligned} \tag{38}$$

which is equivalent to inferring that the transfer function $C_m(sI - A_m)^{-1}B_m$ is strictly positive real.

Boundedness of the Signals and Convergence of the Estimation Error in SFOEM3

The following result states the boundedness of the signals and convergence of the tracking/estimation error e and the output error e_o in SFOEM3.

Theorem 5. *Consider the Switched Fractional-Order Error Model 3 defined by* (32) *and* (33)*, with* (34)–(37)*. Then, θ, e_o, and e remain bounded, and $e_o, e \in \mathcal{L}^2$. If ω is also assumed to be a bounded function, then the error e asymptotically converges to zero, i.e., $\lim_{t\to\infty} e(t) = 0$. Moreover, if $\omega \in \mathrm{PE}(m)$, then $\lim_{t\to\infty}\phi(t) = 0$, i.e., $\lim_{t\to\infty}\theta(t) = \theta^*$.*

Proof of Theorem 5. As in the case of SFOEM1, the proof consists of several claims, as it is detailed in the following:

i. *There exists $T_i \in \mathbb{R}_{\geq 0} \cup \{+\infty\}$ such that the existence and uniqueness of continuous solutions holds on $[t_i, t_i + T_i)$, where t_i is any switching time.*

Proof of Claim i. Since ω is $\mathcal{C}^1$, the right-hand side of (32) and (33) is locally Lipschitz-continuous as a function of e, ϕ, and we can find a small enough T_i, such that the right-hand side is Lipschitz-continuous with respect to e, ϕ and continuous with respect to t. Then, Theorem 1 can be applied to (32) and (33), allowing us to write them in the form (7) and allowing us to conclude that e, ϕ are unique and continuous in $[t_i, t_i + T]$. □

ii. *The fractional mode, i.e., $\alpha = \alpha_0 < 1$, can only be activated on time intervals of finite lengths.*

Proof of Claim ii. As in the proof of Claim ii for SFOEM1, let us analyze what happens in those intervals when the fractional mode is active, e.g., $[t_i, t_i + T_i)$ with $T_i \in \mathbb{R}_{\geq 0} \cup \{+\infty\}$. Since the solution of (32) and (33) was proven to be continuous, the right-hand sides of (32) and (33) are continuous too, and we can take the first-order derivative of $e^T Pe$ and use (32) together with trace property $tr[yx^T] = x^T y$ to establish that

$$\frac{d}{dt}e^T Pe = 2e^T P\dot{e} = e^T(A_m^T P + PA_m)e + 2e_o^T\phi^T\omega = -|e|_Q + 2tr\left[\phi^T\omega e_o^T\right]. \tag{39}$$

In the same way, we can take the α_0 derivative of $tr\left[\phi^T\phi\right]$ and use Theorem 2 together with (33) to establish that

$$\frac{1}{\gamma}D_i^{\alpha_0}tr\left[\phi^T\phi\right] \leq \frac{1}{\gamma}2tr\left[\phi^T D_i^{\alpha_0}\phi\right] = -2tr\left[\phi^T\omega e_o^T\right]. \tag{40}$$

Applying the first-order integral to (39) and the α_0-integral to (40), and using Properties 5 and 6, we obtain the following for any $t \in [t_i, t_i + T_i)$:

$$I_i^{1-\alpha_0}I_i^{\alpha_0}|e|_Q = |e(t_i)|_P - |e(t)|_P + I_i^{1-\alpha_0}I_i^{\alpha_0}2tr\left[\phi^T\omega e_o^T\right]. \tag{41}$$

$$I_i^{\alpha_0}2tr\left[\phi^T\omega e_0^T\right] \leq \frac{1}{\gamma}tr[|\phi(t_i)|] - \frac{1}{\gamma}tr[|\phi(t)|]. \tag{42}$$

Combining (41) and (42), the following inequality can be established:

$$I_i^{1-\alpha_0}\left(I_i^{\alpha_0}|e|_Q - \frac{1}{\gamma}tr[|\phi(t_i)|]\right) \leq |e(t_i)|_P. \tag{43}$$

Since $e_o = C_m e$, we claim that inequality (43) implies the existence of $\bar{t} < \infty, \bar{t} \geq t_i$, such that $\|e_o(\bar{t})\| = \epsilon$. By contradiction and recalling that we are in the fractional mode, if $\|e_o(t)\| > \epsilon$ for all $t \geq t_i$, then there exists $\bar{\epsilon} > 0$, such that $\|e(t)\| > \bar{\epsilon}$ for all $t \geq t_i$, since $\|e_o(t)\| \leq \|C_m\|\|e(t)\|$ by using the induced matrix norm for C_m. This implies that the left-hand side of (43) goes to $+\infty$ because the integrand of $I_i^{1-\alpha_0}$ goes to $+\infty$. This contradicts Inequality (43). Therefore, the existence of $\bar{t} < \infty$, such that $\|e_o(\bar{t})\| = \epsilon$ is guaranteed, which means that the integer mode is triggered some finite time after the fractional mode has started. This proves Claim (ii). □

iii. *There is no finite escape time in each mode of operation.*

Proof of Claim iii. In the integer-order mode, the claim can be easily proved by constructing a Lyapunov function $V = e^T Pe + tr\left[\frac{1}{\gamma_1}\phi^T\phi\right]$ and using (39) and (40) with $\alpha = 1$, leading to $\dot{V} \leq 0$. When the fractional-order mode is active, by Claim (ii), and after finite time, the switching condition $\|e_o(t_{i+1})\| = \epsilon$ must hold, implying that e_o cannot escape at $[t_i, t_{i+1}]$. Then, $e_o \in \mathcal{C}[t_i, t_{i+1}]$, and since $\omega \in \mathcal{C}^1$, the right-hand side of (33) is bounded at $[t_i, t_{i+1}]$. This implies, by α-integration, that ϕ is bounded $[t_i, t_{i+1}]$, i.e., ϕ does not escape. Then, the forcing function in (39) is bounded, and since A_m is stable, e remains bounded at $[t_i, t_{i+1}]$ and the claim is proved. □

iv. *There exists a finite number of switches, after which the mode becomes integer. In particular, there is no Zeno solution.*

Proof of Claim iv. According to Claim (ii), if the number of switches is finite, then the final mode is necessarily integer, and there is no Zeno solution. Then, to prove Claim (iv), it is enough to prove that the number of switches is finite. For this, as in the case of SFOEM1, it is enough to show that $\mathcal{T}$, given by (37), is bounded by a constant that does not depend on i.

As in the proof of Claim (ii), in the fractional mode, Condition $\|e_o\| \geq \epsilon$, the equivalence of norms in $\mathbb{R}^n$, and $e_o = C_m e$ imply the existence of a constant C_1, independent of t_i, such that $|e|_Q \geq C_1$ for all t when fractional-order mode is active. By using the normalization factor in (34) (recall that $\phi(t) = \theta(t) - \theta^*$) and the equivalence of norms in finite dimension spaces, there exists a constant C_2, independent of t_i, such that $\frac{1}{\gamma}tr[|\phi(t_i)|] \leq C_2$. Also, due to the choice of the hysteresis function h and the equivalence of norms, there exists constant $C_3 > 0$, independent of t_i, such that $|e(t_i)| \leq C_3(\epsilon + \delta)$. Using these constants

C_1, C_2, C_3 to bound (43) and solving the resulting integrals, it can be written that for all t when the fractional-order mode is active, it holds that

$$C_1(t - t_i) - \frac{C_2}{(1 - \alpha_0)\Gamma(1 - \alpha_0)}(t - t_i)^{1-\alpha_0} \leq C_3(\epsilon + \delta). \quad (44)$$

Inequality (44) establishes an upper bound for $\mathcal{T}$ (length of the time interval $t - t_i$ when the fractional-order mode is active) that does not depend on i, and the claim follows. □

v. *Statement of Theorem 5 holds.*

Proof of Claim v. From Claims (iii) and (iv), it is enough to prove the statement for integer mode. By considering the positive-definite function V as defined in the proof of Claim (iii), we get

$$\dot{V} = -|e|_Q. \quad (45)$$

This implies $\phi, e, e_o, \theta \in \mathcal{L}_\infty$ and $e, e_o \in \mathcal{L}_1$. By (32), if $\omega \in \mathcal{L}_\infty$, then we also have $\dot{e} \in \mathcal{L}_\infty$. Using the Barbalat Lemma, we conclude that $\lim_{t\to\infty} e(t) = 0$ and $\lim_{t\to\infty} e_o(t) = 0$. By (33), we also have $\lim_{t\to\infty} \dot{\phi}(t) = 0$. Hence, for any $T > 0$, $\phi(t+T) - \phi(t) = \int_t^{t+T} \dot{\phi} d\tau$ goes to zero as $t \to \infty$. This implies that in

$$\int_t^{t+T} |(\phi(\tau) - \phi(t))^T \omega(\tau)| \, d\tau \geq \int_t^{t+T} |\phi(t)^T \omega(\tau)| d\tau - \int_t^{t+T} |\phi(\tau)^T \omega(\tau)| d\tau, \quad (46)$$

the left-hand side goes to zero when $\omega \in \mathcal{L}_\infty$ and $t \to \infty$. By additionally assuming that $\omega \in \mathrm{PE}(m)$, which means that there exists $T_0, \epsilon_1 > 0$ such that $|\int_t^{t+T_0} u^T \omega(\tau) d\tau| > \epsilon_1 |u|$ for any $t \geq 0$ and $u \in \mathbb{R}^m - \{0\}$, we can find $T_0, \epsilon_2 > 0$ such that $|\int_t^{t+T_0} M^T \omega(\tau) d\tau| > \epsilon_2 |M|$ for any $t \geq 0$ and $M \in \mathbb{R}^{m\times m} - \{0\}$, simply by considering each component of vector $M^T\omega$ as the product of a column of M with ω. Therefore, if $\omega \in \mathrm{PE}(m)$, we obtain the following from (46)—for t that is large enough and $\epsilon_0 > 0$:

$$\int_t^{t+T_0} |\phi(\tau)^T \omega(\tau)| d\tau \geq \epsilon_0 |\phi(t)|. \quad (47)$$

On the other hand, integrating (32) on $[t, t+T]$ for arbitrary $T > 0$, we have

$$e(t+T) = e(t) + \int_t^{t+T} A_m e \, d\tau + \int_t^{t+T} B_m \phi^T \omega \, d\tau. \quad (48)$$

By sending $t \to \infty$, and since e goes to zero, it follows from (48) that

$$\int_t^{t+T} \phi^T \omega \, d\tau \to 0. \quad (49)$$

However, if $\omega \in \mathrm{PE}(m)$, a contradiction is obtained with (47) whenever ϕ does not converge to zero. Therefore, if $\omega \in \mathrm{PE}(m)$, then $\lim_{t\to\infty} \phi(t) = 0$ and the claim is proved. □

□

Remark 1. *In practical applications of Theorems 4 and 5, the bounded assumption on ω can be relaxed to ω being a bounded function of e and/or θ, in the sense that the boundedness of e and/or θ imply a bounded ω. For instance, in the adaptive control setup of [8], it holds that $\omega = e + \eta$, with η being a bounded signal. This is because the boundedness of e and θ was established regardless of the boundedness of ω in Theorem 5; hence, if e and/or ϕ are bounded, this implies that ω is bounded. Therefore, Theorem 5 guarantees the convergence of e to zero.*

6. Analysis of Switched Fractional-Order Error Model 4

SFOEM4 arises when neither vector $e(t)$ is accessible nor Condition (38) holds. In such a situation, for the particular case where the parameter error ϕ is a vector ($q = 1$) rather than a matrix, the solution consists of introducing an additional signal e_a into the adaptive scheme, as detailed below.

The structure of SFOEM4 is as follows:

$$e_o(t) = W(s)\phi^T(t)\omega(t) \tag{50}$$

$$e_a(t) = \theta^T(t)W(s)I_m\omega(t) - W(s)\theta^T(t)\omega(t) \tag{51}$$

$$\varepsilon(t) = e_o(t) + e_a(t) \tag{52}$$

$$D_{a(t)}^{\alpha(t)}\phi(t) = -\gamma(t)\ (W(s)I_m\omega(t))\varepsilon(t) \tag{53}$$

$$\gamma(t) = \frac{\gamma_i}{1+\theta^T(t_i)\theta(t_i)}, \quad \gamma(0) = \gamma_0 \tag{54}$$

$$\alpha(t) = \begin{cases} \alpha_0 & if \quad |\varepsilon(t)| > \epsilon \quad \text{and} \quad h(t) \\ 1 & if \qquad \text{othewise} \end{cases} \tag{55}$$

$$h(t) = \begin{cases} \text{False} & if \quad (\ i > C, \quad \|\varepsilon(t)| < \epsilon + \delta \quad \text{and} \quad \mathcal{T}(t) \leq i \cdot \delta\ ) \\ \text{True} & if \qquad \text{otherwise} \end{cases} \tag{56}$$

$$\mathcal{T}(t) := \max_j \{|t_{j+1} - t_j| : t_{j+1} \leq t \quad \& \quad \alpha(t_j) = \alpha_0\}. \tag{57}$$

In Equations (50)–(53), variable $e_o \in \mathbb{R}$ is the output error, $e_a(t) \in \mathbb{R}$ is an auxiliary error, $\varepsilon(t) \in \mathbb{R}$ is the augmented error, and $W(s)$ is the transfer function between $\phi^T\omega$ and e_o, an exponentially stable time-invariant linear filter that operates component-wise when applied to a vector or a matrix. $\phi \in \mathbb{R}^m$ is the parameter error, e.g., $\phi(t) = \theta(t) - \theta^*$, with $\theta^* \in \mathbb{R}^m$ the unknown parameter and $\theta \in \mathbb{R}^m$ the adjustable parameter. On the other hand $\omega \in \mathbb{R}^m$ is a measurable C^1 function, I_m is the identity matrix of order m, and $\gamma \in \mathbb{R}^+$ is adaptive gain. The structure and parameters of (54)–(57) are the same as explained in previous sections.

Boundedness of the Signals and Convergence of the Estimation Error in SFOEM4

The following result states the boundedness of the signals and the convergence of the output error e_o in SFOEM4.

Theorem 6. *Consider the Switched Fractional-Order Error Model 4 defined by* (50)–(53)*, with* (54)–(57)*. Then, θ remains bounded. If ω is also a bounded function, then the error e is bounded. Moreover, if $\dot{\omega}$ is also assumed to be bounded, then e converges asymptotically to zero, i.e.,* $\lim_{t\to\infty} e(t) = 0$*. Moreover, if $\omega \in \mathrm{PE}(m)$, then* $\lim_{t\to\infty}\phi(t) = 0$*, i.e.,* $\lim_{t\to\infty}\theta(t) = \theta^*$.

Proof of Theorem 6. If we recall that $\phi(t) = \theta(t) - \theta^*$, Equation (51) can be written as

$$\begin{aligned} e_a(t) &= \theta^T(t)W(s)I_m\omega(t) + \theta^{*T}W(s)I_m\omega(t) - \theta^{*T}W(s)I_m\omega(t) \\ &\quad - W(s)\theta^T(t)\omega(t) + W(s)\theta^{*T}\omega(t) - W(s)\theta^{*T}\omega(t) \\ &= \phi^T(t)W(s)I_m\omega(t) - W(s)\phi^T(t)\omega(t) + \theta^{*T}W(s)I_m\omega(t) - W(s)\theta^{*T}\omega(t). \end{aligned} \tag{58}$$

Since θ^* is a constant vector and W is an exponentially stable time-invariant linear filter, the term $\eta(t) = \theta^{*T} W(s) I_m \omega(t) - W(s)\theta^{*T}\omega(t)$ is exponentially decaying. By substituting (58) and (50) in (52) and using the notation $\xi = W(s)I_m\omega(t)$ in both (52) and (53), it follows that

$$\varepsilon(t) = \phi^T(t)\xi(t) + \eta(t) \tag{59}$$

$$D^{\alpha(t)}_{a(t)}\phi(t) = -\gamma(t)\,\varepsilon(t)\xi(t). \tag{60}$$

As can be observed, Equations (59) and (60) have the same structure as SFOEM1, which was studied in Section 3, for the case when ϕ is a vector rather than a matrix, but not for the term η. We show that the items of the proof of Theorem 3 still hold in this case, with the following key additional elements. Since W is a time-invariant linear filter, it maps $\mathcal{C}^1$ to $\mathcal{C}^1$ so that Claim (i) holds. Items (ii), (iii), and (iv) follow by using ([51], Theorem 1), recalling that the α-fractional integral of an exponentially decaying function like η is bounded and converges to zero when $\alpha < 1$. Finally, Item (v) follows by applying the result in ([2], p. 282). In this way, we have $\phi \in \mathcal{L}_\infty$. Moreover, if $\omega \in \mathcal{L}_\infty$, then $e_o \in \mathcal{L}_\infty$; if $\dot{\omega} \in \mathcal{L}_\infty$, then $\lim_{t\to\infty} e_o(t) = 0$; and if $\omega \in \mathrm{PE}(m)$, then $\lim_{t\to\infty}\phi(t) = 0$. To obtain these conclusions, note that the boundedness of $\omega, \dot{\omega}$ implies the boundedness of $\xi, \dot{\xi}$. Moreover, note that the boundedness and convergence to zero of ε implies the boundedness and convergence to zero of e_o and, consequently, of e. □

Remark 2. *To address the case where ϕ is a matrix ($q > 1$) rather than a vector, some difficulties arise and the complexity of the resulting adaptive scheme increases dramatically. This case is currently under research and will be addressed in a future work. Still, the particular case where $\phi \in \mathbb{R}^{m\times q}$ and the transfer matrix $W(s) \in \mathbb{R}^{q\times q}$ between $\phi^T\omega$ and $e_0 \in \mathbb{R}^q$ is diagonal can be solved in a way similar to the one stated above. Specifically, the problem can be decomposed as follows:*

$$e_{o,i}(t) = W_i(s)\phi_i^T(t)\omega(t) \tag{61}$$

$$e_{a,i}(t) = \theta_i^T(t)W_i(s)I_m\omega(t) - W_i(s)\theta_i^T(t)\omega(t) \tag{62}$$

$$\varepsilon_i(t) = e_{o,i}(t) + e_{a,i}(t), \tag{63}$$

where $W_i \in \mathbb{R}$ are the diagonal elements of $W(s)$, and ϕ_i is a column vector of matrix ϕ. The proof follows as in the case of Theorem 6.

7. Simulation Example

Although the main purpose of this paper is to offer a general theoretical framework to analyze the boundedness of signals and convergence of errors in switched adaptive schemes, it is worth pointing out some examples for which these results are useful. In that sense, the analytical results obtained in this paper could support MRAC schemes with switched adaptive laws such as, for example, in the case presented in [8]. In addition, they can be potentially useful if adaptive observers are proposed [2], but using switched adaptive laws. The implementation of APBC can also benefit from these results if switched adaptive laws are used in the scheme.

To present a numerical example, let us address the output voltage control in an Automatic Voltage Regulator (AVR) using Model Reference Adaptive Control. In [38], MRAC schemes were proposed for this system, using integer-order adaptive laws (IOMRAC) and using fractional-order adaptive laws (FOMRAC). The AVR model used in [38] corresponds to a fifth-order transfer function with a relative degree of 5, leading to an adaptive scheme whose structure coincides with the classic Error Model 4 in the IOMRAC case [2] and with the Fractional-Order Error Model 4 in the FOMRAC case [40].

As a numerical example, this section presents a MRAC controller for the same system but using switched adaptive laws (SFOMRAC). The IOMRAC and FOMRAC controllers proposed in [38] are used here as a reference for comparison. The structures of IOMRAC, FOMRAC, and SFOMRAC are presented in Table 1. As can be observed, the only difference between these controllers is in the adaptive laws used to estimate the controller parameters.

The design parameters $\gamma \in \mathbb{R}^{11}$ of the IOMRAC and $\alpha_0 \in \mathbb{R}^{11}, \gamma \in \mathbb{R}^{11}$ of the FOMRAC controllers presented in [38] were tuned using genetic algorithm optimization, resulting in different values for every component of these design parameter vectors. The reader can check ([38], Section 4) for details. For the SFOMRAC presented in this section, no optimization process was carried out, but three different sets of design parameters $\alpha_0, \gamma, \epsilon$ were used to illustrate different behaviors. In the first case, $\alpha_0 = 0.85, \gamma = 20, \epsilon = 0.15$ was used, while in the second case, $\alpha_0 = 0.6, \gamma = 0.2, \epsilon = 0.07$ was used. Note that these two cases use the same values for every component of the design parameter vectors $\alpha_0, \gamma, \epsilon$. This selection allows placing the SFOMRAC scheme in Table 1 in the form of SFOEM4 studied in this paper, so the boundedness of the signals and convergence of the control errors can be ensured from the analytical results. The third case simulated here uses the same values γ, α_0 as those of FOMRAC and IOMRAC reported in [38], with $\epsilon = 0.1$ for every component of the adaptive law estimating θ and with $\epsilon = 0.5$ for the adaptive law estimating k_1. Note that this third case cannot be placed in the same form as SFOEM4, but we have included it here to describe potential future directions. Simulations were performed in Matlab/Simulink, using the Ninteger Toolbox [52] to implement fractional-order operators.

Table 1. MRAC implementation details for the classic case [38], fractional-order case [38], and switched case.

Reference model	$G_m(s) = \dfrac{1.2}{s^3 + 5.2s^2 + 7s + 1.2}$		
Control law	$u(t) = \theta(t)^T \omega(t)$ $\theta^T(t) = \begin{bmatrix} k(t) & \theta_1^T(t) & \theta_0(t) & \theta_2^T(t) \end{bmatrix} \in \mathbb{R}^{10}$ $\omega(t) = \begin{bmatrix} r(t) & \omega_1^T(t) & y_p(t) & \omega_2^T(t) \end{bmatrix}^T \in \mathbb{R}^{10}$		
Auxiliary signals	$\dot{\omega}_1(t) = \Lambda \omega_1(t) + l u(t)$ $\dot{\omega}_2(t) = \Lambda \omega_2(t) + l y_p(t)$ $\Lambda = \begin{bmatrix} -1 & 0 & 0 & 0 \\ 0 & -2 & 0 & 0 \\ 0 & 0 & -3 & 0 \\ 0 & 0 & 0 & -4 \end{bmatrix}$ $l = [-1\ 1\ 3\ 4]^T$		
Errors	$e_o(t) = y_p(t) - y_m(t)$ $e_a(t) = \theta^T(t)\bar{\omega}(t) - \bar{u}(t)$ $\varepsilon(t) = e_o(t) + k_1(t) e_a(t)$ $\bar{u}(t) = G_m(s) u(t)$		
	SFOMRAC	FOMRAC	IFOMRAC
Adaptive law	$D^{\alpha(t)} k_1(t) = -\gamma \dfrac{\varepsilon(t) e_a(t)}{1 + \bar{\omega}(t)\bar{\omega}^T(t)}$ $D^{\alpha(t)} \theta(t) = -\gamma \dfrac{\varepsilon(t)\bar{\omega}(t)}{1 + \bar{\omega}(t)\bar{\omega}^T(t)}$ $\bar{\omega}(t) = G_m(s)\omega(t)$ $\alpha(t)$ as in (55), (56), (57)	$\alpha(t) = \alpha_0$	$\alpha(t) = 1$

Table 2 shows the resulting values of several performance indices used in [38] to tune the IOMRAC and FOMRAC controller parameters using optimization. These performance indices correspond to percent overshoot (OS), settling time (T_s), steady state error (E_{ss}), integral of the absolute value of the output error (IAE) as in (64), and the control energy

(ISI) as in (65). The sum of all performance indices J is also shown in Table 2 as an overall measure to evaluate the performance of the controllers.

$$IAE = \int_0^T |y_p(t) - y_m(t)|dt, \tag{64}$$

$$ISI = \int_0^T |u(t)|dt, \tag{65}$$

where T is the simulation time, y_p is the AVR output voltage, y_m is the reference model output, and u is the control signal.

Figure 1 shows the AVR output voltage and the control signal for every adaptive scheme in the upper row, which also allows us to display a visual reference of the performance indices in Table 2. The bottom row shows the evolution of IAE and ISI.

Table 2. Resulting performance indices for the simulated examples.

	IAE	ISI	OS	T_s	E_{ss}	J
IOMRAC	1.55	7.93	0	21.64	0	31.12
FOMRAC	2.82	7.45	0	33.67	0.01	43.95
SFOMRAC case 1	0.98	8.71	0	19.83	0	29.52
SFOMRAC case 2	8.89	6.95	7.39	73.24	0.03	96.5
SFOMRAC case 3	1.75	7.75	0	19.21	0	28.71

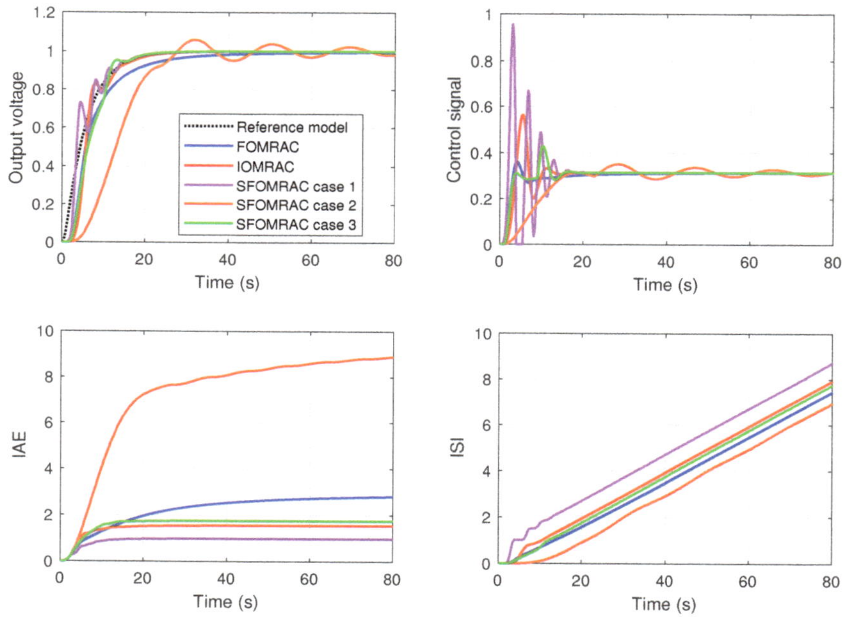

Figure 1. Controlled output, control signal, and performance indices IAE and ISI using IOMRAC [38], FOMRAC [38], and SFOMRAC controllers for the AVR system.

As can be observed from the upper plot of Figure 1, signals in the scheme remain bounded and the AVR output converges to the reference model output, which implies that the output error converges to zero. Since this is a control problem, the convergence of the estimated parameters to their ideal unknown values is not relevant; therefore, no analysis is made here in that regard.

As can be observed from the bottom plots, SFOMRAC in case 1 allows us to obtain a lower IAE than in the optimized IOMRAC and FOMRAC, but at the expense of a higher control energy. SFOMRAC in case 2, on the other hand, allows for an improvement in control energy to be generated but causes the IAE to deteriorate. However, SFOMRAC in case 3 allows us to obtain better balance between IAE and ISI than with non-switched controllers (green line is in-between the red and blue lines). This can also be observed from Table 2, where the values of ISI and IAE for SFOMRAC in case 3 are in-between the values of IOMRAC and FOMRAC. From Table 2, it can also be seen that SFOMRAC in cases 1 and 3 improves the settling time obtained in the non-switched cases. If we check the performance index J from Table 2, it can be observed that the switched schemes in cases 1 and 3 outperform the non-switched schemes, with case 3 being the best, with an improvement of 7. 75% with respect to the IOMRAC controller and an improvement of 34.6% with respect to the FOMRAC controller.

Even when there is no analytical proof available yet for case 3, when different values can be used for every component of SFOMRAC's design parameters $\alpha_0, \gamma, \epsilon$, these results show that the signals are bounded and the error converges to zero. Moreover, improved performance can be obtained with these additional degrees of freedom, and these cases serve as future research directions.

8. Conclusions

In this paper, the analysis of four switched fractional-order error models was presented, allowing us to prove the boundedness of solutions and convergence to zero of the estimation/tracking error in every adaptive scheme that could be placed in one of these forms. The results also state the conditions required to achieve parameter convergence in these schemes.

In contrast to previous works, the four switched error models were analyzed considering the parameter error ϕ as a matrix (multi-variable case) for three out of four error models. It was found that the same excitation condition for the vector case was sufficient to estimate the matrix parameters.

A simulation example included in this paper for an adaptive control problem shows that controller parameters can be found for the switched approach, such that improved overall performance is obtained, compared to the non-switched approaches.

Author Contributions: Conceptualization, N.A.-C. and J.A.G.; methodology, N.A.-C. and J.A.G.; formal analysis, N.A.-C. and J.A.G.; investigation, N.A.-C., J.A.G., J.C.T.-T. and Y.C.; resources, N.A.-C. and Y.C.; writing—original draft preparation, N.A.-C. and J.A.G.; writing—review and editing, N.A.-C., J.A.G., J.C.T.-T. and Y.C.; visualization, N.A.-C.; supervision, N.A.-C.; project administration, N.A.-C.; funding acquisition, N.A.-C., J.A.G., J.C.T.-T. and Y.C. All authors have read and agreed to the published version of the manuscript.

Funding: This research was funded by ANID CHILE under grant numbers Fondecyt 1220168, Fondecyt 3230674, FONDEF IT23I0117, and Basal Project AFB230001.

Data Availability Statement: The original contributions presented in the study are included in the article. The data analyzed in Section 7 is generated automatically at simulation level, using the proposed control strategy and the details included in the manuscript. Further inquiries can be directed to the corresponding author/s.

Conflicts of Interest: The authors declare no conflicts of interest. The funders had no role in the design of this study; in the collection, analyses, or interpretation of data; in the writing of this manuscript; or in the decision to publish the results.

References

1. Gaudio, J.E.; Gibson, T.E.; Annaswamy, A.M.; Bolender, M.A.; Lavretsky, E. Connections Between Adaptive Control and Optimization in Machine Learning. *arXiv* **2019**, arXiv:1904.05856.
2. Narendra, K.; Annaswamy, A. *Stable Adaptive Systems*; Dover Publications: Mineola, NY, USA, 2005.

3. Åström, K.J.; Wittenmark, B. *Adaptive Control*; Dover Publications Inc.: Mineola, NY, USA, 2008.
4. Sastry, K.; Bodson, M. *Adaptive Control: Stability, Convergence and Robustness*; Prentice Hall: West Nyack, NY, USA, 1989.
5. Duarte-Mermoud, M.A.; Castro-Linares, R.; Facuse-Castillo, A. Adaptive Passivity of Nonlinear Systems Using Time-Varying Gains. *Dyn. Control.* **2001**, *11*, 333–351. [CrossRef]
6. Annaswamy, A.M.; Fradkov, A.L. A historical perspective of adaptive control and learning. *Annu. Rev. Control.* **2021**, *52*, 18–41. [CrossRef]
7. Gaudio, J.; Annaswamy, A.; Bolender, M.; Lavretsky, E.; Gibson, T. A Class of High Order Tuners for Adaptive Systems. *IEEE Control. Syst. Lett.* **2021**, *5*, 391–3961. [CrossRef]
8. Aguila-Camacho, N.; Gallegos, J.A. Error-Based Switched Fractional Order Model Reference Adaptive Control for MIMO Linear Time Invariant Systems. *Fractal Fract.* **2024**, *8*, 109. [CrossRef]
9. Kilbas, A.A.; Srivastava, H.M.; Trujillo, J.J. *Theory and Applications of Fractional Differential Equations, Volume 204 (North-Holland Mathematics Studies)*; Elsevier Science Inc.: Amsterdam, The Netherlands, 2006.
10. Shastri, S.V.; Narendra, K.S. Fractional Order Derivatives in Systems Theory. In Proceedings of the American Control Conference, Denver, CO, USA, 1–3 July 2020; pp. 5076–5081.
11. West, B. *Fractional Calculus View of Complexity: Tomorrow's Science*; CRC Press: Boca Raton, FL, USA, 2016; pp. 1–274. [CrossRef]
12. Kaur, A.; Takhar, P.; Smith, D.; Mann, J.; Brashears, M. Fractional differential equations based modeling of microbial survival and growth curves: Model development and experimental validation. *J. Food Sci.* **2008**, *73*, 403–414. [CrossRef]
13. Ozaktas, H.M.; Zalevsky, Z.; Kutay, M.A. *The Fractional Fourier Transform: With Applications in Optics and Signal Processing*; John Wiley & Sons: Chichester, UK, 2001.
14. Assaleh, K.; Ahmad, W.M. Modeling of speech signals using fractional calculus. In Proceedings of the 9th International Symposium on Signal Processing and Its Applications, Sharjah, United Arab Emirates, 12–15 February 2007; pp. 1–4.
15. Dai, Z.; Peng, Y.; Mansy, H.; Sandler, R.; Royston, T. A model of lung parenchyma stress relaxation using fractional viscoelasticity. *Med. Eng. Phys.* **2015**, *37*, 752–758. [CrossRef]
16. Srivastava, H.; Dubey, V.; Kumar, R.; Singh, J.; Kumar, D.; Baleanu, D. An efficient computational approach for a fractional-order biological population model with carrying capacity. *Chaos Solitons Fractals* **2020**, *138*, 109880. [CrossRef]
17. Ugarte, J.P.; Tobón, C. Fractional-order modeling of myocardium structure effects on atrial fibrillation electrograms. *Math. Biosci.* **2024**, *378*, 109331. [CrossRef]
18. Arafa, A.A.M.; Rida, S.Z.; Khalil, M. A fractional-order model of HIV infection: Numerical solution and comparisons with data of patients. *Int. J. Biomath.* **2014**, *7*, 140036. [CrossRef]
19. Amilo, D.; Sadri, K.; Kaymakamzade, B.; Hincal, E. A mathematical model with fractional-order dynamics for the combined treatment of metastatic colorectal cancer. *Commun. Nonlinear Sci. Numer. Simul.* **2024**, *130*, 107756. [CrossRef]
20. Rahman, M.; Karaca, Y.; Agarwale, R.P.; David, S.A. Mathematical modelling with computational fractional order for the unfolding dynamics of the communicable diseases. *Appl. Math. Sci. Eng.* **2024**, *32*, 2300330. [CrossRef]
21. Caetano, C.; Morgado, L.; Lima, P.; Hens, N.; Nunes, B. A fractional order SIR model describing hesitancy to the COVID-19 vaccination. *Appl. Numer. Math.* **2025**, *207*, 608–620. [CrossRef]
22. Tarasov, V.E.; Tarasova, V.V. Long and Short Memory in Economics: Fractional-Order Difference and Differentiation. *arXiv* **2017**, arXiv:1612.07903. [CrossRef]
23. Johansyah, M.D.; Sambas, A.; Qureshi, S.; Zheng, S.; Abed-Elhameed, T.M.; Vaidyanathan, S.; Sulaiman, I.M. Investigation of the hyperchaos and control in the fractional order financial system with profit margin. *Partial. Differ. Equations Appl. Math.* **2024**, *9*, 100612. [CrossRef]
24. Guha, A.; Routh, B.; Prudhvi Velivela, N. Adaptive Extended Kalman Filtering Based State-of-Charge and Voltage Estimation of a Lithium-Ion Battery Using a Fractional-Order Model. *IEEE Sensors J.* **2024**, *24*, 26225–26234. [CrossRef]
25. Zhu, G.; Kong, C.; Wang, J.V.; Kang, J.; Yang, G.; Wang, Q. A fractional-order model of lithium-ion battery considering polarization in electrolyte and thermal effect. *Electrochim. Acta* **2023**, *438*, 141461. [CrossRef]
26. Yu, M.; Li, Y.; Podlubny, I.; Gong, F.; Sun, Y.; Zhang, Q.; Shang, Y.; Duan, B.; Zhang, C. Fractional-order modeling of lithium-ion batteries using additive noise assisted modeling and correlative information criterion. *J. Adv. Res.* **2020**, *25*, 49–56. [CrossRef]
27. Mok, R.; Ahmad, M.A. Smoothed Functional Algorithm with Norm-limited Update Vector for Identification of Continuous-time Fractional-order Hammerstein Models. *IETE J. Res.* **2024**, *70*, 1814–1832. [CrossRef]
28. Wang, Y.; Yang, Z.; Luo, Y.; Yang, R.; Sun, L.; Sapnken, F.E.; Narayanan, G. A novel structural adaptive Caputo fractional order derivative multivariate grey model and its application in China's energy production and consumption prediction. *Energy* **2024**, *312*, 133622. [CrossRef]
29. Pan, J. Fractional-order sliding mode control of manipulator combined with disturbance and state observer. *Robot. Auton. Syst.* **2025**, *183*, 104840. [CrossRef]
30. Chaudhary, K.S.; Kumar, N. Hybrid neural network-based fractional-order sliding mode controller for tracking control problem of reconfigurable robot manipulators using fast terminal type switching law. *Eng. Appl. Artif. Intell.* **2025**, *139*, 109515. [CrossRef]
31. Fan, H.; Wei, H.; Xu, D.; Liu, Y. Closed-Loop Iterative Optimized Fractional-Order PID Current Control of PMSM. *IEEE Trans. Ind. Inform.* **2024**, 1–10. [CrossRef]
32. Wei, W.; Shiwei, C.; Junfang, F. Fractional-order ILC-based adaptive integrated guidance and control of flight vehicles with system lag. *Nonlinear Dyn.* **2024**, *112*, 22039–22056. [CrossRef]

33. Cajo, R.; Mac, T.T.; Plaza, D.; Copot, C.; De Keyser, R.; Ionescu, C. A Survey on Fractional Order Control Techniques for Unmanned Aerial and Ground Vehicles. *IEEE Access* **2019**, *7*, 66864–66878. [CrossRef]
34. Jamil, A.A.; Tu, W.F.; Ali, S.W.; Terriche, Y.; Guerrero, J.M. Fractional-Order PID Controllers for Temperature Control: A Review. *Energies* **2022**, *15*, 3800. [CrossRef]
35. Kumarasamy, V.; Ramasamy, V.; Chandrasekaran, G.; Chinnaraj, G.; Sivalingam, P.; SanjeevKumar, N. A review of integer order PID and fractional order PID controllers using optimization techniques for speed control of brushless DC motor drive. *Int. J. Syst. Assur. Eng. Manag.* **2023**, *14*, 1139–1150. [CrossRef]
36. de Almeida, A.M.; Lenzi, M.K.; Lenzi, E.K. A Survey of Fractional Order Calculus Applications of Multiple-Input, Multiple-Output (MIMO) Process Control. *Fractal Fract.* **2020**, *4*, 22. [CrossRef]
37. Tejado, I.; HosseinNia, S.; Vinagre, B. Adaptive gain-order fractional control for network-based applications. *J. Fract. Calc. Appl. Anal.* **2014**, *17*, 462–482. [CrossRef]
38. Aguila-Camacho, N.; Duarte-mermoud, M.A. Fractional Adaptive Control for an Automatic Voltage Regulator. *ISA Trans.* **2013**, *52*, 807–815. [CrossRef]
39. Aburakhis, M.; Ordóñez, R. Generalization of Direct Adaptive Control Using Fractional Calculus Applied to Nonlinear Systems. *J. Control. Autom. Electr. Syst.* **2024**, *35*, 428–439. [CrossRef]
40. Aguila-Camacho, N.; Duarte-mermoud, M.A. Boundedness of the solutions for certain classes of fractional differential equations with application to adaptive systems. *ISA Trans.* **2016**, *60*, 82–88. [CrossRef]
41. Duarte-mermoud, M.A.; Aguila-Camacho, N. Fractional adaptive systems in the presence of bounded disturbances and parameter variations. *Int. J. Adapt. Control. Signal Process.* **2017**, *31*, 1273–1284. [CrossRef]
42. Aguila-Camacho, N.; Gallegos, J.; Duarte-mermoud, M.A. Analysis of fractional order error models in adaptive systems: Mixed order cases. *J. Fract. Calc. Appl. Anal.* **2019**, *22*, 1113–1132. [CrossRef]
43. Gallegos, J.A.; Aguila-Camacho, N.; Duarte-Mermoud, M.A.; Travieso-Torres, J.C.; Ceballos-Benavides, G.E. Switched systems with changing derivation order: Stability and applications. *J. Frankl. Inst.* **2021**, *358*, 3943–3963. [CrossRef]
44. Gallegos, J.; Aguila-Camacho, N.; Núñez, F. High-Gain Adaptive Control With Switching Derivation Order and Its Application to a Class of Multiagent Systems. *IEEE Trans. Syst. Man, Cybern. Syst.* **2024**, *54*, 3960–3971. [CrossRef]
45. Ortega, R.; Aranovskiy, S.; Pyrkin, A.A.; Astolfi, A.; Bobtsov, A.A. New Results on Parameter Estimation via Dynamic Regressor Extension and Mixing: Continuous and Discrete-Time Cases. *IEEE Trans. Autom. Control.* **2021**, *66*, 2265–2272. [CrossRef]
46. Diethelm, K. *The Analysis of Fractional Differential Equations: An Application-Oriented Exposition Using Operators of Caputo Type*; Springer: Berlin/Heidelberg, Germany, 2004.
47. Gallegos, J.; Aguila-Camacho, N.; Duarte-Mermoud, M. Smooth Solutions to mixed-order fractional differential systems with applications to stability analysis. *J. Integral Equations Appl.* **2019**, *31*, 59–84. [CrossRef]
48. Tuan, H.; Trinh, H. Stability of fractional-order nonlinear systems by Lyapunov direct method. *IET Control. Theory Appl.* **2018**, *12*, 2417–2422. [CrossRef]
49. Lavretsky, E. Combined/Composite Model Reference Adaptive Control. *IEEE Trans. Autom. Control.* **2009**, *54*, 2692–2697. [CrossRef]
50. Annaswamy, A.M. Adaptive Control and Intersections with Reinforcement Learning. *Annu. Rev. Control. Robot. Auton. Syst.* **2023**, *6*, 65–93. [CrossRef]
51. Gallegos, J.A.; Duarte-Mermoud, M.A.; Castro-Linares, R. Mixed order robust adaptive control for general linear time invariant systems. *J. Frankl. Inst.* **2018**, *355*, 3399–3422. [CrossRef]
52. Valerio, D.; Da Costa, J.S. Ninteger: A non-integer control toolbox for MatLab. In Proceedings of the 1st International Conference on Fractional Differentiation and its Applications (ICFDA), Bordeaux, France, 19–21 July 2004.

fractal and fractional

Article

Fractional-Order Control Method Based on Twin-Delayed Deep Deterministic Policy Gradient Algorithm

Guangxin Jiao [1], Zhengcai An [1], Shuyi Shao [1,*] and Dong Sun [2]

[1] College of Automation Engineering, Nanjing University of Aeronautics and Astronautics (NUAA), Nanjing 211106, China; 17863960870@163.com (G.J.); azc0802@163.com (Z.A.)
[2] Jincheng Nanjing Engineering Institute of Aircraft Systems, Nanjing 211199, China; dongsun_njust@126.com
* Correspondence: shaosy@nuaa.edu.cn

Abstract: In this paper, a fractional-order control method based on the twin-delayed deep deterministic policy gradient (TD3) algorithm in reinforcement learning is proposed. A fractional-order disturbance observer is designed to estimate the disturbances, and the radial basis function network is selected to approximate system uncertainties in the system. Then, a fractional-order sliding-mode controller is constructed to control the system, and the parameters of the controller are tuned using the TD3 algorithm, which can optimize the control effect. The results show that the fractional-order control method based on the TD3 algorithm can not only improve the closed-loop system performance under different operating conditions but also enhance the signal tracking capability.

Keywords: FODOB; FOSMC; radial basis function network; TD3 algorithm

1. Introduction

Fractional-order calculus stands out for its flexible description of the behavior of nonlocal and non-Markovian dynamics, providing a rich mathematical tool for modeling complex systems. Its main applications include the modeling of nonlinear and nonsmooth system dynamics [1], the simulation of multiscale complex systems [2], the analysis of non-Markovian processes [3], as well as in the fields of signal processing, control systems and financial modeling [4]. In the field of control, fractional-order control has been combined with many traditional control schemes, such as fractional-order PID control [5], fractional-order robust control [6], and fractional-order sliding-mode control [7]; these fractional-order control methods have been developed in depth at the theoretical level to form a sound theoretical system and have achieved extensive and powerful results in practical applications.

As a traditional control strategy, fractional-order sliding-mode control is able to cope with the nonlinearity and uncertainty of the system more flexibly [8–10]. However, with the increase in system complexity and nonlinearity, choosing the optimal values of controller parameters often becomes a great difficulty [11]. Faced with this problem, different scholars have used a variety of optimization algorithms over the years, such as early adaptive control [12], the particle swarm algorithm [13], the genetic algorithm [14], and the wolf-pack algorithm [15]. These algorithms have had some success, but as technology advances, they struggle to handle increasingly complex problems.

In recent years, with the continuous development of artificial intelligence technology, more and more advanced algorithms have been applied to the control field [16]. Among them, reinforcement learning algorithms based on neural networks are increasingly becoming a focus of research because their flexibility and adaptability make them more adaptable to complex environments [17–23]. Reinforcement learning efficiently masters system dynamics by learning through the interaction of an agent with the environment [24]. This approach transforms the system control problem into a process in which the agent learns the optimal control strategy through continuous trial and error [25], and it can also learn without prior knowledge of the system model [26,27], which offers a flexible approach for real-time controller parameter optimization. Specifically, under the framework of reinforcement learning, the intelligent body adjusts the control strategy according to feedback signals by interacting with the environment and gradually optimizes the controller parameters [28,29]. This learning approach is more adaptive, especially in the face of a high-order system complexity, significant nonlinearities, and rapid changes in dynamic characteristics, where traditional static parameter optimization methods may appear to be inadequate [30]. Therefore, reinforcement learning provides a more intelligent solution for controller parameter optimization, which is expected to show more significant performance improvements in practical applications.

In this paper, a fractional-order control method based on TD3 reinforcement learning is proposed to optimize the parameters. A fractional-order disturbance observer is designed for estimating the disturbance signals present in the system, while an RBF network is designed for approximating the possible uncertainties in the system, and finally, a fractional-order sliding-mode controller is designed for controlling the system. For the parameter optimization part, it is performed by the TD3 reinforcement learning algorithm, which consists of six deep neural network networks designed to inhibit the bootstrap phenomenon in the reinforcement learning algorithm [31], so that the output of the network can converge to the optimal solution quickly. On this theoretical basis, a valve-controlled hydraulic system is selected for design and simulation in Matlab/Simulink to verify the effectiveness of the method proposed in this paper, and at the same time, in order to reflect the advantages and disadvantages of the proposed method, a series of comparative experiments are designed.

In summary, the main work of this paper can be succinctly summarized in the following three areas:

i. The fractional-order disturbance observer is designed to estimate the system disturbance signal, and the RBF network is selected to approximate the uncertainties of the system; then, a fractional-order sliding-mode controller is designed to control the system according to the estimated value of the disturbance observer;
ii. The TD3 algorithm is introduced to optimize the parameters of the controller, and an improved loss function is designed to improve the learning performance of the algorithm, accelerate the convergence of the network output, and optimize the control effect;
iii. It is verified through simulation that not only the control effect of the proposed method is better than the selected comparison control method, but that its also has a great robustness and generalization capability.

This paper is organized as follows. Section 2 introduces the proposed fractional-order control method, including the system state equation generalization and the designed fractional-order disturbance observer with an RBF network structure, based on which the fractional-order sliding-mode controller is designed. Section 3 introduces the fundamentals of the TD3 algorithm [32] and defines the reward function and loss function used. Section 4 demonstrates the stability of the designed method to verify its theoretical correctness. In Section 5, the proposed method is simulated in Matlab/Simulink using a valve-controlled hydraulic system as a model, and a series of control simulations are performed to verify the practical feasibility of the proposed method. Finally, conclusions and future work are presented.

2. Design of Fractional-Order Control Method

In this section, the fractional-order control part of the proposed method is accomplished for a system generalization including a fractional-order disturbance observer, an RBF network, and a fractional-order sliding-mode controller.

For a common system, the equation of state can be expressed as

$$\begin{cases} \dot{x} = Ax + Bu + f(x) + d \\ y = Cx \end{cases} \tag{1}$$

where x is the system state, and $x \in R^n$; A is the system parameter matrix, $A \in R^{n \times n}$, B is the control matrix, and $B \in R^{n \times q}$; u is the control signal, $u \in R^q$, $C \in R^{p \times n}$ is the output matrix, $d \in R^{n \times 1}$ is the disturbance signal, and $f(x) \in R^{n \times 1}$ is the uncertainty of system.

In this paper, the Riemann–Liouville fractional-order calculus formula is defined as [33]:

$$D^{\alpha} g(x) = \frac{1}{\Gamma(n-\alpha)} \frac{d^n}{x^n} \int_a^x \frac{g(t)}{(x-t)^{\alpha+1-n}} dt \tag{2}$$

where α represents the differential, $0 < \alpha < 1$, $\Gamma(\cdot)$ represents the Gamma function, n represents the smallest integer larger than α, usually taken as 1, and a, x represents the lower and upper limits of the integral.

For the disturbance signal, the fractional-order disturbance observer is given by

$$\begin{cases} D^{\lambda} z = -L\left(D^{\lambda-1} z + L D^{\lambda-1} x\right) - L\left(A D^{\lambda-1} x + B D^{\lambda-1} u + D^{\lambda-1} \hat{f}(x)\right) \\ \hat{d} = z + Lx \end{cases} \tag{3}$$

where z is the disturbance observer auxiliary vector, L is the gain matrix, $-L$ is the Hurwitz matrix with n distinct eigenvalues, $\hat{d}$ is the estimate of the disturbance, and $\hat{f}(x)$ is an estimate of the model uncertainty.

Theorem 1. *Assume that the disturbance is bounded and its derivative is also bounded, which means that* $\|d(t)\| \leq \varsigma$, $\|\dot{d}(t)\| \leq \zeta$. *Define the estimation error of the disturbance* $e_d(t) = d(t) - \hat{d}(t)$. *Based on the structure of the designed disturbance observer, the estimation error of the disturbance is bounded when the system uncertainty estimation error* $\tilde{f}(x)$ *is bounded, which means that* $\|e_d(t)\| \leq \xi$, *where* ξ *is a very small constant greater than zero.*

Proof. Define $M = -L$; then, M is a Hurwitz matrix with n distinct eigenvalues, so there exists an invertible matrix X such that $X^{-1} M X = diag(\lambda_1, \lambda_2, \cdots, \lambda_n)$. Therefore, there exists a positive constant σ such that $\|e^{Mt}\| \leq \sigma e^{\lambda_{\max}(M)t}$, where $\sigma = \|X^{-1}\| \|X\|$ [34]. Differentiating the estimation error $e_d(t)$ for the disturbance yields the following equation:

$$\begin{aligned} \frac{d(e_d(t))}{dt} &= \dot{d}(t) - \dot{\hat{d}}(t) = \dot{d}(t) - \frac{d}{dt}(z + Lx) \\ &= \dot{d}(t) + L(z + Lx) + L\left(Ax + Bu + \hat{f}(x)\right) - L(Ax + Bu + f(x) + d) \\ &= \dot{d}(t) + L\hat{d}(t) - Ld(t) + L\hat{f}(x) - Lf(x) \\ &= \dot{d}(t) - Le_d(t) - L\tilde{f}(x) \end{aligned}$$

It can be obtained that $\dot{e}_d(t) = \dot{d}(t) + Me_d(t) + M\tilde{f}(x)$. The subsequent proof is given in the stability analysis on $e_d(t)$. □

For the model uncertainty that may exist in the system, a radial basis function (RBF) network is used for approximation. The network structure of an RBF has the ability for high-speed learning, the ability to approximate a nonlinear function, which improves the model's fitting ability, and it can adapt to a variety of complex input–output mapping, so it is very flexible in practical applications. Moreover, compared with other types of neural networks, the RBF network structure is relatively simple, which makes it easy to realize and adjust [35]. The structure of the RBF network used in this paper is shown in Figure 1.

The input–output relationship for each layer can be expressed as an input layer, $\theta_i = x_i, (i = 1, 2, \cdots, n)$, an implicit layer, $h_j = e^{net_j}, net_j = -\sum_{j=1}^{m} \frac{\|\theta_i - c_j\|^2}{b_j^2}, (j = 1, 2, \cdots m)$, and an output layer, $Y_i = W_i{}^T H = \omega_{i1} h_1 + \omega_{i2} h_2 + \cdots + \omega_{im} h_m$, where $\|\cdot\|$ is the Euclidean paradigm, H is an implicit layer function vector $H = [h_1 h_2 \cdots h_m]^T$, c_j is the center vector, b_j is the width of the radial basis function, and W_i is the ith set of column vectors of the weight matrix W from the implicit layer to the output layer, which means that

$W = \begin{bmatrix} \omega_{11} & \omega_{12} & \cdots & \omega_{1m} \\ \omega_{21} & \omega_{22} & \cdots & \omega_{2m} \\ \vdots & \vdots & \ddots & \vdots \\ \omega_{n1} & \omega_{n2} & \cdots & \omega_{nm} \end{bmatrix}^T$, where i is the number of neurons in the input and output layer, and j is the number of neurons in the hidden layer.

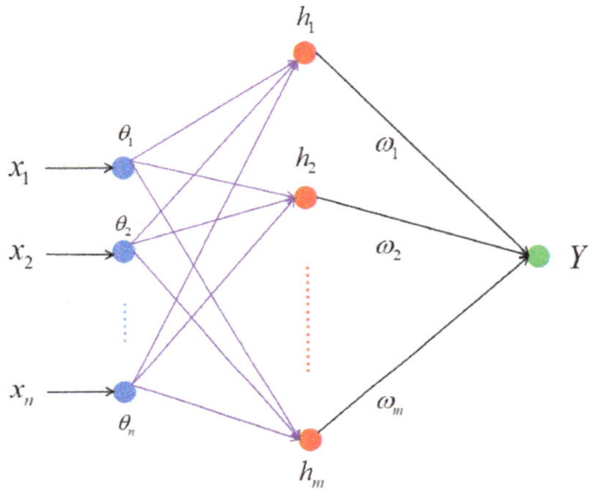

Figure 1. Structure of the RBF network.

Defining the optimal weights vector as W^*, the system uncertainty part function $f(x)$ can be expressed as [36]

$$f(x) = W^{*T}H + \varepsilon \tag{4}$$

where $\varepsilon = \begin{bmatrix} \varepsilon_1 & \varepsilon_2 & \cdots & \varepsilon_n \end{bmatrix}^T$ is the smallest approximation error of the RBF network, $H = \aleph(x, c, b)$, $\aleph(\cdot)$ is the arithmetic function from the input layer to the output of the implicit layer; since the system uncertainty $f(x)$ is bounded, there exists an upper bound W_{max} for W^*, which means that $\|W^*\| \leq W_{\max}$.

Defining the estimate of the RBF network for the system uncertainty $f(x)$ as $\hat{f}(x) = \hat{W}^T H$, the estimation error can be written as [36]

$$\begin{aligned} \tilde{f}(x) &= f(x) - \hat{f}(x) \\ &= W^{*T}H - \hat{W}^T H + \varepsilon \\ &= \left(W^{*T} - \hat{W}^T\right)H + \varepsilon \\ &= \tilde{W}^T H + \varepsilon \end{aligned} \tag{5}$$

where ε satisfies $\|\varepsilon\| \leq \varepsilon_{\max}$, with $\varepsilon_{\max}$ a bounded constant.

Based on the fractional-order disturbance observer's estimate of the disturbance signal $\hat{d}$ and the RBF network's estimate of the system's uncertainty $\hat{f}(x)$, the sliding-mode surface of the fractional-order sliding-mode controller is given by

$$s = C_1 e + C_2 D^{\lambda-1} e \tag{6}$$

where e is the system tracking error, $e = y_d - y$, determined by the desired output y_d and the actual output y of the system, C_1 and C_2 are positive real numbers, and $0 < \lambda < 1$ is the order of the fractional-order differentiation.

Deriving the sliding-mode surface and taking the sliding-mode convergence law as $-ks - k_s\text{sgn}(s)$, where k_s and k are the sliding-mode convergence law parameters and $k_s, k > 0$, and substituting into (1), the control law can be obtained by

$$u = (CB)^{-1}\left(\dot{y}_d + \frac{C_2}{C_1}D^{\lambda}e + \frac{ks + k_s\text{sgn}(s)}{C_1}\right) - B^{-1}Ax - B^{-1}d - B^{-1}f(x) \quad (7)$$

Substituting the disturbance signal estimator of (3) with the system uncertainty estimator $\hat{f}(x)$ yields the following final control law:

$$u = (CB)^{-1}\left(\dot{y}_d + \frac{C_2}{C_1}D^{\lambda}e + \frac{ks + k_s\text{sgn}(s)}{C_1}\right) - B^{-1}Ax - B^{-1}\hat{d} - B^{-1}\hat{f}(x) \quad (8)$$

3. TD3 Algorithm Based on Fractional-Order Control Method

In this section, a TD3 algorithm based on the designed fractional-order control method is proposed, and we design the reward function and loss function of the TD3 algorithm based on the proposed fractional-order control method, which in turn makes it effective for the optimization of the parameters of the fractional-order control method.

As a kind of reinforcement learning algorithm, the basic principle of the TD3 algorithm is also composed of a critic and an actor, two kinds of networks of the agent and the environment for the interactive operation. More specifically, Figure 2 shows the agent interacting with the environment to obtain the state s_t and rewards r_t, the output of the action a_t at moment t, and the update of the two kinds of networks within the agent is based on the loss function.

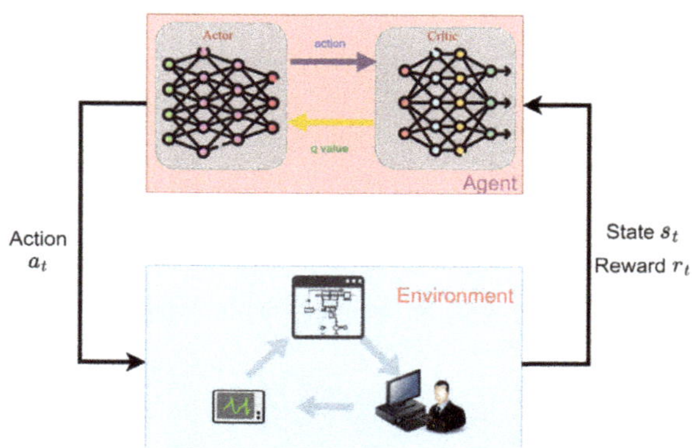

Figure 2. The relationship between environment and agent.

In this paper, the state of the environment s_t is set as the error signal for each state of the system, which means that $s_t = \begin{bmatrix} e_1(t) & e_2(t) & \cdots & e_n(t) \end{bmatrix}^T$, and the reward signal r_t is set to the negative of the weighted sum of the absolute values of the error signals for each state of the system, which means $r_t = -\sum_{i=1}^{n} \eta_i |e_i(t)|$, where $\eta_i > 0$ denotes the weight of the ith error signal in the reward function, which needs to be set according to the actual situation; the action a_t of the actor network's output are the parameters of the fractional-order controller which should be optimized.

In the reinforcement learning algorithm, the Bellman equation is defined as [37]

$$U_t = r_t + \gamma U_{t+1} \quad (9)$$

where r_t denotes the reward acquired by the intelligence in the environment, $U_t = \sum_{k=1}^{n} \gamma^{k-1} r_{t+k-1}$ is a value function that represents the sum of the rewards that the intelligence expects to acquire in a continuous process after moment t, and where $0 < \gamma < 1$ denotes the discount rate.

For both sides of (9), the expectation based on the state S and action A at a moment can be obtained, and the value function of the action at that moment can be obtained by

$$E_\pi[U_t|S_t = s_t, A_t = a_t] = E_\pi(r(s_t, a_t)) + \gamma E_\pi[U_{t+1}|S_{t+1} = s_{t+1}, A_{t+1} = a_{t+1}] \tag{10}$$

In the reinforcement learning algorithm, the estimation of the value function of the action is carried out through the critic network, so there is a certain bias, at which point the Bellman equation's equal sign does not hold. Defining the difference between the left and right sides of (10) as the temporal-difference error (TD error) $\delta_{TD}(t)$, it can be written as [38]

$$\delta_{TD}(t) = Q_\pi(s_t, a_t|\varpi_t) - (r(s_t, a_t) + \gamma Q_\pi(s_{t+1}, a_{t+1}|\varpi_t)) \tag{11}$$

where ϖ_t is the parameter of the critic network at moment t.

In order to solve the bootstrapping of the overestimation in traditional reinforcement learning algorithms, the TD3 algorithm adopts two sets of critic networks and target critic networks with identical structure and parameters; the corresponding network outputs are called Critic1 network $Q_{1,\pi}(s, a|\varpi_1)$, Critic2 network $Q_{2,\pi}(s, a|\varpi_2)$, Target Critic1 network $Q'_{1,\pi}(s, a|\varpi'_1)$, and Target Critic2 network $Q'_{2,\pi}(s, a|\varpi'_2)$. The algorithm adopts the strategy of smoothing regularization to reduce the estimation error, adding a noise signal $\epsilon \sim clip(\mathcal{N}(0, \sigma); -c, c)$ to the action output from the actor network, so that the target value $y_{target} = r(s_t, a_t) + \gamma \cdot Q_\pi(s_{t+1}, a_{t+1})$ can be expressed as [31]

$$y_{target} = r(s_t, a_t) + \gamma \min_{i=1,2}(Q'_{1,\pi}(s_{t+1}, (a'_{t+1} + \epsilon)|\varpi'_j)) \tag{12}$$

The agent in the TD3 algorithm stores each state transfer pair (s, a, r, s') in the replay buffer during each interaction with the environment, and since in the process of learning by the intelligent body, the selection of the experience should be considered to prioritize the learning value, in this paper, a nonuniform sampling was designed as

$$P(X = r_i) = e^{\alpha(|r_i|)} \left(\sum_{i=1}^{n} e^{\alpha(|r_i|)} \right)^{-1} \tag{13}$$

where $P(X = r_i)$ denotes the probability of drawing a state transfer pair corresponding to reward r_i, α is a parameter that controls the shape of the distribution, and n denotes the total number of samples in the replay buffer.

Based on the sampling probability in (13), N samples are taken from the replay buffer and the loss function of the critic network is constructed using the mean squared error (MSE), which can be written as

$$J_\varpi(t) = \sum_{j=1}^{2} J_{\varpi_j}(t) = -N^{-1} \sum_{j=1}^{2} \sum_{i=1}^{N} \left(Q_{j,\pi}(s, a|\varpi)_i - y_{target_{j,i}} \right)^2 \tag{14}$$

For the actor network, in order to maximize the future expected return of the current strategy, which means that $\pi(a|\theta) = \arg\max_\theta E_{s_t, a_t} \left(\sum_{i=t}^{\infty} \gamma^{i-t} R_i \right)$, and in order to improve the explorability of the action space, the loss function was designed as

$$J_\theta(t) = -\left[N^{-1} \sum_{j=1}^{N} E_{s_{t_j}, a_{t_j}} \left(\sum_{i=t_j}^{\infty} \gamma^{i-t_j} R_i \right) + \alpha H_\theta(\pi) \right] \tag{15}$$

where θ is the parameter of the actor network, $H_\theta(\pi) = \sum_a \pi(s|\theta) \log\pi(s|\theta)$ is a measure of uncertainty in the distribution of strategies, and α is the weight parameter for that item.

For both critic network and actor network parameters ϖ, θ are updated using the gradient descent method, which can be expressed as [31]

$$\begin{aligned} \varpi_{t+1} &= \varpi_t - \alpha \nabla_\varpi J_\varpi(t) \\ \theta_{t+1} &= \theta_t - \beta \nabla_\theta J_\theta(t) \\ &= \theta_t - \beta\left(N^{-1} \sum \nabla_a Q_\pi(s, a|\omega) \nabla_\theta \pi(s|\theta) + \alpha \nabla_\theta H_\theta(\pi)\right) \end{aligned} \tag{16}$$

where α and β are the learning rate of the critic network and actor network.

For the target network parameter, a soft update is performed by introducing the update shift τ of the network parameter, which means [31]:

$$\begin{aligned} \varpi'_{i,t+1} &= \tau\varpi_{i,t+1} + (1-\tau)\varpi'_{i,t} \qquad (i = 1,2) \\ \theta'_{t+1} &= \tau\theta_{t+1} + (1-\tau)\theta'_t \end{aligned} \tag{17}$$

Finally, the TD3 algorithm uses lagged updating for the actor network, which means that the actor network is updated once when the critic network is updated multiple times. Based on the above description, a structure of a fractional-order control method based on the TD3 algorithm is proposed, and the corresponding pseudocode is shown in Algorithm 1.

Algorithm 1 FOCS pseudocode based on the TD3 algorithm

```
Initialize critic networks Q_{1,π}(s,a|ϖ1) and Q_{2,π}(s,a|ϖ2) and actor network π_θ with
random parameters ϖ1, ϖ2, θ
Initialize target networks ϖ'1 ← ϖ1, ϖ'2 ← ϖ2, θ' ← θ
Initialize replay buffer B
for t = t0 to t = tf do
    Output action based on current parameters a_t ~ π_θ(s_t) + ε, ε ~ clip(N(0,σ); −c, c),
    Take a_t for FOCS, observe the error e(t), calculate r, and observe new state s_{t+1}
    Store the transfer pair (s, a, r, s') in B
    if n ≥ N \\ n is the number of transfer pair in replay buffer B.
        Sample N transfer pairs and calculate the loss function of critic network J_ϖ(t)
        Update ϖ by ϖ_{t+1} ← ϖ_t − α∇_ϖ J_ϖ(t)
            if Critic network after d updates
                Calculate the loss function of critic network J_θ(t)
                Updateθ by θ_{t+1} ← θ_t − β∇_θ J_θ(t)
                Update target network by
                        ϖ'_{t+1} ← τϖ_{t+1} + (1 − τ)ϖ'_t
                        θ'_{t+1} ← τθ_{t+1} + (1 − τ)θ'_t
            end if
    end if
end for
```

4. Stability Analysis

In this section, the stability of the proposed control method is substantiated through a detailed proof, emphasizing the theoretical feasibility of the proposed approach.

For the proposed fractional-order disturbance observer, the Lyapunov function is defined as

$$V_d = \frac{1}{2} e_d^T e_d \tag{18}$$

The derivation of (18) and substitution into (1), (3), and (5) yield

$$\begin{aligned}
\dot{V}_d &= e_d^T \dot{e}_d = e_d^T \left(\dot{d} - \dot{\hat{d}}\right) = e_d^T (\dot{d} - (\dot{z} + L\dot{x})) \\
&= e_d^T (\dot{d} - (D^{-\lambda+1} D^{\lambda} z + L(Ax + Bu + f(x) + d))) \\
&= e_d^T (\dot{d} - Le_d - L\tilde{f}(x)) = e_d^T M e_d + e_d^T M \tilde{f}(x) + e_d^T \dot{d} \\
&\leq \lambda_{\max}(M) e_d^T e_d + \lambda_{\max}(M) e_d^T (\tilde{W}^T H) + \lambda_{\max}(M) e_d^T \varepsilon + \tfrac{1}{2}(e_d^T e_d + \dot{d}^T \dot{d}) \\
&\leq \left(2\lambda_{\max}(M) + \tfrac{1}{2}\right) e_d^T e_d + \tfrac{\lambda_{\max}(M)}{2} tr(H^T H) tr(\tilde{W}^T \tilde{W}) + \tfrac{1}{2}(\lambda_{\max}(M) \varepsilon^T \varepsilon + \dot{d}^T \dot{d}) \\
&\leq \left(2\lambda_{\max}(M) + \tfrac{1}{2}\right) e_d^T e_d + \tfrac{\lambda_{\max}(M) m}{2} tr(\tilde{W}^T \tilde{W}) + \tfrac{1}{2}(\lambda_{\max}(M) \varepsilon_{\max}^2 + \zeta^2)
\end{aligned} \tag{19}$$

where $\lambda_{\max}(M)$ is the largest eigenvalue of M which is a Hurwitz matrix so it can be obtained that $\lambda_{\max}(M) < 0$. In order to make the estimation error of the fractional-order disturbance observer for the disturbance signal stable, M should satisfy $\lambda_{\max}(M) < -1/4$.

In order to keep the approximation error of the designed RBF network for the system uncertainty within a certain range, the adaptive law of the RBF network weights and the implicit layer function update law are taken as

$$\dot{\hat{W}} = \eta^{-1}(-\mathcal{K}\hat{W} - C_1 H s^T C) \tag{20}$$

where $\eta \in R^{m \times m}$ is the designed positive-definite diagonal matrix, $\mathcal{K} > 0$ are the designed parameters of the adaptive law, s and C_1 are the sliding-mode surface and the parameter which was designed in (6), and H is the implicit layer function vector.

For the proposed RBF network, the Lyapunov function is defined as

$$V_f = \frac{1}{2} tr(\tilde{W}^T \eta \tilde{W}) \tag{21}$$

Derive Equation (21), and substitute into the RBF network's adaptive law (20):

$$\begin{aligned}
\dot{V}_f &= tr\left(\tilde{W}^T \eta \dot{\tilde{W}}\right) = tr\left[\tilde{W}^T \eta \left(\dot{W}^* - \dot{\hat{W}}\right)\right] \\
&= -tr\left[\tilde{W}^T \eta (\eta^{-1}(-\mathcal{K}\hat{W} - C_1 H s^T C))\right] \\
&= tr(\tilde{W}^T \mathcal{K} \hat{W}) + tr(\tilde{W}^T C_1 H s^T C) \\
&= \mathcal{K} tr\left[\tilde{W}^T (W^* - \tilde{W})\right] + tr(\tilde{W}^T C_1 H s^T C) \\
&\leq -\tfrac{\mathcal{K}}{2} tr(\tilde{W}^T \tilde{W}) + \tfrac{\mathcal{K}}{2} \|W^*\|^2 + tr(\tilde{W}^T C_1 H s^T C) \\
&\leq -\tfrac{\mathcal{K}}{2} tr(\tilde{W}^T \tilde{W}) + \tfrac{\mathcal{K}}{2} W_{\max}^2 + tr(\tilde{W}^T C_1 H s^T C)
\end{aligned} \tag{22}$$

Based on the designed sliding-mode surface function, define the Lyapunov function

$$V_s = \frac{1}{2} s^T s \tag{23}$$

Deriving Equation (23) and substituting the system state equation shown in (1) with the control law shown in (8) yield

$$\begin{aligned}
\dot{V}_s &= s^T \dot{s} = s^T (C_1 \dot{e} + C_2 D^{\lambda} e) = s^T (C_1 (\dot{y}_d - C\dot{x}) + C_2 D^{\lambda} e) \\
&= s^T (C_1 (\dot{y}_d - C(Ax + Bu + f(x) + d)) + C_2 D^{\lambda} e) \\
&= s^T (C_1 \left(\dot{y}_d - C\left(Ax + B\left((CB)^{-1}\left(\dot{y}_d + \tfrac{ks + k_s \mathrm{sgn}(s) + C_2 D^{\lambda} e}{C_1}\right)\right.\right.\right. \\
&\qquad \left.\left.\left. - B^{-1} Ax - B^{-1}\hat{d} - B^{-1}\hat{f}(x)\right) + f(x) + d\right)\right) + C_2 D^{\lambda} e) \\
&= s^T (-ks - k_s \mathrm{sgn}(s) - C_1 C \tilde{f}(x) - C_1 C e_d) \\
&= -k s^T s - s^T (k_s \mathrm{sgn}(s) + C_1 C \tilde{f}(x) + C_1 C e_d) \\
&\leq \left(-k + \tfrac{k_s}{2} + \tfrac{C_1 \|C\|}{2}\right) s^T s + \tfrac{C_1 \|C\|}{2} e_d^T e_d + \tfrac{k_s}{2} - s^T C_1 C (\tilde{W}^T H + \varepsilon)
\end{aligned} \tag{24}$$

To prove the overall stability of the original system with the addition of a fractional-order control method, define the Lyapunov function

$$V = V_d + V_f + V_s \tag{25}$$

A derivation of Equation (25) and a substitution of the Lyapunov derivation of (19), (22), and (24) lead to

$$\begin{aligned}
\dot{V} &= \dot{V}_d + \dot{V}_f + \dot{V}_s \\
&\le \left(2\lambda_{\max}(M) + \tfrac{C_1\|C\|+1}{2}\right)e_d^T e_d + \left(-k + \tfrac{k_s}{2} + \tfrac{C_1\|C\|}{2}\right)s^T s - \left(\tfrac{\mathcal{K}}{2} - \tfrac{m\lambda_{\max}(M)}{2}\right)tr(\tilde{W}^T\tilde{W}) \\
&\quad + \tfrac{C_1\|C\|}{2}s^T s + \tfrac{C_1\|C\|}{2}\varepsilon_{\max}^2 + \tfrac{\mathcal{K}}{2}W_{\max}^2 + \tfrac{1}{2}(\zeta^2 + \lambda_{\max}(M)\varepsilon_{\max}^2) + \tfrac{k_s}{2} \\
&\le \left(-k + \tfrac{k_s}{2} + C_1\|C\|\right)s^T s + \left(2\lambda_{\max}(M) + \tfrac{C_1\|C\|+1}{2}\right)e_d^T e_d \\
&\quad - \left(\tfrac{\mathcal{K}}{2} - \tfrac{m\lambda_{\max}(M)}{2}\right)tr(\tilde{W}^T\tilde{W}) + \tfrac{1}{2}(\zeta^2 + \lambda_{\max}(M)\varepsilon_{\max}^2 + C_1\|C\|\varepsilon_{\max}^2 + \mathcal{K}W_{\max}^2 + k_s) \\
&\le -\left(k - \tfrac{k_s}{2} - C_1\|C\|\right)s^T s - \left(-2\lambda_{\max}(M) - \tfrac{C_1\|C\|+1}{2}\right)e_d^T e_d \\
&\quad - \left(\tfrac{\mathcal{K}}{2} - \tfrac{m\lambda_{\max}(M)}{2}\right)tr(\tilde{W}^T\tilde{W}) + \delta \\
&\le -\kappa V + \delta
\end{aligned} \tag{26}$$

where $\kappa = \min\left(\left(k - \frac{k_s}{2} - C_1\|C\|\right), \left(-2\lambda_{\max}(M) - \frac{C_1\|C\|+1}{2}\right), \left(\frac{\mathcal{K}}{2} - \frac{m\lambda_{\max}(M)}{2}\right)\right)$ to satisfy the system stability, the control system parameter $0 < C_1 < \frac{-4\lambda_{\max}(M)-1}{\|C\|}$, $k - \frac{k_s}{2} > C_1\|C\|$, $k_s > 0$ should be selected, and $\delta = \frac{1}{2}(\zeta^2 + \lambda_{\max}(M)\varepsilon_{\max}^2 + C_1\|C\|\varepsilon_{\max}^2 + \mathcal{K}W_{\max}^2 + k_s)$. At this point, the system is boundedly stable.

5. Simulation Result

In this section, in order to validate the effectiveness of the proposed methodology, a valve-controlled hydraulic system is selected as the object of study to design a series of simulations. Firstly, to verify the online learnability of the TD3 algorithm, three rounds of training states during the training process are selected for the comparison. Secondly, the control effect is verified for the optimized fractional-order control method of the TD3 algorithm by employing the prescribed performance fractional-order sliding-mode controller (PPC-FOSMC), an unoptimized fractional-order sliding-mode controller (FOSMC), and a sliding-mode controller (SMC). Furthermore, the antidisturbance is proved for the proposed method, and the simulation verification is carried out for different input signals and disturbance signals. The results show that the fractional-order control method based on the TD3 algorithm not only has good online learning ability and generalization ability but also has a better control effect than the traditional FOSMC, SMC, and PPC-FOSM. Finally, in order to verify that the designed method can still maintain a better control effect under noise disturbance, a Gaussian noise is selected as the disturbance signal to simulate using the online learning method.

5.1. Simulation Results of System Online Learning

A valve-controlled hydraulic system is a third-order system [39] which can be described approximately as

$$\begin{cases} \dot{x} = Ax + Bu + B_d d + f(x) \\ y = Cx \end{cases} \tag{27}$$

where $x = [x_v, x_s, x_a]^T$ denotes the position, velocity, and acceleration, respectively, u denotes the control input, and y denotes the output of the position. The state matrix, input matrix, disturbance matrix, and output matrix A, B, B_d, and C are, respectively,

$$\begin{gathered}
A = \begin{bmatrix} 0 & 1 & 0 \\ 0 & 0 & 1 \\ -R_1 & -R_2 & -R_3 \end{bmatrix}, B = \begin{bmatrix} 0 \\ 0 \\ b \end{bmatrix}, \\
B_d = \begin{bmatrix} 0 & 0 & 1 \end{bmatrix}^T, \quad C = \begin{bmatrix} 1 & 0 & 0 \end{bmatrix},
\end{gathered} \tag{28}$$

where $R_1 = \frac{4kk_{ce}\beta}{mV_t}$, $R_2 = \frac{4B_p k_{ce}\beta}{mV_t} + \frac{k}{m}$, $R_3 = \frac{4k_{ce}\beta}{V_t} + \frac{B_p}{m}$, $b = \frac{4k_v k_{sv} k_q \beta A_p}{mV_t}$, and the values of the parameters of the valve-controlled hydraulic system are shown in Table 1.

Table 1. Parameters of the valve-controlled hydraulic system.

Name	Symbol	Numerical Value
Total mass of piston and load converted to piston	m	5 kg
Piston area of hydraulic cylinder	A_p	0.4734 m^2
Springiness	k	1
Flow-pressure amplification factor	k_{ce}	3.005×10^{-13}
Modulus of elasticity of hydraulic fluid	β	0.005
Total volume of hydraulic cylinder	V_t	1.5 m^3
Viscous damping coefficient of piston and load	B_p	5
Servo amplifier gain	k_v	1
Servo valve gain	k_{sv}	4.733×10^{-3}
Flow gain of slide valve	k_q	1

The disturbance signal during the TD3 training was set to be $d = 0.5sin(5\pi t)$, the system uncertainty was $f(x) = sin(2x_a)cos(3x_a)$, and the system expected the output position signal to be $x_{pd} = sin(t)$; the training parameters are shown in Table 2.

Table 2. Parameters of the TD3 algorithm.

Name	Numerical Value
Episode	1300
Episode step	100
Sampling time	0.1 s
Final time	10 s
Learning rate	0.001
Replay buffer	1×10^6
Minimum batch size	128
Number of actor network layers	4
Number of critic network layers	9

Learning to train the system with the above parameters, the episode rewards and average rewards obtained after 1900 episodes are shown in Figure 3a, and Figure 3b–d show the state of the system for the selected episode 1, episode 500, and episode 1826. It can be noticed that in the first 420 episodes, the episode reward fluctuates a lot, and after episode 420, the episode reward and the average reward stabilize for the first time, but at that point, the network still cannot achieve a particularly good training effect. From episode 630 to episode 1300, the agent further explores the action space, and after episode 1300, the episode reward and average reward converge to more optimal values, corresponding to the system state. It can be found that the position and velocity signals of the system corresponding to episode 1 have the worst tracking performance with both large errors and desired signals, while the acceleration signals have the worst tracking performance with more noise. Compared to episode 1, episode 500's systematic position signal can already track the desired signal at the 0.8th second but with a small error, the velocity signal can track the desired signal after 2 s but with a small noise in between, and the acceleration signal can only track the desired signal after the 4th second. Finally, compared to the previous two, episode 1826 is definitely the best performer, with the position signal tracking the desired signal in the first 0.1 s with very little error, and the velocity signal tracking the desired signal with little error after that, despite oscillating in the first 0.1 s, and with the acceleration signal tracking the desired signal after oscillating in the first 0.4 s with an acceptable error.

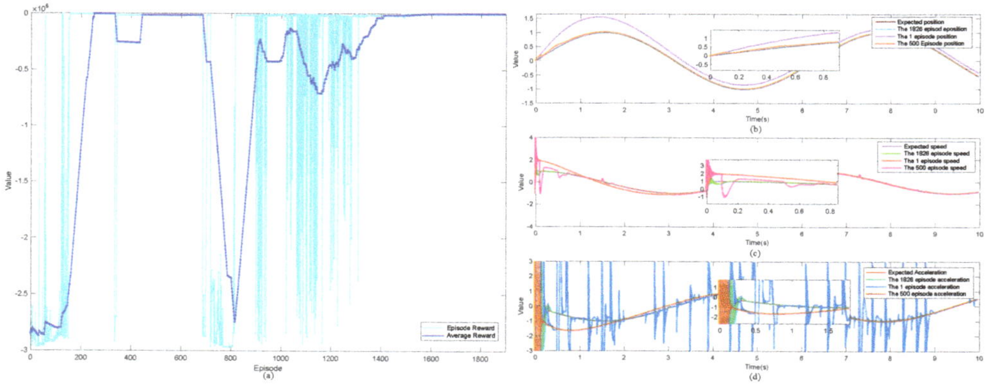

Figure 3. Episode reward and average reward with the training results of the 1826th episode: (**a**) episode reward and average reward of 1900 episodes, (**b**) the 1st, 500th, and 1826th episode's position and expected position signals; (**c**) the 1st, 500th, and 1826th episode's speed and expected signals; (**d**) the 1st, 500th, 1826th episode's acceleration and expected signals.

In order to better visually represent the training state of set 1826, the system's individual signals and desired signals and their errors at that point in time are shown in Figure 4. It can be seen that the position, velocity, and acceleration signals of the system have good tracking performance, and the errors are all within a small range. The same is true for the estimation of disturbance signals.

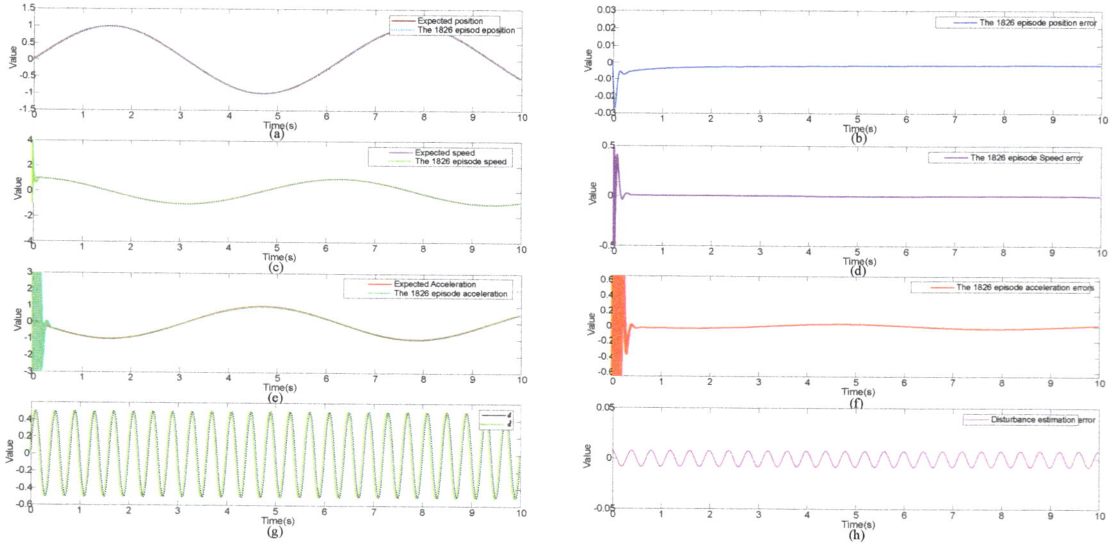

Figure 4. The 1st, 500th, and 1826th episode's system signal and expected signals with the error: (**a**) position signal and expected signal, (**b**) position tracking error, (**c**) speed signal and expected signal, (**d**) speed tracking error, (**e**) acceleration signal and expected signal, (**f**) acceleration tracking error, (**g**) estimation of the disturbance signal and the actual signal, (**h**) estimation error of the disturbance signal.

5.2. Simulation Results of TD3-FOCS

For the online learning results of the valve-controlled hydraulic fractional-order control method, the training results of the 1826th episode are shown in Figure 5 with $C_1 = 0.001$, $C_2 = 290$, $k = 945, k_s = 43, \lambda = 0.55$ selected as the controller parameters, and the parameters of FODOB and the RBF network were set to $L = 157$, $\mathcal{K} = 7$, and $\eta = I$, which is a single-unit diagonal matrix for the subsequent comparative simulation experiments; the simulation time was set to 100 s, and the final results are shown in Figure 6.

As can be seen from Figure 6, for the position signal, the proposed TD3-FOSMC had the fastest convergence speed (first 0.03 s) and maintained the final steady-state error in the range of (−0.001, +0.001), which was the best among the experimental results. For the velocity signal, it also had the fastest convergence speed (first 0.25 s) and maintained the final steady-state error in the range of (−0.005, +0.005), which was the best performance among all the experimental methods. For acceleration signals, although the prescribed performance of the fractional-order sliding-mode controller ended up with a smaller error range, the convergence time was greater than the proposed TD3-FOSMC, which converged in the first 0.3 s. Combining all the above control effects, it can be concluded that the proposed TD3-FOSMC has a better speed and stability.

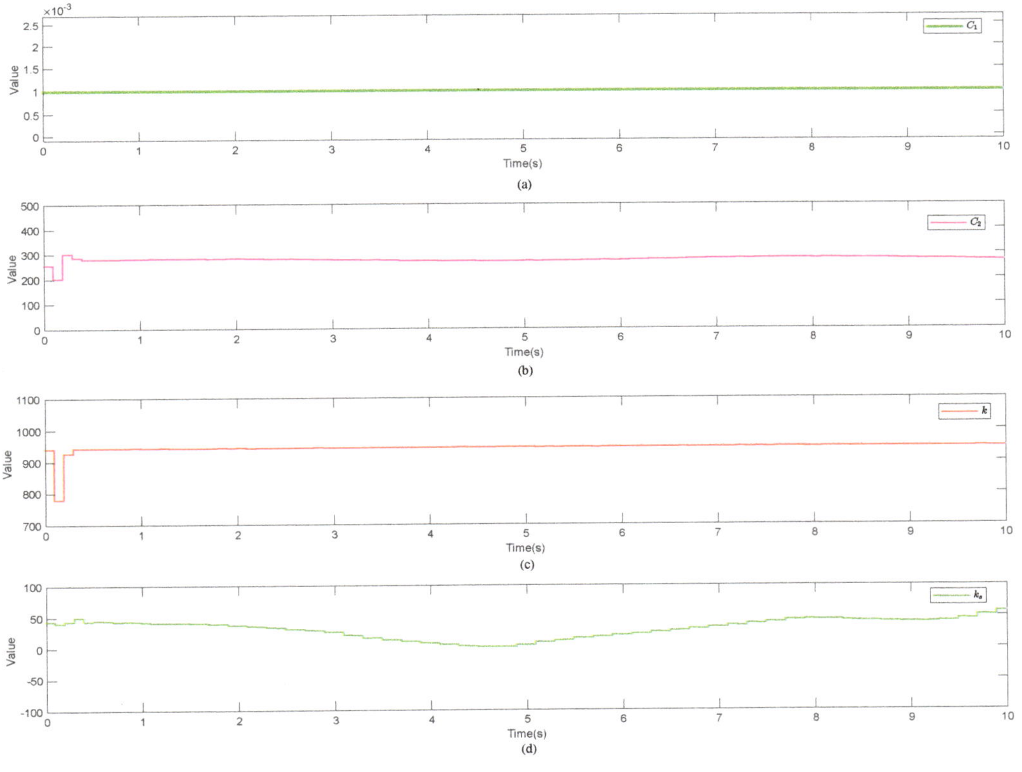

Figure 5. Training results of the 1826th episode: (**a**) parameter C_1, (**b**) parameter C_2, (**c**) parameter k, (**d**) parameter k_s.

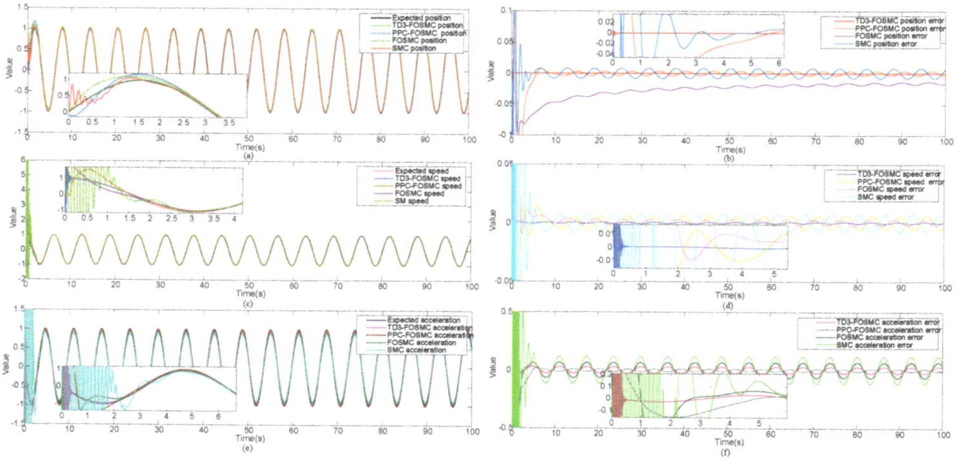

Figure 6. TD3-optimized fractional-order control vs. unoptimized control method's system signals and expected signals and their errors: (**a**) position signal and expected signal, (**b**) position tracking error, (**c**) speed signal and expected signal, (**d**) speed tracking error, (**e**) acceleration signal and expected signal, (**f**) acceleration tracking error.

5.3. Simulation Results of Different Control Signals and Disturbance Signals

In order to demonstrate the robustness and versatility of the method, two control signals (a triangular wave composite signal and a step signal) and two disturbance signals (a strong sine wave signal disturbance and a triangular wave composite signal) were selected for the simulation verification. This allowed a more comprehensive assessment of the method's antidisturbance performance under different control and disturbance conditions, thus demonstrating the effectiveness of the method in mitigating the effects of an external disturbance in a variety of practical situations.

As observed in Figure 7, the simulation results revealed the capability of accurately tracking the desired signal under the influence of a triangular wave composite signal and a strong sine wave signal disturbance. Specifically, the position signal achieved tracking within a remarkably short duration of 0.07 s, with very small overshoots (less than 1%) in the first 0.02 s, and the final steady-state error confined to a narrow range of $(-0.001, 0.001)$. In the first 0.3 s, the velocity signal had large to small fluctuations in the range of $(-3.7, 6)$; after that, it achieved synchronization with the desired signal and kept the steady-state error within the range of $(-0.01, 0.01)$. On the other hand, the acceleration signal fluctuated in the range of $(-12, 12)$ for the first 0.5 s and was after synchronized with the desired signal and kept the steady-state error within the range of $(-0.25, 0.25)$. The results show that despite the large overshoot in the system response for the different input signals and disturbance signals, the system still had good stability and fast performance.

Analyzing the results shown in Figure 8, when subjected to a step signal and a strong triangular wave composite signal disturbance during the simulation, the obtained results exhibited a marginally reduced performance compared to the previous scenario. Taking the simulation time of 10 s as an example, it can be found that for the position, velocity, and acceleration signals, there were large fluctuations in the first 1 s, and their stabilization time was slowing down sequentially. For the position signal, there were fluctuations within the range of (5, 15) within the first 0.3 s; after that, it could track the desired signal, and the final steady-state error was limited to the range of (0, 0.004). The velocity signal exhibited fluctuations within the range of $(-10, 10)$ during the first 0.5 s, after which it synchronized with the desired signal, and the final steady-state error was constrained to $(-0.002, 0)$. For the acceleration signal, fluctuations within the range of $(-20, 20)$ occurred during the initial 0.8 s, following which it converged to the desired value. The final steady-state error remained confined to a smaller range of (−0.0001, 0.0001). Despite a slight decrease in performance, the results underscore the method's ability to maintain effective control even in the presence of challenging input signals and disturbance conditions.

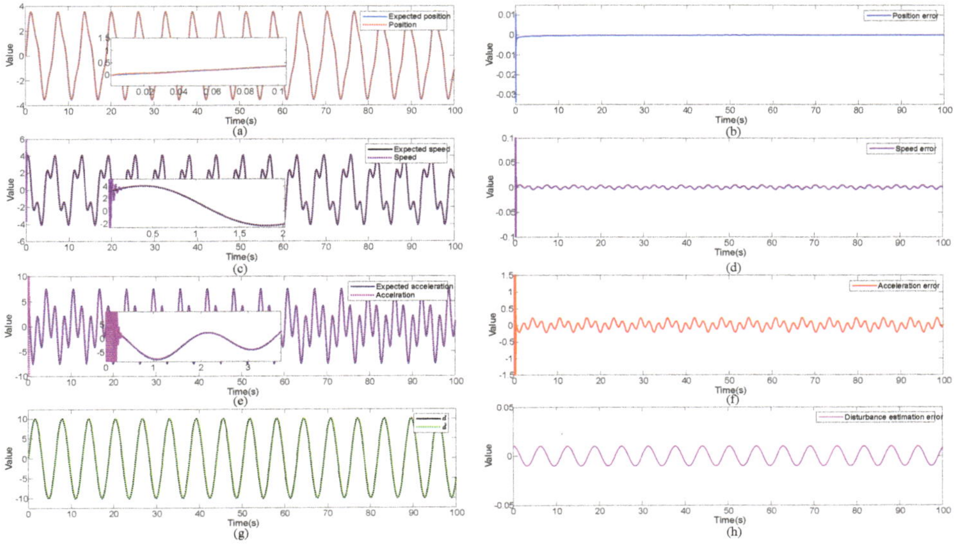

Figure 7. Simulation results of triangular wave composite signals and strongly interfering sine wave signals: (**a**) position signal and expected signal, (**b**) position tracking error, (**c**) speed signal and expected signal, (**d**) speed tracking error, (**e**) acceleration signal and expected signal, (**f**) acceleration tracking error, (**g**) estimation of the disturbance signal and actual signal, (**h**) estimation error of the disturbance signal.

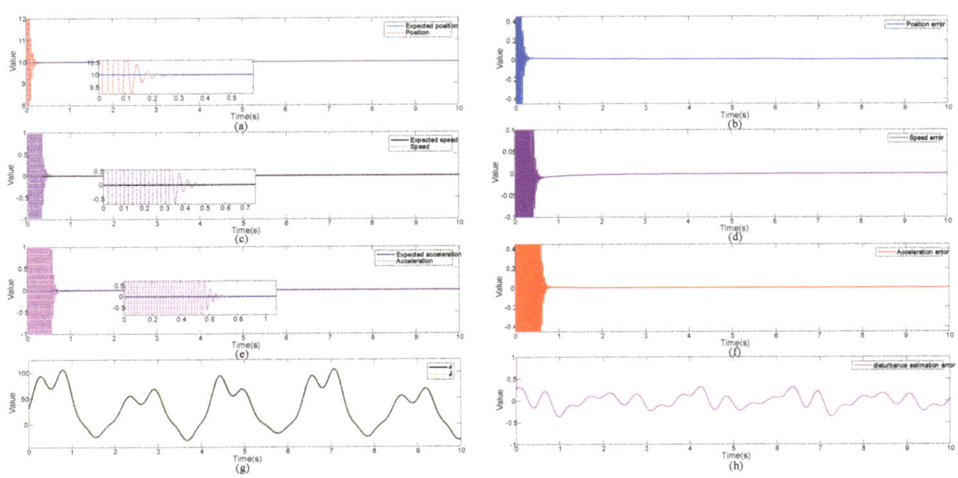

Figure 8. Simulation results of step signals and triangular wave composite signal: (**a**) position signal and expected signal, (**b**) position tracking error, (**c**) speed signal and expected signal, (**d**) speed tracking error, (**e**) acceleration signal and expected signal, (**f**) acceleration tracking error, (**g**) estimation of the disturbance signal and actual signal, (**h**) estimation error of the disturbance signal.

5.4. Simulation Results under Noise Disturbance

The control performance is illustrated for the method under noise disturbance. The noise disturbance signal was selected with a mean value of zero, a variance of five, and a frequency of 10 Hz, and the simulation verification was carried out by using the online learning of the agent. The results are shown in Figure 9.

Analyzing Figure 9, it can be seen that the position signal exhibited minimal susceptibility form the noise disturbance signal, with an overshoot of less than 3% observed only within the initial 0.03 s. Subsequently, the position signal could track the desired signal, and the final steady-state error could be maintained within the range of $(-0.001, 0)$. The velocity signal demonstrated minimal susceptibility to noise disturbances, with oscillations confined to the range of $(-1.6, 4.7)$ within the initial 0.08 s. Following that period, the signal could adeptly track the desired trajectory, and the final steady-state error could satisfy the narrow bounds of $(-0.0015, 0.0015)$. Comparing with the earlier online learning outcomes, the performance of the acceleration signal was degraded under the noise signal. The acceleration signal was oscillatory within the range of $(-6, 6)$ during the initial 0.15 s. Despite these challenges, it could approximately track the desired signal. However, due to the influence of noise, it maintained a final steady-state error within the range of $(-0.13, 0.13)$. Thus, based on the analysis above, the proposed control method could achieve the control of system within the bounded range under the noise signal.

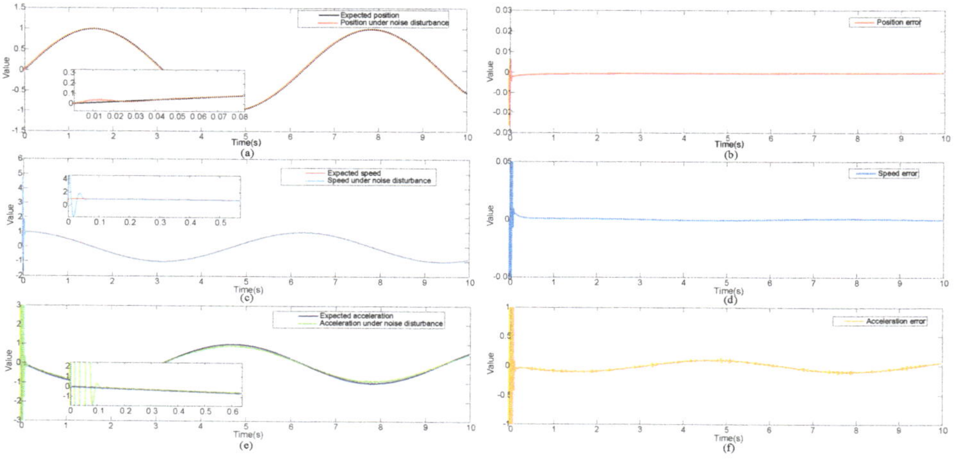

Figure 9. Simulation results under noise disturbance: (**a**) position signal and expected signal, (**b**) position tracking error, (**c**) speed signal and expected signal, (**d**) speed tracking error, (**e**) acceleration signal and expected signal, (**f**) acceleration tracking error.

6. Conclusions

In this paper, a fractional-order control method based on the TD3 algorithm was introduced. A fractional-order disturbance observer was designed to estimate the system's disturbance signal, and an RBF network was selected to approximate the uncertainties in the system, and the fractional-order sliding-mode control method was also adopted to design the controller. A valve-controlled hydraulic system was simulated and validated in Matlab/Simulink using the agent online learning and the optimization parameters. Different control signals and disturbance signals were used in the optimized fractional-order control system. The results showed that the limitations of this method mainly lay in the difficulty of setting up the training environment comprehensively. Despite some shortcomings, the proposed method was generally fast and had good antidisturbance ability.

Author Contributions: Conceptualization, G.J. and S.S.; methodology, G.J. and Z.A.; software, G.J. and D.S.; validation, G.J. and D.S.; formal analysis, G.J. and Z.A.; writing—original draft preparation, G.J.; writing—review and editing, G.J. and S.S.; supervision, S.S.; funding acquisition, S.S. All authors have read and agreed to the published version of the manuscript.

Funding: This research was funded by the Fund of Science and Technology on Space Intelligent Control Laboratory Foundation (grant number HTKJ2023KL502002); the China Postdoctoral Science Foundation (grant number 2020M681587); the Jiangsu Province Postdoctoral Science Foundation (grant number 2020Z112).

Data Availability Statement: Data are contained within the article.

Conflicts of Interest: The authors declare no conflicts of interest.

References

1. Spanos, P.D.; Malara, G. Nonlinear vibrations of beams and plates with fractional derivative elements subject to combined harmonic and random excitations. *Probabilistic Eng. Mech.* **2020**, *59*, 142–149. [CrossRef]
2. Magin, R.L. Fractional calculus models of complex dynamics in biological tissues. *Comput. Math. Appl.* **2010**, *36*, 1586–1593. [CrossRef]
3. Atangana, A. Non validity of index law in fractional calculus: A fractional differential operator with Markovian and non-Markovian properties. *Phys. A Stat. Mech. Its Appl.* **2018**, *505*, 688–706. [CrossRef]
4. Gonzalez, E.A.; Petráš, I. Advances in fractional calculus: Control and signal processing applications. In Proceedings of the 2015 16th International Carpathian Control Conference (ICCC), Szilvasvarad, Hungary, 27–30 May 2015; pp. 147–152. . [CrossRef]
5. Warrier, P.; Shah, P. Optimal Fractional PID Controller for Buck Converter Using Cohort Intelligent Algorithm. *Appl. Syst. Innov.* **2021**, *4*, 50. [CrossRef]
6. Razzaghian, A. A fuzzy neural network-based fractional-order Lyapunov-based robust control strategy for exoskeleton robots: Application in upper-limb rehabilitation. *Math. Comput. Simul.* **2022**, *193*, 567–583. [CrossRef]
7. Fei, J.; Wang, Z.; Pan, Q. Self-Constructing Fuzzy Neural Fractional-Order Sliding Mode Control of Active Power Filter. *IEEE Trans. Neural Netw. Learn. Syst.* **2023**, *34*, 10600–10611. [CrossRef]
8. Jakovljević, B.; Pisano, A.; Rapaić, M.R.; Usai, E. On the sliding-mode control of fractional-order nonlinear uncertain dynamics. *Int. J. Robust Nonlinear Control* **2016**, *24*, 782–798. [CrossRef]
9. Mirrezapour, S.Z.; Zare, A.; Hallaji, M. A new fractional sliding mode controller based on nonlinear fractional-order proportional integral derivative controller structure to synchronize fractional-order chaotic systems with uncertainty and disturbances. *J. Vib. Control* **2022**, *28*, 773–785. [CrossRef]
10. Deepika, D. Hyperbolic uncertainty estimator based fractional-order sliding mode control framework for uncertain fractional-order chaos stabilization and synchronization. *ISA Trans.* **2022**, *123*, 76–86. [CrossRef] [PubMed]
11. Delavari, H.; Ghaderi, R.; Ranjbar, A.; Momani, S. Fuzzy fractional-order sliding mode controller for nonlinear systems. *Commun. Nonlinear Sci. Numer. Simul.* **2010**, *15*, 963–978. [CrossRef]
12. Sun, G.; Ma, Z. Practical tracking control of linear motor with adaptive fractional-order terminal sliding mode control. *IEEE/ASME Trans. Mechatron.* **2017**, *22*, 2643–2653. [CrossRef]
13. Djari, A.; Bouden, T.; Boulkroune, A. Design of fractional-order sliding mode controller (FSMC) for a class of fractional-order non-linear commensurate systems using a particle swarm optimization (PSO) Algorithm. *J. Control Eng. Appl. Inform.* **2014**, *16*, 46–55.
14. Karthikeyan, A.; Rajagopal, K. Chaos control in fractional-order smart grid with adaptive sliding mode control and genetically optimized PID control and its FPGA implementation. *Complexity* **2017**, *2017*, 3815146. [CrossRef]
15. Han, S. Modified grey-wolf algorithm optimized fractional-order sliding mode control for unknown manipulators with a fractional-order disturbance observer. *IEEE Access* **2020**, *8*, 18337–18349. [CrossRef]
16. Salman, G.A.; Jafar, A.S.; Ismael, A.I. Application of artificial intelligence techniques for LFC and AVR systems using PID controller. *Int. J. Power Electron. Drive Syst.* **2019**, *10*, 1694. [CrossRef]
17. Perrusquía, A.; Yu, W. Identification and optimal control of nonlinear systems using recurrent neural networks and reinforcement learning: An overview. *Neurocomputing* **2021**, *438*, 145–154. [CrossRef]
18. Liu, Y.J.; Zhao, W.; Liu, L.; Li, D.; Tong, S.; Chen, C.P. Adaptive neural network control for a class of nonlinear systems with function constraints on states. *IEEE Trans. Neural Netw. Learn. Syst.* **2023**, *34*, 2732–2741. [CrossRef] [PubMed]
19. Chen, Z.; Niu, B.; Zhang, L.; Zhao, J.; Ahmad, A.M.; Alassafi, M.O. Command filtering-based adaptive neural network control for uncertain switched nonlinear systems using event-triggered communication. *Int. J. Robust Nonlinear Control* **2022**, *32*, 6507–6522. [CrossRef]
20. Katz, S.M.; Corso, A.L.; Strong, C.A.; ; Kochenderfer, M.J. Verification of image-based neural network controllers using generative models. *J. Aerosp. Inf. Syst.* **2022**, *19*, 574–584. [CrossRef]
21. Kiumarsi, B.; Vamvoudakis, K.G.; Modares, H.; Lewis, F.L. Optimal and autonomous control using reinforcement learning: A survey. *IEEE Trans. Neural Netw. Learn. Syst.* **2017**, *29*, 2042–2062. [CrossRef]

22. Duraisamy, P.; Nagarajan, Santhanakrishnan, M.; Rengarajan, A. Design of deep reinforcement learning controller through data-assisted model for robotic fish speed tracking. *J. Bionic Eng.* **2023**, *20*, 953–966. [CrossRef]
23. Wei, C.; Xiong, Y.; Chen, Q.; Xu, D. On adaptive attitude tracking control of spacecraft: A reinforcement learning based gain tuning way with guaranteed performance. *Adv. Space Res.* **2023**, *71*, 4534–4548. [CrossRef]
24. Barto, A.G. Reinforcement learning. In *Neural Systems for Control*; Academic Press: Cambridge, MA, USA, 1997; pp. 7–30.
25. Nian, R.; Liu, J.; Huang, B. A review on reinforcement learning: Introduction and applications in industrial process control. *Comput. Chem. Eng.* **2020**, *139*, 106886. [CrossRef]
26. Zhu, Y.; Zhao, D.; Li, X. Using reinforcement learning techniques to solve continuous-time non-linear optimal tracking problem without system dynamics. *IET Control Theory Appl.* **2016**, *10*, 1339–1347. [CrossRef]
27. Syafiie, S.; Tadeo, F.; Martinez, E. Model-free learning control of neutralization processes using reinforcement learning. *Eng. Appl. Artif. Intell.* **2007**, *20*,767–782. [CrossRef]
28. Dogru, O.; Velswamy, K.; Ibrahim, F.; Wu, Y.; Sundaramoorthy, A.S.; Huang, B.; Xu, S.; Nixon, M.; Bell, N. Reinforcement learning approach to autonomous PID tuning. *Comput. Chem. Eng.* **2022**, *161*, 107760. [CrossRef]
29. Taherian, S.; Kuutti, S.; Visca, M.; Fallah, S. Self-adaptive torque vectoring controller using reinforcement learning. In Proceedings of the 2021 IEEE International Intelligent Transportation Systems Conference (ITSC), Indianapolis, IN, USA, 19–22 September 2021; IEEE: Piscataway, NJ, USA, 2021; pp 172–179.
30. Yan, L.; Webber, J.L.; Mehbodniya, A.; Moorthy, B.; Sivamani, S.; Nazir, S.; Shabaz, M. Distributed optimization of heterogeneous UAV cluster PID controller based on machine learning. *Comput. Electr. Eng.* **2022**, *101*, 108059. [CrossRef]
31. Wu, J.; Wu, Q.J.; Chen, S.; Pourpanah, F.; Huang, D. A-TD3: An Adaptive Asynchronous Twin Delayed Deep Deterministic for Continuous Action Spaces. *IEEE Access* **2022**, *10*,128077–128089. [CrossRef]
32. Dankwa, S.; Zheng, W. Twin-delayed ddpg: A deep reinforcement learning technique to model a continuous movement of an intelligent robot agent. In Proceedings of the 3rd International Conference on Vision, Image and Signal Processing, Vancouver, BC, Canada, 26–28 August 2019; pp. 1–5.
33. Li, C.; Qian, D.; Chen, Y.Q. On Riemann-Liouville and caputo derivatives. *Discret. Dyn. Nat. Soc.* **2011**, *2011*, 562494. [CrossRef]
34. Walsh, G.C.; Ye, H.; Bushnell, L.G. Stability analysis of networked control systems. *IEEETrans. Control Syst. Technol.* **2002**, *10*, 438–446. [CrossRef]
35. Er M.J.; Wu, S.; Lu, J.; Toh, H.L. Face recognition with radial basis function (RBF) neural networks. *IEEE Trans. Neural Netw.* **2002**, *13*, 697–710. [PubMed]
36. Fei, J.; Wang, H.; Fang, Y. Novel neural network fractional-order sliding-mode control with application to active power filter. *IEEE Trans. Syst. Man, Cybern. Syst.* **2021**, *52*, 3508–3518. [CrossRef]
37. Fei, Y.; Yang, Z.; Chen, Y.; Wang, Z. Exponential bellman equation and improved regret bounds for risk-sensitive reinforcement learning. *Adv. Neural Inf. Process. Syst.* **2021**, *34*, 20436–20446.
38. Doya, K. Reinforcement learning in continuous time and space. *Neural Comput.* **2000**, *12*, 219–245. [CrossRef]
39. Duan, Z.; Sun, C.; Li, J.; Tan, Y. Research on servo valve-controlled hydraulic motor system based on active disturbance rejection control. *Meas. Control.* **2024**, *57*, 113–123. [CrossRef]

fractal and fractional

MDPI

Article

Vibration Suppression of the Vehicle Mechatronic ISD Suspension Using the Fractional-Order Biquadratic Electrical Network

Yujie Shen [1,*], Zhaowei Li [1], Xiang Tian [1,2], Kai Ji [1] and Xiaofeng Yang [3]

1. Automotive Engineering Research Institute, Jiangsu University, Zhenjiang 212013, China; lizhaowei272154@163.com (Z.L.); auzn0009@163.com (X.T.); jk11036@163.com (K.J.)
2. Chery New Energy Automobile Co., Ltd., Wuhu 241000, China
3. School of Automotive and Traffic Engineering, Jiangsu University, Zhenjiang 212013, China; yangxf18@ujs.edu.cn

* Correspondence: shenyujie@ujs.edu.cn

Abstract: In order to break the bottleneck of the integer-order transfer function in vehicle ISD (inerter-spring-damper) suspension design, a positive real synthesis design method of vehicle mechatronic ISD suspension based on the fractional-order biquadratic transfer function is proposed. The emergence of the fractional-order components disrupts the equivalence relationship between the passivity of components and the positive realness of integer-order transfer functions in traditional networks. In this paper, the positive real condition of the fractional-order biquadratic transfer function is given. Then, a quarter dynamic model of the vehicle mechatronic ISD suspension is established, and the parameters of the fractional-order biquadratic transfer function and vehicle suspension are obtained by an NSGA-II multi-objective genetic algorithm. Moreover, the structure of the external circuit and the parameters of the electrical components are obtained by the fractional-order passive network synthesis theory. The simulation results show that under the condition of random road input and vehicle speed of 20 m/s, the root-mean-square (RMS) value of the vehicle body acceleration and the dynamic tire load of the fractional-order ISD suspension are reduced by 7.98% and 18.75% compared with the traditional passive suspension, while under the same condition, the integer-order ISD suspension can only reduce by 5.34% and 16.07%, respectively. The results show that employing a fractional-order biquadratic electrical network in the vehicle mechatronic ISD suspension enhances vibration isolation performance compared with the suspension using an integer-order biquadratic electrical network.

Keywords: vehicle suspension; mechatronic inerter; positive real synthesis; fractional-order biquadratic transfer function

1. Introduction

The vehicle suspension system is an essential component intended to support the weight of the vehicle body while mitigating the effects of uneven road surfaces. It plays a vital role in minimizing vehicle vibrations. Suspension performance has a major impact on vehicle driving safety, operating stability, and riding comfort. The evolution of suspension systems has progressed from traditional passive configurations, consisting of springs and dampers, to semi-active [1–3] and active suspensions [4–6]. Nonetheless, both semi-active and active suspension systems necessitate precise state and parameter estimation, often employing filter algorithms [7,8]. This complexity increases the challenges associated with their design and introduces drawbacks such as high energy consumption and elevated costs. By comparison, passive suspension still plays an irreplaceable role because of its simple and reliable structure. Traditional passive suspension systems face the challenge of structural solidification, limiting their ability to meet the increasingly demanding performance requirements of vehicle suspension, and the proposal of "inerter" [9] makes the suspension composed of "inerter-spring-damper" attract attention in the realm of engineering.

The application of inerter in fields such as aerospace engineering [10], civil engineering [11], and railway engineering [12] has proved the vibration isolation performance improvement and the application value of the device. Based on the previous research, Professor Smith conducted a further study on vehicle suspensions and found that the application of an inerter can effectively reduce body acceleration and dynamic tire load [13]. Nie [14] realized the passive realization of skyhook damping suspension system by means of the method of reduced-order ISD suspension. Zhang [15] combined a hydro-pneumatic spring with an inerter to reduce the vehicle body and wheel vibration effectively. Wang [16] investigated the influence of an inerter on the multi-directional vibrations. Du [17] explored 21 engineering feasible vehicle ISD suspension structures, identifying 12 configurations that exhibited superior vibration isolation performance compared to traditional passive suspensions. Shen [18] researched a nonlinear fluid inerter-based hydro-pneumatic suspension, and the key parameters of the suspension were optimized by the non-dominated sorting genetic algorithm II (NSGA-II), which was proposed by Srinivas and Deb [19] and has been applied in many fields [20–22]. Yang [23] considered an energy-harvesting vehicle suspension with an inerter. The above methods are all from the perspective of structural methods to carry on the structural design of suspension systems. The work is huge, and the suspension structure with excellent vibration isolation performance is easily missed. From the perspective of the impedance method to design the suspension system, we can gain inspiration from passive network synthesis in electricity. This used to be a key issue in circuit theory, and many scholars have studied it [24,25]. However, with the development of integrated circuits, the heat of related research has declined. The birth of the inerter makes the new electromechanical similarity theory correspond perfectly and promotes the development of passive mechanical network synthesis theory. The previous research on the network synthesis theory of electrical systems can be leveraged for the reverse impedance design of suspension systems.

Papageorgiou first proposed the impedance method for passive ISD suspension design [26]. Due to the passive network synthesis for low order transfer function being relatively simple, numerous scholars [27–32] mainly focus on the simplest realization of biquadratic and bicubic transfer function. According to the Bott–Duffin circuit synthesis [25], a biquadratic transfer function can be implemented passively using no more than nine components, while a bicubic transfer function requires no more than ten components. Jiang investigated the passive implementations of the biquadratic transfer function using at least six and five components [27,28]. Wang [29] extended this research by exploring sufficient and necessary conditions for passive implementation of the biquadratic transfer function with at least four and three components, respectively. Zhang [30] addressed the network synthesis problem, demonstrating the passive implementation of the bicubic transfer function with at least seven components. Wang and Chen further examined the conditions under which the bicubic transfer function can be passively implemented using at least six and five components [31,32]. However, with the reduction of the number of components in the synthetic passive implementation of the passive network, the constraints of biquadratic and bicubic transfer functions become more and more stringent. Many studies [33–35] have found that fractional order can be used to describe more complex system dynamics. Chen [36] presented a design methodology to achieve critical damping by integrating fractional dampers and inerters. Fractional-order mechanical components exhibit more intricate vibration transmission characteristics compared to their integer-order counterparts, offering new design perspectives for mechanical systems. However, implementing fractional-order mechanical elements remains challenging in engineering. At the same time, fractional-order electrical elements have been widely researched in recent years [37,38]. By designing the external circuit of the mechatronic inerter [39], complex impedance outputs can be easily achieved. Shen [40] investigated a shock mitigation system with a three-element electrical network structure comprising a fractional inductor, a fractional capacitor, and a resistor. Subsequently, Shen [41] analyzed the effects of parametric perturbations, including the numerical values and orders of electrical components. However, the exploration of the fractional-order transfer functions remains limited. The emergence of fractional order necessitates a departure from network synthesis conditions tailored for integer order transfer functions, as they are not fully applicable to fractional-order transfer functions. Therefore, this paper concentrates on investigating the influence of the fractional-order biquadratic transfer functions implemented in mechatronic inerters on the vibration suppression effectiveness of vehicle suspension systems. The paper is structured as follows:

In Section 2, the construction of vehicle mechatronic ISD suspension with a fractional-order biquadratic electrical network is given. In Section 3, the NSGA-II multi-objective genetic algorithm is employed to optimize the critical suspension parameters, and the passive network positive real synthesis of fractional-order biquadratic transfer function is given. Section 4 is the dynamic analysis of the suspension system. Some conclusions and discussions are drawn in Section 5.

2. Model of Suspension Dynamics

2.1. Model of Vehicle Mechatronic ISD Suspension

The 2-DOF (degrees of freedom) dynamic model is established to evaluate the vertical vibration of vehicle suspension in Figure 1.

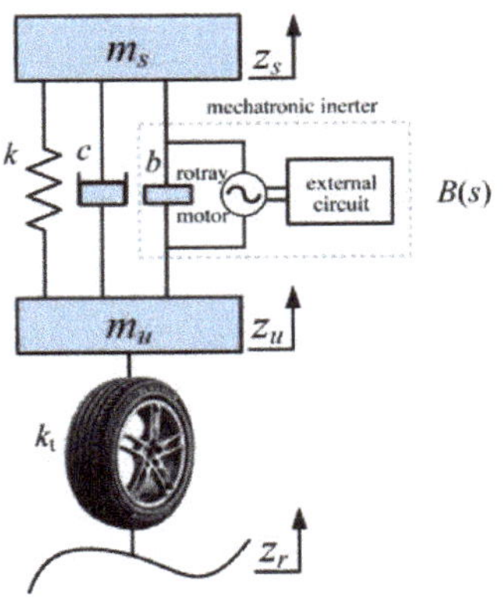

Figure 1. Model of vehicle suspension.

In Figure 1, m_s and m_u are the sprung mass and unsprung mass of the vehicle, respectively, z_s and z_u are the corresponding vertical displacements, k is the stiffness of the suspension support spring, c is the damper coefficient of the suspension, b is the inertance coefficient of inerter, $B(s)$ is the impedance expression of the mechatronic inerter, k_t is the equivalent spring stiffness of the tire, and z_r is the pavement input. The Laplacian equation of the suspension is obtained:

$$\begin{cases} m_s s^2 Z_s + [k + cs + sB(s)](Z_s - Z_u) = 0 \\ m_u s^2 Z_u - [k + cs + sB(s)](Z_s - Z_u) + k_t(Z_u - Z_r) = 0 \end{cases} \tag{1}$$

In the Equation (1), Z_s, Z_u, and Z_r represent the Laplacian transforms of z_s, z_u, and z_r, respectively.

The constructed suspension parameters shown in Table 1 are given according to reference [42].

Table 1. Parameters of vehicle suspension.

Parameters	Values
Sprung Mass m_s	675/kg
Unsprung Mass m_u	62.5/kg
Tire Stiffness k_t	290,000/N·m^{-1}
Spring Stiffness k	53,100/N·m^{-1}

2.2. Fractional-Order Biquadratic Electrical Network

The schematic diagram of the ball-screw mechatronic inerter [40] used in this paper is shown in Figure 2.

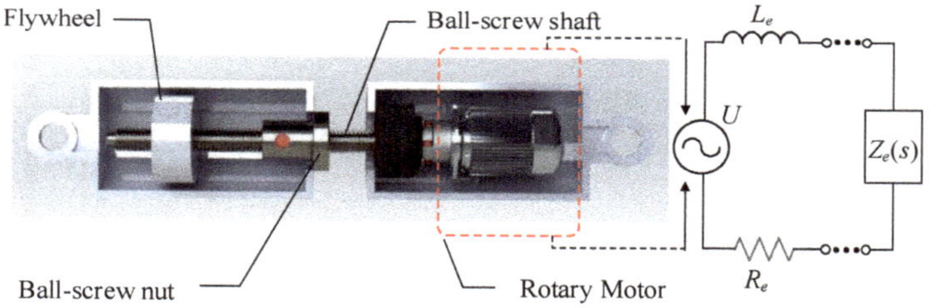

Figure 2. Schematic diagram of the mechatronic inerter.

Ignoring the equivalent inductance L_e and resistance R_e of the rotating motor, we can obtain the impedance function expression $B(s)$ of the mechanical inerter:

$$B(s) = bs + (\frac{2\pi}{P})k_t k_e Y_e(s) \tag{2}$$

where b represents the inertance coefficient of the inerter, s denotes the Laplacian operator, P signifies the lead of ball screw, k_e and k_t are the induced electromotive force coefficient and the thrust coefficient of the rotary motor, severally, and $Y_e(s)$ is the admittance expression of the external circuitry of the rotating motor. The admittance expression of the external circuit is as follows:

$$Y_e(s) = \frac{As^{\alpha+\beta}+Bs^{\alpha}+Cs^{\beta}+D}{Es^{\alpha+\beta}+Fs^{\alpha}+Gs^{\beta}+H} \tag{3}$$

In the equation, A, B, C, D, E, F, G, and H are the coefficients of the biquadratic admittance expression, s is the Laplacian operator, and α and β are the order of the Laplacian operator. According to articles [43,44], when the order of fractional-order inductors and capacitors meet $0 < \alpha, \beta < 1$, passive networks can be formed with passive components.

3. Parameter Optimization

3.1. NSGA-II Multi-Objective Genetic Algorithm

NSGA (non-dominated sorting genetic algorithm) is a genetic algorithm based on the Pareto optimization concept. This algorithm improves the selection regeneration method on the basis of the genetic algorithm; each individual is stratified according to its dominant and non-dominant relationship, and then the selection operation is performed. However, the NSGA has some problems, such as high complexity of non-dominated sorting time and lack of elite retention strategy to improve the performance of the algorithm. NSGA-II, therefore, uses an elitism-based non-dominated sorting method for ranking and sorting each individual and uses a crowding distance approach in its section operator for keeping the diversity among the obtained Pareto optimal solutions [45]. The algorithm flow chart is shown in Figure 3.

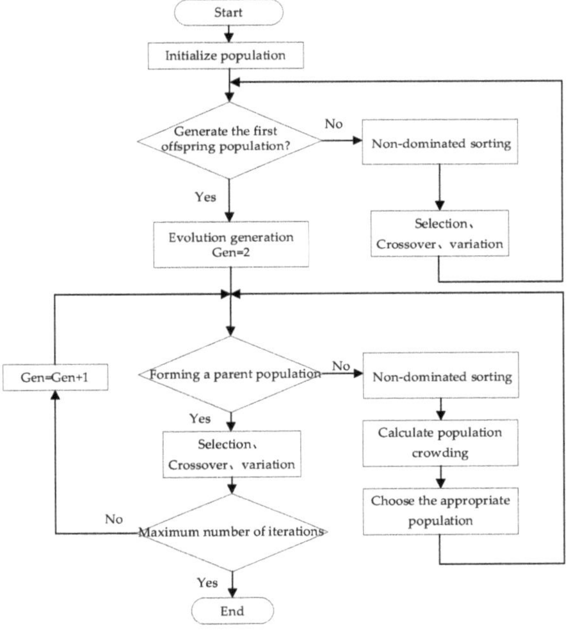

Figure 3. Flowchart of NSGA-II.

This research mainly focuses on the riding comfort and operating stability of the vehicle. In addition, we still set a hard limit on the suspension working space: ± 0.08 m [46]. Since the single-objective optimization algorithm has limitations in determining weight coefficients and handling non-uniform objective dimensions, we adopt a multi-objective optimization algorithm, which exhibits strong global search capability and robustness. The Pareto front generated by this algorithm provides multiple feasible solutions for decision-making. More information can be obtained from the Pareto front for the analysis of the correlation between the targets, and the suspension structure with the best performance can be selected according to the focus. With this in mind, we can consider both the minimum vehicle body acceleration and the dynamic tire load through Equation (4).

$$\begin{cases} J_1 = minBA(X) \\ J_3 = minDTL(X) \end{cases} \tag{4}$$

where J_1 is the minimum RMS value of the vehicle body acceleration to be optimized; J_3 is the minimum RMS value of the dynamic tire load.

A kind of passive suspension with a parallel structure composed of the parallel spring and damper elements has been elected as the contrast suspension in this paper. We improved the efficiency of the optimization algorithm by setting E as 1. The optimization variables were selected as follows: damping coefficient $c/\text{N}\cdot\text{s}\cdot\text{m}^{-1}$, inertance b/kg of the mechatronic inerter, and coefficient A, B, C, D, F, G, H of the fractional-order transfer function and order α, β. In the optimization process, the range of variables is set as follows:

$$\begin{cases} 0 < c < 5000 \\ 0 < b < 1000 \\ 0 \leq A, B, C, D, F, G, H < inf \\ 0 < \alpha,\ \beta < 1 \end{cases} \tag{5}$$

In addition, the above variables must satisfy both the nonlinear inequality constraints of theorem 1 and the nonlinear equality constraints of theorem 2 in Appendix A. Theorem 1 makes the fractional-order biquadratic function a positive real function, and theorem 2 makes the numerator and denominator of the fractional-order biquadratic function have common factors.

In the following, we will refer to the fractional-order ISD suspension as FO-ISD suspension and the integer-order ISD suspension as IO-ISD suspension in the text description section. Figure 4 shows the optimized Pareto front. It represents the curve formed by all the Pareto optimal solutions obtained in multi-objective optimization problems in the graph. This curve shows the trade-off relationship between different objectives. Choosing different optimal solutions on the curve means that the corresponding parameters can change the performance tendency of the suspension system. In Figure 4, the optimal parameter solutions for the FO-ISD suspension are represented by red triangles, the IO-ISD suspension is indicated by blue dots, and the black squares represent the optimal parameter solutions for the traditional passive suspension.

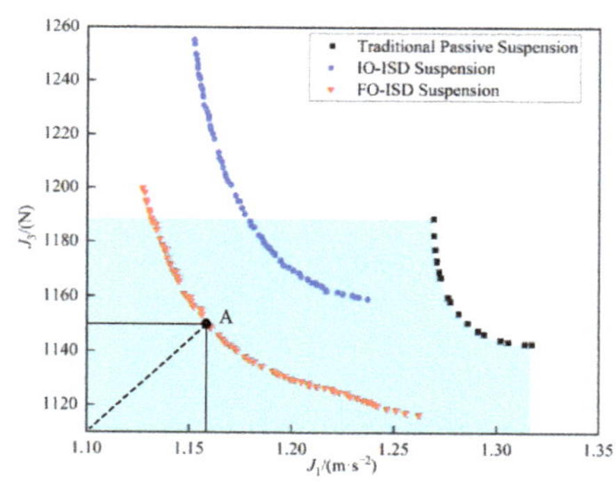

Figure 4. Pareto front.

It can be seen from the figure that compared with the traditional passive suspension, both the FO-ISD suspension and the IO-ISD suspension have a certain degree of reduction in the two indicators. However, the reduction of J_1 in the IO-ISD suspension will lead to a more serious deterioration of J_3, which results in fewer available parameters for the IO-ISD suspension. In contrast, the curve formed by the optimal parameter solutions of the FO-ISD suspension is closer to the origin of the coordinate axis, which means that the two performance indicators of the FO-ISD suspension are better than those of the IO-ISD suspension, and there are more available points better than the traditional passive suspension. Therefore, the FO-ISD suspension has more and better parameter options. It can reduce J_1 while also reducing J_3, thereby comprehensively improving the vehicle's suspension performance. The optimal normalized weighted sum of the J_1 and J_3 is selected as the final optimization variable so as to balance and comprehensively improve the performance of the vehicle suspension. Its formula is as follows:

$$f = 0.5 \times \frac{J_1}{J_{1pas}} + 0.5 \times \frac{J_3}{J_{3pas}} \tag{6}$$

Through comparative calculations, it can be found that the f-value at point A is the smallest, at 0.9479. Therefore, the parameters of the FO-ISD suspension corresponding to point A are selected, as shown in Table 2.

Table 2. Optimized parameters.

Parameters	Values
Damper coefficient c	$1000/\mathrm{N{\cdot}s{\cdot}m^{-1}}$
Inertance b	10/kg
A	0.21
B	3.70
C	0.01
D	0.20
F	0.52
G	2.88
H	49.13
Fractional order α_1	0.82
Fractional order β_1	0.88

3.2. Positive Real Synthetic Passive Implementation of Fractional-Order Biquadratic Transfer

The positive realness of the transfer function is very important for the determination of passive networks. However, with the arrival of fractional-order components, the equivalence relationship between component passivity and positive real transfer function in traditional networks is broken. Nevertheless, from the study of articles [47–49], it is found that the transfer function obtained after appropriate variable substitution for fractional-order network satisfies the positive realness condition.

According to the positive real synthetic passive realization condition of the fractional-order transfer function, the simplest realization condition of two, three, four, and five electrical-element network components is tested according to the simplest principle [50]. In Appendix B, based on the positive real synthetic passive realization conditions of the fractional-order biquadratic transfer function given in the literature, it can be obtained that the A, B, C, D, E, F, G, H optimized in this paper meet the fourteenth condition of the five-element implementation conditions of the transfer function $Y_e(s)$, that is, $AH + DE - BG - CF = 0$, $AD - BC = 0$, $BE > AF$ and satisfy $BH > DF$ or $DF > BH$. Thus, the optimized fractional-order biquadratic transfer function $Y_e(s)$ meets the conditions of positive real function and satisfies inequalities (7); it can be implemented passively with a five-element structure, as shown in Figure 5, which is composed of three resistors, a fractional-order capacitor, and a fractional-order inductor.

$$\begin{cases} BH - DF > 0 \\ BE - AF > 0 \\ AD - BC = 0 \\ (AH + DE) - (BG + CF) = 0 \end{cases} \tag{7}$$

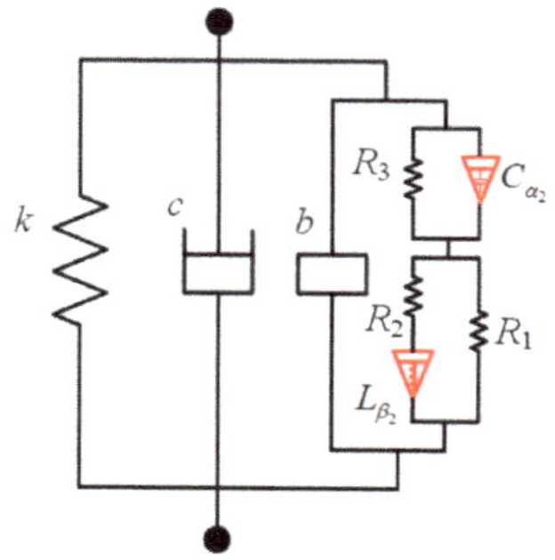

Figure 5. Optimal electrical structure.

The expression of each parameter component in Figure 5 is as follows:

$$\begin{cases} R_1 = \frac{E}{A} \\ R_2 = \frac{EF}{BE-AF} \\ R_3 = \frac{BH-DF}{BD} \\ C_{\alpha_2} = \frac{B^2}{BH-DF} \\ L_{\beta_2} = \frac{E^2}{BE-AF} \end{cases} \tag{8}$$

Table 3 shows that the transfer function coefficients obtained by optimization are substituted into the component expressions of each parameter.

Table 3. Parameters of the external electrical circuit.

Parameters	Values
Resistance R_1	4.69/Ω
Resistance R_2	0.15/Ω
Resistance R_3	246.86/Ω
Fractional-order capacitance C_α	0.08/F
Fractional-order of capacitance α_2	0.82
Fractional-order inductance L_β	0.28/H
Fractional-order of inductance β_2	0.88

4. Simulation Analysis

4.1. Random Pavement Input

The expression for random pavement input [51] is as follows:

$$\dot{z}_r = -0.111[uz_r(t) + 40\sqrt{G_0 u w_0(t)}] \tag{9}$$

where u is the velocity of vehicle, G_0 is the road roughness coefficient, and $w(t)$ represents the Gaussian white noise. Table 4 presents a comparison of performance indexes for each suspension at three different speeds. Figures 6a, 7a and 8a show the comparison of RMS values of performance indicators at speeds ranging from 10 m/s to 30 m/s. Figures 6b, 7b and 8b show the comparison of performance index at the speed of 20 m/s. The formula for calculating the RMS of a performance indicator is as follows:

$$\text{BA}_{\text{RMS}} = \sqrt{\frac{1}{n}\sum_{i=1}^{n} a_i^2} \tag{10}$$

$$\mathrm{DTL_{RMS}} = \sqrt{\frac{1}{n}\sum_{i=1}^{n} F_i^2} \tag{11}$$

where a_i is the acceleration value at each point in time; F_i is the dynamic tire load at each time point; and n is the total number of time points.

Table 4. Evaluation indexes at various velocities.

Suspension	Velocity u/(m/s)	RMS of Vehicle Body Acceleration/(m·s^{-2})	RMS of Suspension Working Space/(m)	RMS of Dynamic Tire Load/(N)
Traditional passive suspension	10	0.9062	0.0081	844.0
	20	1.2641	0.0112	1184.5
	30	1.5178	0.0133	1435.1
IO-ISD suspension	10	0.8550	0.0068	832.8
	20	1.1966	0.0094	1171.4
	30	1.4438	0.0111	1423.8
FO-ISD suspension	10	0.8304	0.0066	821.4
	20	1.1632	0.0091	1155.9
	30	1.4054	0.0108	1406.2

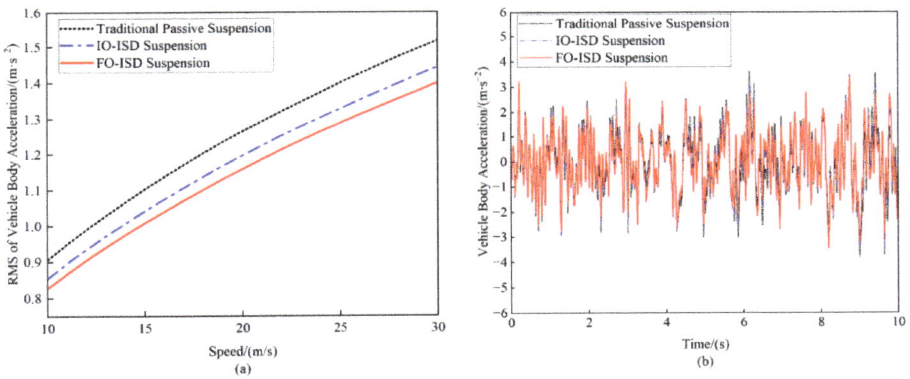

Figure 6. (**a**) Comparison of RMS values of vehicle body acceleration at speeds ranging from 10 m/s to 30 m/s. (**b**) Comparison of vehicle body acceleration at the speed of 20 m/s.

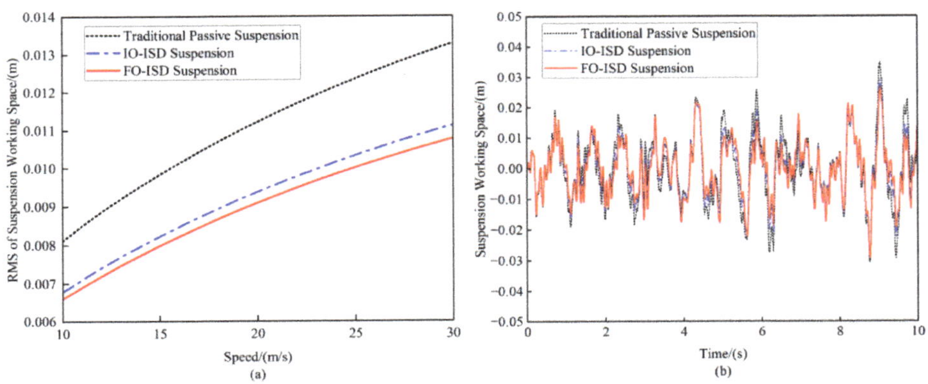

Figure 7. (**a**) Comparison of RMS values of suspension working space at speeds ranging from 10 m/s to 30 m/s. (**b**) Comparison of suspension working space at the speed of 20 m/s.

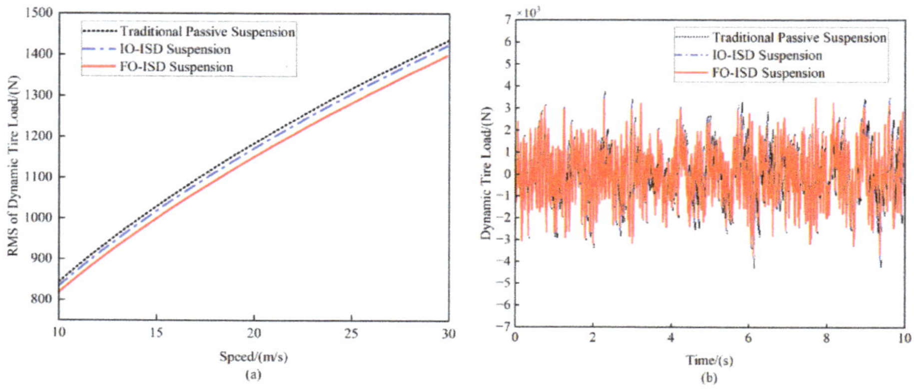

Figure 8. (**a**) Comparison of RMS values of dynamic tire load at speeds ranging from 10 m/s to 30 m/s. (**b**) Comparison of dynamic tire load at the speed of 20 m/s.

It can be seen that the FO-ISD suspension shows excellent advantages compared to the other two suspensions in the three performance indexes. Specifically, for the vehicle body acceleration, the RMS values of IO-ISD suspension decreased by 5.65%, 5.34%, and 4.88% at speeds of 10 m/s, 20 m/s, and 30 m/s, respectively. Meanwhile, the FO-ISD suspension decreased by 8.36%, 7.98%, and 7.41% at the corresponding speeds. Regarding the suspension working space, the IO-ISD suspension reduced by 16.05%, 16.07%, and 16.54% at different speeds, whereas the FO-ISD suspension achieved reductions of 18.52%, 18.75%, and 18.80%. Although the improvement in dynamic tire load was relatively small, IO-ISD suspension showed optimization margins of 1.32%, 1.11%, and 0.79%, while FO-ISD suspension achieved 2.68%, 2.39%, and 2.01%.

The results indicate superior performance of the FO-ISD suspension over the IO-ISD suspension across three key performance metrics. This superiority is attributed to the more intricate vibration transmission characteristics inherent in FO-ISD suspension, which enhances its vibration isolation capabilities.

4.2. Frequency Responses Analysis

For the frequency response to the suspension indexes, sinusoidal road input is given in different frequency domains, and the output amplitude is divided by the input amplitude to obtain the gain value of the corresponding index. Sinusoidal road input is as follows:

$$z_r(t) = Asin(2\pi ft) \tag{12}$$

In Formula (12), A represents the amplitude of sinusoidal road input, which is 0.01 m; f represents the frequency of the sinusoidal road input, with values ranging from [0.01,15].

The frequency responses of suspension indexes are shown in Figures 9–12.

The vehicle body acceleration gain is shown in Figure 9. In the figure, the red line represents the FO-ISD suspension, the blue line represents the IO-ISD suspension, and the black line represents the traditional passive suspension. It can be seen from the figure that in the low-frequency band (0–5 Hz), although the low-frequency peak of the FO-ISD suspension exceeds that of the IO-ISD suspension, it is still better than the traditional passive suspension. In the high-frequency band (10–15 Hz), the differences in the high-frequency peaks of the three types of suspensions are very small. However, the performance of the FO-ISD suspension is always better than that of the traditional passive suspension and the IO-ISD suspension. This indicates that the fractional-order design enhances the energy dissipation capacity of the inerter through the complex impedance characteristics, thereby more effectively suppressing the body vibration.

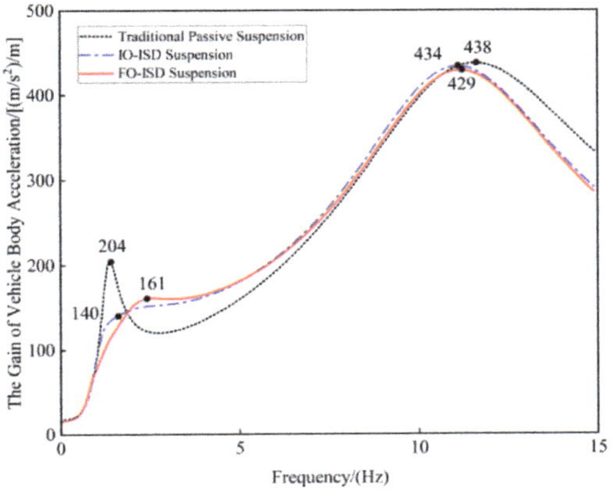

Figure 9. Frequency responses of vehicle body acceleration.

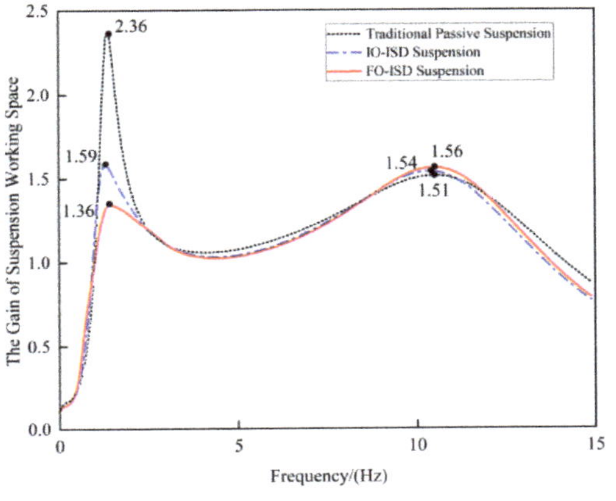

Figure 10. Frequency responses of suspension working space.

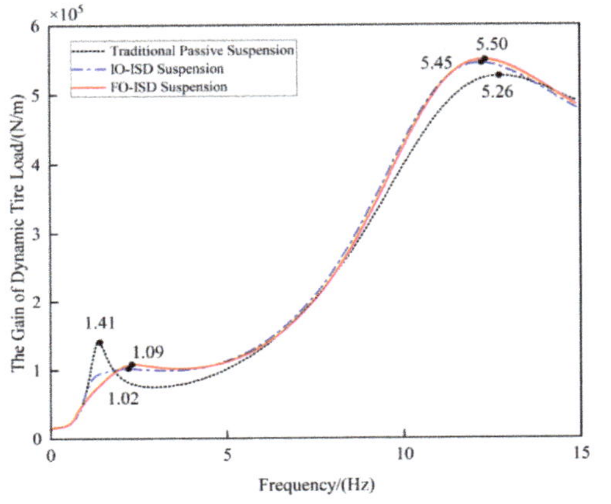

Figure 11. Frequency responses of dynamic tire load.

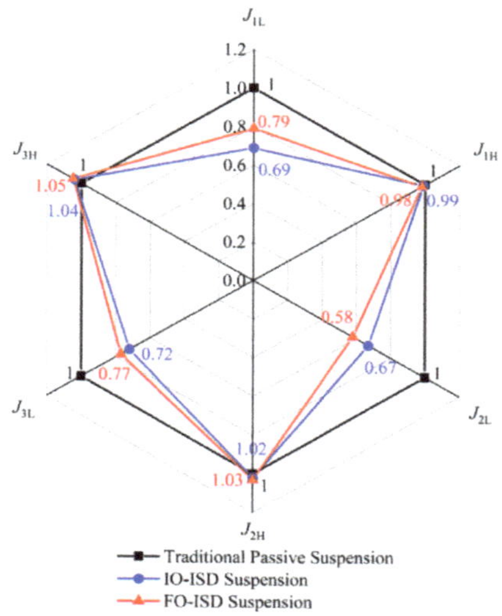

Figure 12. Radar diagram of low and high-frequency peak of suspensions.

As can be seen from Figure 10, although both types of ISD suspensions are slightly worse at the high-frequency peak, the low-frequency gain of the suspension working space of the FO-ISD suspension is significantly better than that of the other two types of suspensions, effectively improving the working space of the vehicle suspension when working in the low-frequency band. In Figure 11, although the dynamic tire load at the low-frequency peak of the FO-ISD suspension exceeds that of the IO-ISD suspension, it is still better than the traditional passive suspension. At the high-frequency resonance peak, the FO-ISD and IO-ISD suspensions deteriorate slightly, respectively. Overall, the FO-ISD suspension is slightly worse at the high-frequency peak, but the overall trend is similar to that of the IO-ISD suspension. In the low-frequency range, due to the complex impedance, the FO-ISD suspension fully exerts the advantage of the inerter. Compared with the IO-ISD suspension, the low-frequency amplitude suppression effect is further enhanced in most frequencies.

In Figure 12, the comparison of various indicators can be seen more intuitively. Among them, J_{1L} denotes the low-frequency peak of vehicle body acceleration, while J_{1H} represents its high-frequency counterpart. Similarly, J_{2L} and J_{2H} denote the low and high-frequency peaks of suspension working space, respectively. Meanwhile, J_{3L} and J_{3H} signify the low and high-frequency peaks of the dynamic tire load. In terms of body acceleration, although the J_{1L} of the FO-ISD suspension in the low-frequency band is slightly higher than that of the IO-ISD suspension, it is still 21% lower than that of the traditional passive suspension. The low-frequency peak of the IO-ISD suspension is reduced by 31%. In the case of J_{1H} in the high-frequency band, the performance of the FO-ISD and IO-ISD suspensions is similar. However, the FO-ISD suspension is still slightly better than the IO-ISD suspension and can further reduce body acceleration. In terms of suspension working space, the J_{2L} of the FO-ISD suspension working space is reduced by 42% compared with that of the traditional passive suspension, and the IO-ISD suspension working space is reduced by 33%. This shows that the FO-ISD suspension can maintain the optimal suspension working space when working at low frequency. In the case of J_{3L} in the low-frequency band, although the FO-ISD suspension is slightly higher than the IO-ISD suspension, it is still 23% lower than the traditional passive suspension. This effectively reduces the dynamic tire load of the vehicle and improves driving stability. In the case of J_{2H} and J_{3H} in the high-frequency band, the FO-ISD and IO-ISD suspensions exhibit slight deterioration. However, they are still within the acceptable range.

4.3. Impulse Pavement Input

The vehicle is set to drive through a speed reducer with a height of 0.05 m and a width of 0.3 m at a speed of 10 m/s [52]. The impulse response of dynamic performance indexes is illustrated in Figures 13–15; Table 5 shows the corresponding peak-to-peak (PTP) values.

Table 5. PTP values comparison between different suspensions.

	Traditional Passive Suspension	IO-ISD Suspension	FO-ISD Suspension
PTP value of the vehicle body acceleration/($m{\cdot}s^{-2}$)	13.6815	13.1320	12.8691
PTP value of the suspension working space/(m)	0.0428	0.0412	0.0416
PTP value of the dynamic tire load/(N)	21,488	21,298	21,314

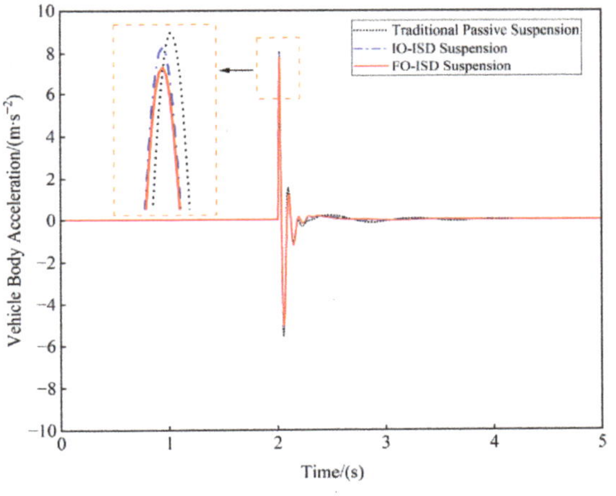

Figure 13. The impulse response of vehicle body acceleration.

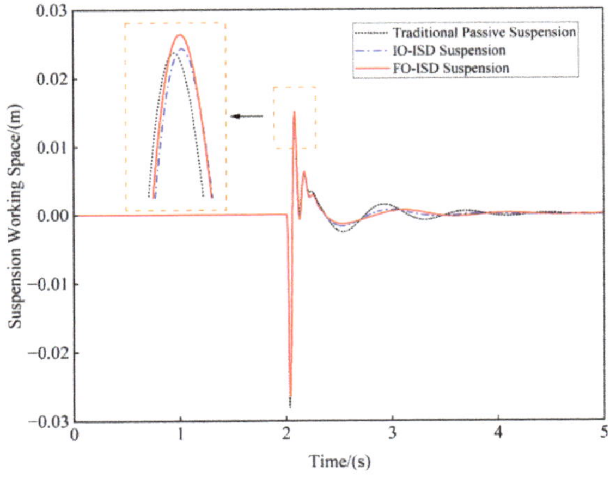

Figure 14. The impulse response of suspension working space.

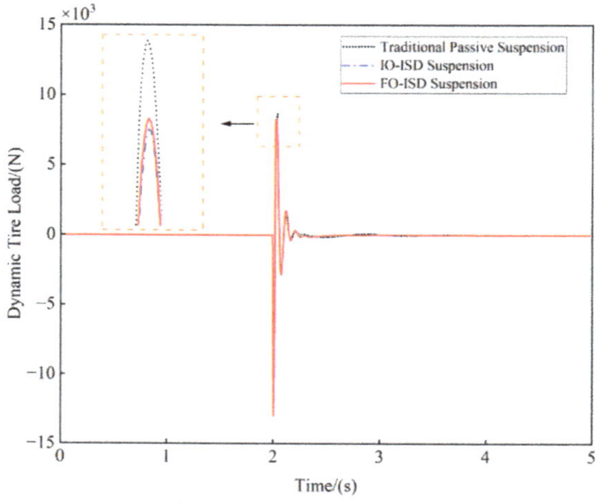

Figure 15. The impulse response of dynamic tire load.

The FO-ISD suspension optimizes the PTP values of the three suspension performance indexes compared to the traditional passive suspension. Specifically, the vehicle body acceleration shows a reduction of 5.94%, the suspension working space reduces by 2.80%, and the dynamic tire load exhibits marginal improvement. In comparison with IO-ISD suspension, the vehicle body acceleration is optimized by 2.00%, whereas the other two performance indexes exhibit a slight worsening in PTP values.

5. Discussion

This paper employs the fractional-order calculus theory to propose a design methodology for vehicle mechatronic ISD suspension utilizing a fractional-order biquadratic electrical network. A dynamic model of the vehicle suspension is formulated, with suspension parameters to be optimized using NSGA-II. The external network structure is synthesized using the fractional-order passive network design principles. Simulation results validate that the proposed fractional-order ISD suspension exhibits superior vibration isolation effectiveness compared to both integer-order ISD suspension and traditional passive suspension.

It is worth mentioning that the positive real synthetic passive implementation of the fractional-order transfer function can be applied to other vibration isolation fields to obtain better vibration isolation performance. In the future, we will focus on the simplest implementation of fractional-order transfer functions and the engineering implementation of fractional-order electrical components.

Author Contributions: Conceptualization, Y.S.; methodology, Y.S.; software, Z.L.; investigation, X.T.; data curation, K.J.; writing—original draft preparation, Z.L. and K.J.; writing—review and editing, X.T. and X.Y.; funding acquisition, Y.S. and X.Y. All authors have read and agreed to the published version of the manuscript.

Funding: This work was supported by the National Natural Science Foundation (Grant No. 52072157), the Postgraduate Research and Practice Innovation Program of Jiangsu Province (Grant No. SJCX23_2084), the Young Elite Scientists Sponsorship Program by CAST (Grant No. 2022QNRC001), and the China Postdoctoral Science Foundation (No. 2024T171048, 2024M753653, 2024M761549).

Data Availability Statement: All the pertinent data are available in the article.

Conflicts of Interest: Author Xiang Tian was employed by the company Chery New Energy Automobile Co., Ltd. The remaining authors declare that the research was conducted in the absence of any commercial or financial relationships that could be construed as a potential conflict of interest.

Abbreviations

The following abbreviations are used in this manuscript:

RMS	Root-mean-square
ISD	Inerter-spring-damper
DOF	Degrees of freedom
NSGA	Non-dominated sorting genetic algorithms
FO	Fractional-order
IO	Integer-order
PTP	Peak-to-peak

Appendix A

The positive real condition of the fractional-order biquadratic transfer function is given by the following theorem [51]:

Theorem 1. *Consider the fractional biquadratic transfer function $Y_e(s)$ of the Formula (3), where $A, B, C, D, E, F, G, H \geq 0$ and $0 < \alpha, \beta < 1$. Then, the necessary and sufficient condition for a positive real function of fractional order $Y_e(s)$:*

$$\text{when } \alpha = \beta,\ \left(\sqrt{AH} - \sqrt{DE}\right)^2 < (B + C)(F + G); \tag{A1}$$

$$\text{when } \alpha \neq \beta,\ (AH + DE) - (BG + CF) \leq 2\left(\sqrt{BCFG} + \sqrt{ADEH}\right) \tag{A2}$$

Theorem 2. *For the fractional-order biquadratic transfer function represented by Formula (3), a necessary and sufficient condition for common factors to exist between its numerator and denominator is met when one of the following conditions holds:*

$$AG - CE = 0 \text{ and } AH + BG - CF - DE = 0 \text{ and } BH - DF = 0 \tag{A3}$$

$$AF - BE = 0 \text{ and } AH + CF - BG - DE = 0 \text{ and } CH - DG = 0 \tag{A4}$$

Appendix B

The following are the positive real synthetic passive realization conditions of the fractional-order biquadratic transfer function [51]. According to the principle of simplicity, the simplest realization conditions of the two, three, four, and five electrical network components are tested in turn. The specific flow chart of its network synthesis process is shown in Figure A1.

Binary implementation conditions of the transfer function $Y_e(s)$ (5 coefficients are 0):

1. $B > 0$, $C > 0$ and $E > 0$ or $H > 0$, the remaining coefficients are 0;
2. $F > 0$, $G > 0$ and $A > 0$ or $D > 0$, the remaining coefficients are 0;
3. $A > 0$, $D > 0$ and $F > 0$, the remaining coefficients are 0;
4. $B > 0$, $E > 0$ and $H > 0$, the remaining coefficients are 0.

Three-element implementation conditions of the transfer function $Y_e(s)$ (in conditions 1–4, 4 coefficients are 0, and in conditions 5–10, 3 coefficients are 0):

1. $B > 0$, $C > 0$ and $A, E > 0$ or $D, H > 0$, the remaining coefficients are 0;
2. $F > 0$, $G > 0$ and $A, E > 0$ or $D, H > 0$, the remaining coefficients are 0;
3. $A > 0$, $B > 0$, $D > 0$, $F > 0$, the remaining coefficients are 0;
4. $B > 0$, $E > 0$, $F > 0$, $H > 0$, the remaining coefficients are 0;
5. $BG = CF$, and $A > 0$ or $D > 0$ or $E > 0$ or $H > 0$, the remaining coefficients are 0;
6. $CF = DE$, and $B > 0$ or $G > 0$, the remaining coefficients are 0;
7. $AH = BG$, and $C > 0$ or $F > 0$, the remaining coefficients are 0;
8. $BG = DE$, and $A > 0$ or $H > 0$, the remaining coefficients are 0;
9. $AH = CF$, and $D > 0$ or $E > 0$, the remaining coefficients are 0;
10. $AH = DE$, and $C > 0$ or $G > 0$, the remaining coefficients are 0.

Four-element implementation condition of the transfer function $Y_e(s)$ (in conditions 1–7, 2 coefficients are 0, and in conditions 8–15, 1 coefficient is 0);

1. $BG = CF$, $A = E = 0$, and satisfy $BH > DF$ or $DF > BH$ or $CH > DG$ or $DG > CH$;
2. $BG = CF$, $D = H = 0$, and satisfy $AG > CE$ or $CE > AG$ or $AF > BE$ or $BE > AF$;
3. $AH = CF$, $B = D = 0$, and satisfy $AG > CE$ or $CE > AG$;
4. $AH = CF$, $E = G = 0$, and satisfy $DF > BH$ or $BH > DF$;
5. $BG = DE$, $F = H = 0$, and satisfy $AG > CE$ or $CE > AG$;
6. $BG = DE$, $A = C = 0$, and satisfy $DF > BH$ or $BH > DF$;
7. $AH = DE$, $C = G = 0$, and satisfy $DF > BH$ or $BH > DF$;
8. $BG = CF + DE$, and satisfy $AG = CE$, $H = 0$ or $DF = BH$, $A = 0$;
9. $CF = AH + BG$, and satisfy $AG = CE$, $D = 0$ or $DF = BH$, $E = 0$;
10. $AH = BG + CF$, and satisfy $EH + FG$, $D = 0$ or $AD = BC$, $E = 0$;
11. $DE = BG + CF$, and satisfy $AD = BC$, $H = 0$ or $EH = FG$, $A = 0$;
12. $DE = AH + CF$, and satisfy $AF = BE$, $DF > BH$, $G = 0$ or $CH = DG$, $CE > AG$, $B = 0$;
13. $AH = BG + DE$, and satisfy $AF = BE$, $BH > DF$, $C = 0$ or $CH = DG$, $AG > CE$, $F = 0$;
14. $CF = AH + DE$, $EH = GF$, $B = 0$;
15. $BG = AH + DE$, $AD = BC$, $F = 0$.

Five-element implementation conditions of the transfer function $Y_e(s)$ (in conditions 1–6, 1 coefficient is 0, and in conditions 7–14, all coefficients are > 0):

1. $4BC(FG - EH) = (DE - CF - BG)^2$, $A = 0$;
2. $4BC(FG - EH) = (AH - CF - BG)^2$, $D = 0$;
3. $4FG(BC - AD) = (DE - CF - BG)^2$, $H = 0$;
4. $4FG(BC - AD) = (AH - CF - BG)^2$, $E = 0$;
5. $4AG(BH - DF) = (DE - BG - AH)^2$, $C = 0$;
6. $4CE(DF - BH) = (AH - CF - DE)^2$, $G = 0$;
7. $AH + BG - CF - DE = 0$, $BH - DF = 0$, $CE > AG$, and satisfy $DE > AH$ or $AH > DE$;
8. $AH + BG - CF - DE = 0$, $BH - DF = 0$, $AG > CE$, and satisfy $\mathrm{DE} > \mathrm{AH}$ or $\mathrm{AH} > \mathrm{DE}$;
9. $AH + BG - CF - DE = 0$, $AG - CE = 0$, $DE > AH$, and satisfy $BH > DF$ or $DF > BH$;
10. $AH + BG - CF - DE = 0$, $AG - CE = 0$, $AH > DE$, and satisfy $BH > DF$ or $DF > BH$;
11. $AH + DE - BG - CF = 0$, $EH - FG = 0$, $DG > CH$ and satisfy $AG > CE$ or $CE > AG$;
12. $AH + DE - BG - CF = 0$, $EH - FG = 0$, $AF > BE$ and satisfy $BH > DF$ or $DF > BH$;
13. $AH + DE - BG - CF = 0$, $AD - BC = 0$, $CH > DG$ and satisfy $AG > CE$ or $CE > AG$;
14. $AH + DE - BG - CF = 0$, $AD - BC = 0$, $BE > AF$ and satisfy $BH > DF$ or $DF > BH$.

In order to follow the simplest principle of passive implementation of the transfer function, the implementation conditions of six elements and more are not given here.

When all coefficients A, B, C, D, E, F, G, H are positive, a network comprising five series-parallel elements with two fractional-order reactance elements can be implemented.

If at least one of the coefficients A, B, C, D, E, F, G, H is zero, the network can be realized using up to four elements.

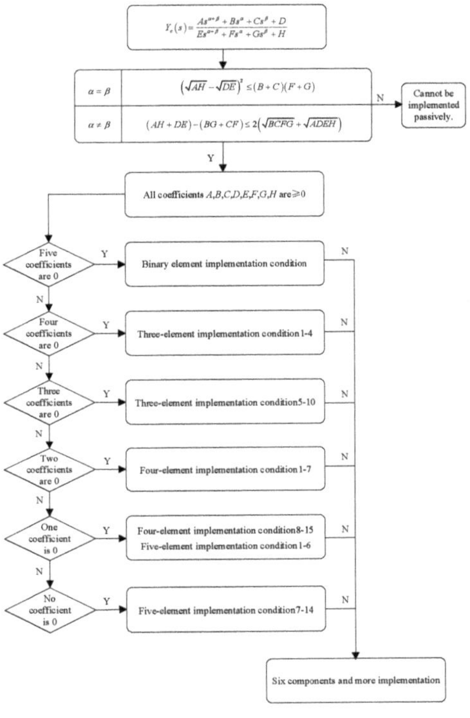

Figure A1. References.

References

1. Sankaranarayanan, V.; Emekli, M.E.; Gilvenc, B.A.; Guvenc, L.; Ozturk, E.S.; Ersolmaz, E.S.; Sinal, M. Semiactive suspension control of a light commercial vehicle. *IEEE/ASME Trans. Mechatron.* **2008**, *13*, 598–604. [CrossRef]
2. Liu, C.N.; Chen, L.; Yang, X.F.; Zhang, X.L.; Yang, Y. General theory of skyhook control and its application to semi-active suspension control strategy design. *IEEE Access* **2019**, *7*, 101552–101560. [CrossRef]
3. Shen, Y.J.; Chen, A.; Du, F.; Yang, X.F.; Liu, Y.L.; Chen, L. Performance enhancements of semi-active vehicle air ISD suspension. *Proc. Inst. Mech. Eng. Part D J. Automob. Eng.* **2024**, 09544070241233024. [CrossRef]
4. Shen, Y.J.; Jia, M.Q.; Yang, X.F.; Liu, Y.L.; Chen, L. Vibration suppression using a mechatronic PDD-ISD-combined vehicle suspension system. *Int. J. Mech. Sci.* **2023**, *250*, 108277. [CrossRef]
5. Nguyen, D.N.; Nguyen, T.A.; Dang, N.D. A novel sliding mode control algorithm for an active suspension system considering with the hydraulic actuator. *Lat. Am. J. Solids Struct.* **2022**, *19*, e424. [CrossRef]
6. Yu, Y.; Li, Z.; Zhou, Y.; Wang, X. A nonlinear model predictive control for air suspension in hub motor electric vehicle. *Proc. Inst. Mech. Eng. Part D J. Automob. Eng.* **2025**, *239*, 2–3. [CrossRef]
7. Yin, Y.; Li, S.E.; Tang, K.; Cao, W.; Wu, W.; Li, H. Approximate optimal filter design for vehicle system through actor-critic reinforcement learning. *Automot. Innov.* **2022**, *5*, 415–426. [CrossRef]
8. Zhong, S.; Zhao, Y.; Ge, L.; Shan, Z.; Ma, F. Vehicle state and bias estimation based on unscented kalman filter with vehicle hybrid kinematics and dynamics models. *Automot. Innov.* **2023**, *6*, 571–585. [CrossRef]
9. Smith, M.C. Synthesis of mechanical networks: The inerter. *IEEE Trans. Autom. Control* **2002**, *47*, 1648–1662. [CrossRef]
10. Zhang, Y.; Xu, K.; Zang, J.; Ni, Z.; Zhu, Y.; Chen, L. Dynamic design of a nonlinear energy sink with NiTiNOL-steel wire ropes based on nonlinear output frequency response functions. *Appl. Math. Mech.* **2019**, *40*, 1791–1804. [CrossRef]
11. Shen, W.; Niyitangamahoro, A.; Feng, Z.; Zhu, H. Tuned inerter dampers for civil structures subjected to earthquake ground motions: Optimum design and seismic performance. *Eng. Struct.* **2019**, *198*, 109470. [CrossRef]

12. Qiu, Z.; Han, S.; Na, J.; Wang, C. Vertical suspension optimization for a high-speed train with PSO intelligent method. *Comput. Intell. Neurosci.* **2021**, *2021*, 1526792. [CrossRef]
13. Smith, M.C.; Wang, F.C. Performance benefits in passive vehicle suspensions employing inerters. *Veh. Syst. Dyn.* **2004**, *42*, 235–257. [CrossRef]
14. Nie, J.; Yang, Y.; Jiang, T.; Zhang, H. Passive skyhook suspension reduction for improvement of ride comfort in an off-road vehicle. *IEEE Access* **2019**, *7*, 150710–150719. [CrossRef]
15. Zhang, X.L.; Liu, J.; Nie, J.; Wei, H.; Chen, L. Simulation analysis and experiment research on hydro-pneumatic isd suspension. *Shock Vib.* **2021**, *2021*, 2095350. [CrossRef]
16. Wang, Y.; Wang, P.; Meng, H.; Chen, L.Q. Nonlinear vibration and dynamic performance analysis of the inerter-based multi-directional vibration isolator. *Arch. Appl. Mech.* **2022**, *92*, 3597–3629. [CrossRef]
17. Fu, D.; Ming, M.; Yijie, C.; Yajun, W.; Yafeng, Z. Structure design and performance analysis of ISD suspension based on dynamic model and parameter optimization. *Shock Vib.* **2014**, *33*, 59–65. Available online: https://jvs.sjtu.edu.cn/CN/Y2014/V33/I6/59 (accessed on 25 January 2024).
18. Shen, Y.; Qiu, D.; Yang, X.; Chen, J.; Guo, Y.; Zhang, T. Vibration isolation performance analysis of a nonlinear fluid inerter-based hydro-pneumatic suspension. *Int. J. Struct. Stab. Dyn.* **2024**. [CrossRef]
19. Srinivas, N.; Deb, K. Muiltiobjective optimization using nondominated sorting in genetic algorithms. *Evol. Comput.* **1994**, *2*, 221–248. [CrossRef]
20. Bu, J.G.; Lan, X.D.; Zhou, M.; Lv, K.X. Performance optimization of flywheel motor by using NSGA-2 and AKMMP. *IEEE Trans. Magn.* **2018**, *54*, 8103707. [CrossRef]
21. Wang, X.; Fan, Y.; Lu, X.; Chen, Q.; Lee, C.H. Multiobjective optimization of a dual stator brushless hybrid excitation motor based on response surface model and NSGA 2. *IEEE Trans. Ind. Appl.* **2022**, *58*, 6105–6114. [CrossRef]
22. Wang, P.; Zang, T.; Xiao, W.; Zhang, Y.; He, J.; Men, H.; Li, J. An optimized resource allocation approach in limited emergency situations using NSGA-II. In Proceedings of the 2023 9th International Conference on Big Data and Information Analytics (Big DIA), Sorrento, Italy, 15–18 December 2023; pp. 258–264. [CrossRef]
23. Yang, X.; Zhang, T.; Shen, Y.; Liu, Y.; Bui, V.; Qiu, D. Tradeoff analysis of the energy-harvesting vehicle suspension system employing inerter element. *Energy* **2024**, *308*, 132841. [CrossRef]
24. Brune, O. Synthesis of a Finite Two-Terminal Whose Driving-Point Impedance Is a Function of Frequency. Ph.D. Thesis, Massachusetts Institute of Technology, Cambridge, MA, USA, 1931. [CrossRef]
25. Bott, R.; Duffin, R.J. Impedance synthesis without use of transformers. *J. Appl. Phys.* **1949**, *20*, 816. [CrossRef]
26. Papageorgiou, C.; Smith, M.C. Positive real synthesis using matrix inequalities for mechanical networks: Application to vehicle suspension. *IEEE Trans. Control Syst. Technol.* **2006**, *14*, 423–435. [CrossRef]
27. Jiang, J.Z.; Smith, M.C. Series-parallel six-element synthesis of biquadratic impedances. *IEEE Trans. Circuits Syst. I Regul. Pap.* **2012**, *59*, 2543–2554. [CrossRef]
28. Jiang, J.Z.; Smith, M.C. Regular positive-real functions and five-element network synthesis for electrical and mechanical networks. *IEEE Trans. Autom. Control* **2010**, *56*, 1275–1290. [CrossRef]
29. Wang, K. Implementation of Biquadratic Function in Passive Network Synthesis. Master's Thesis, Nanjing University of Science and Technology, Nanjing, China, 2013.
30. Zhang, S.Y.; Jiang, J.Z.; Wang, H.L.; Neild, S. Synthesis of essential-regular bicubic impedances. *Int. J. Circuit Theory Appl.* **2017**, *45*, 1482–1496. [CrossRef]
31. Wang, K.; Chen, M.Z.; Liu, F. Series-parallel mechanical circuit synthesis of a positive-real third-order admittance using at most six passive elements for inerter-based control. *J. Frankl. Inst.* **2023**, *360*, 5442–5480. [CrossRef]
32. Wang, K.; Chen, M.Z. Passive mechanical realizations of bicubic impedances with no more than five elements for inerter-based control design. *J. Frankl. Inst.* **2021**, *358*, 5353–5385. [CrossRef]
33. Gutierrez, R.E.; Rosário, J.M.; Tenreiro Machado, J. Fractional order calculus: Basic concepts and engineering applications. *Math. Probl. Eng.* **2010**, *2010*, 375858. [CrossRef]
34. Caponetto, R.; Graziani, S.; Pappalardo, F.L.; Sapuppo, F. Experimental characterization of ionic polymer metal composite as a novel fractional order element. *Adv. Math. Phys.* **2013**, *2013*, 953695. [CrossRef]
35. Nasreen, N.; Lu, D.; Zhang, Z.; Akgül, A.; Younas, U.; Nasreen, S.; Al-Ahmadi, A.N. Propagation of optical pulses in fiber optics modelled by coupled space-time fractional dynamical system. *Alex. Eng. J.* **2023**, *73*, 173–187. [CrossRef]
36. Chen, Y.; Xu, J.; Tai, Y.; Xu, X.; Chen, N. Critical damping design method of vibration isolation system with both fractional-order inerter and damper. *Mech. Adv. Mater. Struct.* **2022**, *29*, 1348–1359. [CrossRef]
37. Zhang, L.; Kartci, A.; Elwakil, A.; Bagci, H.; Salama, K.N. Fractional-order inductor: Design, simulation, and implementation. *IEEE Access* **2021**, *9*, 73695–73702. [CrossRef]
38. Tsirimokou, G.; Psychalinos, C.; Elwakil, A.S.; Salama, K.N. Experimental behavior evaluation of series and parallel connected constant phase elements. *AEU-Int. J. Electron. Commun.* **2017**, *74*, 5–12. [CrossRef]

39. Wang, F.C.; Chan, H.A. Vehicle suspensions with a mechatronic network strut. *Veh. Syst. Dyn.* **2011**, *49*, 811–830. [CrossRef]
40. Shen, Y.J.; Hua, J.; Fan, W.; Liu, Y.L.; Yang, X.F.; Chen, L. Optimal design and dynamic performance analysis of a fractional-order electrical network-based vehicle mechatronic ISD suspension. *Mech. Syst. Signal Process.* **2023**, *184*, 109718. [CrossRef]
41. Shen, Y.J.; Hua, J.; Hou, Q.H.; Xia, X.H.; Liu, Y.L.; Yang, X.F. Performance analysis of the fractional-order vehicle mechatronic ISD suspension with parameter perturbation. *J. Theor. Appl. Mech.* **2022**, *60*, 141–152. [CrossRef] [PubMed]
42. Feng, Z.; Yu, M.; Evangelou, S.A.; Jaimoukha, I.M.; Dini, D. Mu-synthesis PID control of full-car with parallel active link suspension under variable payload. *IEEE Trans. Veh. Technol.* **2022**, *72*, 176–189. [CrossRef]
43. Liang, G.; Ma, L. Multivariate theory-based passivity criteria for linear fractional networks. *Int. J. Circuit Theory Appl.* **2018**, *46*, 1358–1371. [CrossRef]
44. Liang, G.; Huo, X. Synthesis of fractional-order biquadratic immittance functions. *J. Circuits Syst. Comput.* **2019**, *28*, 1950187. [CrossRef]
45. Gadhvi, B.; Savsani, V.; Patel, V. Multi-objective optimization of vehicle passive suspension system using NSGA-II, SPEA2 and PESA-II. *Procedia Technol.* **2016**, *23*, 361–368. [CrossRef]
46. Onat, C.; Kucukdemiral, I.B.; Sivrioglu, S.; Yuksek, I.; Cansever, G. LPV gain-scheduling controller design for a non-linear quarter-vehicle active suspension system. *Trans. Inst. Meas. Control* **2009**, *31*, 71–95. [CrossRef]
47. Ramachandran, V.; Rao, A.S. The real part of a multivariable positive real function and some applications. *IEEE Trans. Circuits Syst.* **1974**, *21*, 598–605. [CrossRef]
48. Chen, M.Z.; Smith, M.C. A note on tests for positive-real functions. *IEEE Trans. Autom. Control* **2009**, *54*, 390–393. [CrossRef]
49. Bose, N.K. A criterion to determine if two multivariable polynomials are relatively prime. *Proc. IEEE* **1972**, *60*, 134–135. [CrossRef]
50. Huo, X. Passive Synthesis of Fractional Order Biquadratic Impedance Functions. Master's Thesis, North China Electric Power University, Beijing, China, 2019.
51. Loizos, A.; Plati, C. An alternative approach to pavement roughness evaluation. *Int. J. Pavement Eng.* **2008**, *9*, 69–78. [CrossRef]
52. Shao, X.; Naghdy, F.; Du, H.; Qin, Y. Coupling effect between road excitation and an in-wheel switched reluctance motor on vehicle ride comfort and active suspension control. *J. Sound Vib.* **2019**, *443*, 683–702. [CrossRef]

fractal and fractional

MDPI

Article

Fractional-Order Controller for the Course Tracking of Underactuated Surface Vessels Based on Dynamic Neural Fuzzy Model

Guangyu Li [1], Yanxin Li [1,*], Xiang Li [2], Mutong Liu [2,3], Xuesong Zhang [1] and Hua Jin [1]

1 School of Railway Intelligent Engineering, Dalian Jiaotong University, Dalian 116028, China; ligyu@djtu.edu.cn (G.L.)
2 School of Electronic Information and Automation, Civil Aviation University of China, Tianjin 300300, China
3 State Key Laboratory of Rail Transit Vehicle System, Southwest Jiaotong University, Chengdu 610031, China
* Correspondence: leeyx@djtu.edu.cn

Abstract: Aiming at the uncertainty problem caused by the time-varying modeling parameters associated with ship speed in the course tracking control of underactuated surface vessels (USVs), this paper proposes a control algorithm based on the dynamic neural fuzzy model (DNFM). The DNFM simultaneously adjusts the structure and parameters during learning and fully approximates the inverse dynamics of ships. Online identification and modeling lays the model foundation for ship motion control. The trained DNFM, serving as an inverse controller, is connected in parallel with the fractional-order $PI^{\lambda}D^{\mu}$ controller to be used for the tracking control of the ship's course. Moreover, the weights of the model can be further adjusted during the course tracking. Taking the actual ship data of a 5446 TEU large container ship, simulation experiments are conducted, respectively, for course tracking, course tracking under wind and wave interferences, and comparison with five different controllers. This proposed controller can overcome the influence of the uncertainty of modeling parameters, tracking the desired course quickly and effectively.

Keywords: dynamic neural fuzzy model; system identification; fractional-order $PI^{\lambda}D^{\mu}$ controller; course tracking; inverse dynamic characteristics; underactuated surface vessels

1. Introduction

When sailing in the ocean, changes in navigation conditions such as ship speed or loading capacity cause changes in various hydrodynamic derivatives. The ship modeling parameters used for controller design are precisely complex nonlinear functions that are related to these hydrodynamic derivatives. For course tracking control of a ship, the uncertainty caused by ship modeling parameters due to the ship speed has always been very difficult for researchers [1–4], and it is particularly important to find advanced solutions. Ship dynamics are characterized by nonlinearity, large inertia, and longtime delay. The issue of uncertainty in ship motion control has always received much attention [5–7]. In recent years, various intelligent control technologies have begun to flourish, such as particle swarm optimization [8], neural networks [9], and genetic algorithms [10]. Among them, fuzzy control and neural networks for ships are two core technologies of intelligent control, which have successfully solved many practical control problems [11–20]. In 1990, Narendra et al. [21] proposed using neural networks for the identification and control of nonlinear systems. The model adopted a connection mode of combining multilayer networks and recurrent networks. Meanwhile, static and dynamic backpropagation methods for parameter adjustment were discussed in the paper. Some new concepts and methods were introduced in the paper, which pointed out the research direction for the application of neural networks in system identification in the future. Lin and Lee [22] proposed a neural network model for decision systems and fuzzy logic control. Based on a feedforward multilayer network, this model combines the learning ability of neural networks with fuzzy control theory to form a neural-network-based fuzzy logic control and decision system. Its learning efficiency is higher than that of the backpropagation learning algorithm. Sastry et al. [23] proposed using memory neuron networks for the identification of nonlinear dynamic systems. On this basis, model reference adaptive control was explored. The structure of this model was obtained by adding trainable temporal elements to the feedforward network. It can identify dynamic systems without having to be explicitly fed with past inputs and outputs. This model can identify systems with unknown delays or systems whose order is unknown. Barada and Singh [24] presented an optimal adaptive fuzzy neural model structure based on I/O. This model uses the fuzzy inference system of an adaptive network to train the Takagi–Sugeno–Kang fuzzy model. Lee and Teng [25] adopted a recurrent fuzzy neural network for the identification and control of nonlinear dynamic systems. Essentially, this model is a recurrent multilayer connectionist network that realizes fuzzy inference by using dynamic fuzzy rules. Abiyev and Kaynak [26] introduced a structure of the Type 2 Takagi–Sugeno–Kang fuzzy neural system. The update rule for the model's parameters adopts the gradient learning algorithm and fuzzy clustering. This model is used for the identification and control of uncertain systems and has achieved good control effects. Feng and Chen [27] introduced the Takagi–Sugeno (TS) fuzzy system into a broad learning system (BLS) to construct a new type of neuro-fuzzy system. This model replaces the feature nodes of the BLS with a group of TS fuzzy subsystems to preserve the characteristics of the inputs. The model output is obtained by combining the outputs of the enhancement layer with the defuzzification outputs of all the fuzzy subsystems [28–30]. However, most of these intelligent algorithms require an accurate ship model, which directly affects the performance of the controller. Therefore, it is of crucial importance to explore advanced control algorithms.

Many scholars have conducted a great deal of research on mathematical modeling. Fuzzy rules and neural networks are powerful tools to solve this problem. In the early stage of the combination of these two intelligent algorithms, most of them first determine the network structure or give the fuzzy rules in advance, and then use the generalization error to evaluate the modeling accuracy, and then directly use it for course control. As it relies on the trial-and-error method or the expert's knowledge to determine the structure or rules, this kind of algorithm itself has greater uncertainty and limitations [31–39]. In order to design an automatic fuzzy controller, some scholars have proposed a method based on fuzzy neural networks [40–47], by using various optimization algorithms to train the relevant parameters in the controller. However, a disadvantage of this method lies in its excessively long training process and relatively low reliability. In addition, a method based on an indirect adaptive fuzzy controller is also applied to course tracking. However, in this method, it is very difficult to generate adaptive rules. Therefore, during the implementation process, there is still a relatively high dependence on expert knowledge [48]. In recent years, with the development of AI technology, due to the complementary working mechanisms and equivalent functions of fuzzy systems and neural networks, neural fuzzy systems have achieved certain results in the field of ship course tracking and dynamic positioning [49–55]. In [49], the ANFIS controller for training real ship trial data to obtain a mathematical model with the actual ship motion is proposed. And it is combined with a nonlinear feedback method for ship course control. In [51], the authors proposed a method of using the neural fuzzy algorithm in a ship dynamic positioning control system. A controller with self-tuning functionality that relies on the neural fuzzy algorithm is employed to regulate the outboard thruster speed so as to optimize the course and adjust the ship position and path tracking. The six-degree-of-freedom ship motion with a dynamic positioning system is simulated in the time domain by using the fourth-order Runge–Kutta method. The results show that the ship path and course deviation are within an acceptable range. In [53], three adaptive critic-based neuro-fuzzy controllers were put forward by the authors to control the attitude and position of ships. In [54], the authors proposed an adaptive neuro-fuzzy model for identifying the inverse dynamic characteristics in ship course control. This model is capable of automatically dividing the input space, figuring out the number of fuzzy rules and membership functions. Sawtooth and sine waves are, respectively, used as course inputs to obtain two different inverse dynamic models in ship course control. The trained two sets of adaptive neuro-fuzzy models are employed as inverse controllers and are connected with the $PI^{\lambda}D^{\mu}$ controller for ship course control, respectively. The author also points out that since the inverse dynamic model is identified offline, the control conditions should be the same as the training conditions, which is idealized. The controller has limitations.

In ship motion control, the digital PID autopilot is usually used for course operation. High-frequency interference is likely to cause the digital PID autopilot to operate the rudder frequently, making it difficult for the ship's course to reach the preset course [56]. Podlubny I initially put forward the $PI^{\lambda}D^{\mu}$ controller in 1999 [57]. Fractional calculus is introduced to PID controllers. The integral order $\lambda > 0$ and differential order $\mu > 0$ are the orders of PID and can take any real number. As λ and μ are arbitrary positive real numbers, $PI^{\lambda}D^{\mu}$ can be more flexible and robust [58–63]. In [58], the authors proposed a design scheme of a fractional-order $PI^{\lambda}D^{\mu}$ controller combined with an ANFIS for controlling a class of first-order delay systems. The intelligent ANFIS system is employed to dynamically adjust λ and μ of $PI^{\lambda}D^{\mu}$ so as to attain a more superior controlling result. In [59], the authors proposed a method for $PI^{\lambda}D^{\mu}$ controllers based on a fractional-order actor–critic algorithm. The proposed FOPID-FOAC scheme makes use of the $PI^{\lambda}D^{\mu}$ controller and the performance of nonlinear systems is enhanced by applying the FOAC method. In [61], the authors introduced the modeling and formal verification of $PI^{\lambda}D^{\mu}$ controller systems in higher-order logic (HOL). It fully compares $PI^{\lambda}D^{\mu}$ controllers and PID controllers and proves the superiority of the former. Based on the Grünwald–Letnikov definition of fractional-order grids proved by higher-order logic theorems, the relationship between $PI^{\lambda}D^{\mu}$ controllers and PID controllers is proved. Finally, the stability of the system is proved. In [62], the authors made use of a backpropagation neural network and the recursive least squares method to identify the ship motion model, which was affected by external factors. Then, according to the dynamically identified motion model, the $PI^{\lambda}D^{\mu}$ controller based on particle swarm optimization was used to control the ship course.

Taking a 5446TEU large container ship as the research object, this paper proposes the structure of a dynamic neural fuzzy model and its corresponding learning algorithm. The model is a five-layer forward network structure. The DNFM can automatically generate fuzzy rules, adjust the structure and parameters simultaneously during the learning process, and automatically partition the input space, determining the number of membership functions and the number of fuzzy rules. In the structure of the DNFM, the number of rules will directly determine the generalization ability of the system. This model adopts a learning algorithm that can automatically determine the number of rules and adjust the parameters and structure simultaneously, which is mainly divided into two aspects: the determination of fuzzy rules and weights. The generation of fuzzy rules is determined by using the hierarchical learning method. The error descent rate method is used to prune the fuzzy rules, and the basis for pruning is the importance of the fuzzy rules. The DNFM has good generalization ability and can be effectively applied to the learning of the inverse model of ships with uncertainties. Meanwhile, the DNFM can generate appropriate compensation signals both in the modeling and control processes, effectively overcoming the influence of uncertain factors. Next, the fractional-order $PI^{\lambda}D^{\mu}$ controller is used to replace the traditional digital PID autopilot controller. Due to the introduction of the integral order λ and the differential order μ (where $\lambda > 0$ and $\mu > 0$ are arbitrary real orders), the controller can be more flexible and have stronger robustness. Finally, the trained DNFM is used as an inverse controller and connected in parallel with the fractional-order $PI^{\lambda}D^{\mu}$ controller to construct an online DNFM-FOPID controller. The DNFM acts as a feedforward inverse controller to generate compensation signals, while the introduction of the fractional-order $PI^{\lambda}D^{\mu}$ controller is to ensure reliable course tracking control. The simulation results show that the DNFM-FOPID controller can overcome the influence of the uncertainty of ship modeling parameters caused by the change in ship speed, quickly and effectively tracking the desired course.

The principal contributions can be summarized as follows:

i. For the uncertainty problem caused by time-varying modeling parameters associated with ship speed, this paper creates a novel identification method based on the DNFM to identify the inverse dynamic characteristics of ship motion. The DNFM can make adjustments to its structure and parameters simultaneously during learning. It provides a full approximation of the inverse dynamics of ship motion.

ii. Regarding the problem that the digital PID autopilot in ship course control is overly sensitive to high-frequency interference and ship models, this paper uses the $PI^{\lambda}D^{\mu}$ controller to replace the digital PID autopilot. The integral order λ and differential order μ are arbitrary positive real numbers, thus making the controller more flexible and robust.

iii. The trained DNFM, serving as an inverse controller, is connected in parallel with the $PI^{\lambda}D^{\mu}$ controller to be used for the tracking control of the ship's course. The weights of the dynamic neural fuzzy model can be further adjusted online, thus improving the accuracy of the controller.

iv. In order to verify the algorithm performance, a comparison experiment among five different ship course controllers is conducted, namely the ANFM-FOPID controller [54], the PID controller [64], the PSO-PID controller [65], the $PI^{\lambda}D^{\mu}$ controller based on the evolutionary algorithm [66], and the controller proposed in this paper.

The rest of this paper is as follows. In Section 2, the motion model of a USV and the Norrbin nonlinear model of ship course control studied in this paper are presented. In Section 3, the DNFM structure and its learning algorithm are presented. Based on this, it is employed for the online identification of the ship course inverse controller model proposed in this paper. In Section 4, fractional calculus and $PI^{\lambda}D^{\mu}$ controllers are briefly introduced. Based on this, in this study, the $PI^{\lambda}D^{\mu}$ controller is connected in parallel with the ship course inverse controller to achieve online ship course control. In Section 5, the proposed controller is validated. Through Matlab R2016b and relying on the real ship data of 5446 TEU large container ships, simulations of course tracking control, course control under wind and wave interference, and comparison simulations with five different controllers are, respectively, conducted to verify the effectiveness of the algorithm. In Section 6, the conclusion and the direction of future work are presented.

2. USV Mathematical Model

2.1. Motion Model

A ship may be treated as a rigid body. Its actual motion encompasses surge, sway, pitch, heave, roll, and yaw, as shown in Figure 1.

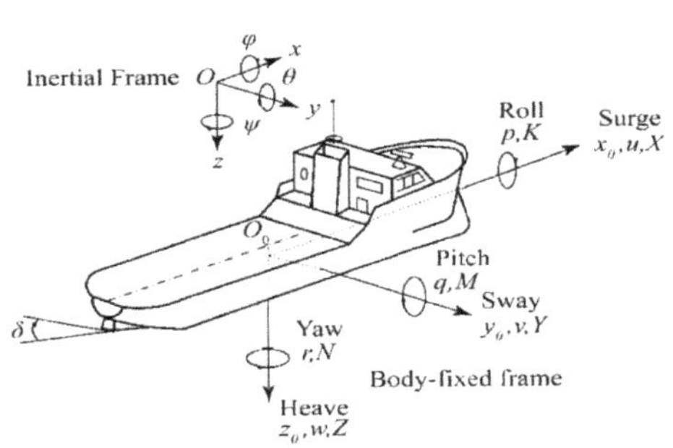

Figure 1. Motion coordinate system.

Ship maneuvering usually only contemplates the movement of the ship in the horizontal plane. For most ship motion and its control problems, the motions of pitch, heave, and roll can be ignored. Only surge, sway, and yaw motions are considered. As a result, the ship motion problem is simplified to planar motion with merely three degrees of freedom. This paper also solely considers the horizontal plane motion control of a USV. The models [67] are

$$\begin{cases} \dot{x} = u\cos\psi - v\sin\psi \\ \dot{y} = u\sin\psi + v\cos\psi \\ \dot{\psi} = r \end{cases} \tag{1}$$

$$\begin{cases} \dot{u} = \frac{m_{22}}{m_{11}}vr - \frac{d_{11}}{m_{11}}u + \frac{1}{m_{11}}\tau_u \\ \dot{v} = -\frac{m_{11}}{m_{22}}ur - \frac{d_{22}}{m_{22}}v \\ \dot{r} = \frac{m_{11}-m_{22}}{m_{33}}uv - \frac{d_{33}}{m_{33}}r + \frac{1}{m_{33}}\tau_r \end{cases} \tag{2}$$

In the equations, x, y, and ψ, respectively, denote the longitudinal position coordinate, transverse position coordinate, and course angle. u, v, and r, respectively, stand for longitudinal, transverse, and yaw angular velocity. m_{11}, m_{22}, and m_{33} represent inherent added mass and are uncertain terms. d_{11}, d_{22}, and d_{33} represent the hydrodynamic damping in transverse, longitudinal, and yawing directions and are uncertain terms. τ_u and τ_r represent the longitudinal propulsion force and ship-turning moment of the propeller.

2.2. The Norrbin Nonlinear Model for Course Control

During the design process of the course controller, the ship is generally considered as a dynamic system. After omitting the transverse drift velocity, it grasps the principal essence of the ship's dynamics from $\delta \to \dot{\psi} \to \psi$. The interference of wind and waves can be turned into a specific rudder angle for input signal δ_D. Along with the actual rudder angle δ, it enters the ship model, as illustrated in Figure 2.

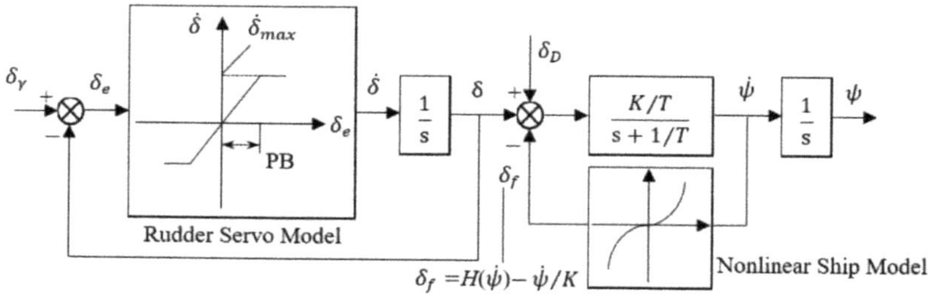

Figure 2. Corresponding nonlinear ship model.

The second-order Nomoto model is as follows:

$$\ddot{\psi} + \frac{1}{T}\dot{\psi} = \frac{K}{T}\delta \tag{3}$$

In order to improve the description accuracy, especially for some statically unstable ships, the second term $\dot{\psi}/T$ on the left side of Equation (3) must be replaced by a nonlinear term $(K/T)H(\dot{\psi})$, and

$$H\left(\dot{\psi}\right) = \alpha\dot{\psi} + \beta\dot{\psi}^3 \tag{4}$$

where α and β are nonlinear coefficients. Thus, the nonlinear second-order ship motion response model becomes

$$\ddot{\psi} + \frac{K}{T}H\left(\dot{\psi}\right) = \frac{K}{T}\delta \tag{5}$$

Obviously, in the linear case, in order to make Equation (3) consistent with Equation (5), it is necessary to have $\alpha = 1/K, \beta = 0$.

Equation (5) can be described by a block diagram, which is the nonlinear ship model on the right part of Figure 2. There exists a nonlinear feedback part from $\dot{\psi} \to \delta_f$ in this model, and the input–output relationship of the latter is represented by the following equation.

$$\dot{\psi} \to \delta_f : \quad \delta_f = H\left(\dot{\psi}\right) - \frac{1}{K}\dot{\psi} = \left(\alpha - \frac{1}{K}\right)\dot{\psi} + \beta\dot{\psi}^3 \tag{6}$$

If a linear ship model is used, obviously $\delta_f = 0$, that is, the feedback part in the figure will disappear automatically.

In the nonlinear mathematical models of Equations (3) and (4), the parameters K, T, α, and β are all related to the ship's speed, and their values have a significant impact on the accuracy of the simulation model. In this section, a general parameter calculation method is presented by using the curve fitting method [68].

Considering the operating characteristics of the ship, when a constant rudder angle δ is applied to the ship, after reaching the steady state, the ship will turn at a constant angular velocity, which is the turning angular velocity $\dot{\psi} = r =$ constant. In Equation (3), if δ and $\dot{\psi}$ are taken as constants, then $\ddot{\psi} = 0$, and we can obtain

$$\frac{K}{T}H\left(\dot{\psi}\right) = \frac{K}{T}\delta \tag{7}$$

That is

$$\delta = H(\dot{\psi}) = a\dot{\psi} + \beta\dot{\psi}^3 \tag{8}$$

Equation (8) is the steady relationship $\delta \sim \dot{\psi}$.

For the ship state-space model equation, considering the nonlinear hydrodynamic forces, there is

$$I'_{(2)}\dot{X}_{(2)} = P'_{(2)}X_{(2)} + Q'_{(2)}U + F'_{NON} \tag{9}$$

$$F'_{NON} = \begin{bmatrix} Y'_{NON} \\ N'_{NON} \end{bmatrix} = \begin{bmatrix} Cf_Y(v,r) \\ Cf_N(v,r) \end{bmatrix} \tag{10}$$

The nonlinear force $f_Y(v, r)$ and the nonlinear torque $f_N(v, r)$ are calculated according to Norrbin's formula:

$$f_Y(v,r) = \begin{cases} T \cdot r|r|\left[-\frac{1}{12} - \frac{1}{L^2}\left(\frac{v}{r}\right)^2\right], -\infty < -\frac{1}{L}\frac{v}{r} < -\frac{1}{2} \\ T \cdot r|r|\left[-\frac{1}{2}\frac{1}{L}\frac{v}{r} - \frac{2}{3}\frac{1}{L^3}\left(\frac{v}{r}\right)^3\right], -\frac{1}{2} < -\frac{1}{L}\frac{v}{r} < \frac{1}{2} \\ T \cdot r|r|\left[-\frac{1}{12} + \frac{1}{L^2}\left(\frac{v}{r}\right)^2\right], \frac{1}{2} < -\frac{1}{L}\frac{v}{r} < \infty \end{cases} \tag{11}$$

$$f_N(v,r) = \begin{cases} T \cdot r|r|\left[-\frac{1}{6}\left(\frac{v}{r}\right)\right], -\infty < -\frac{1}{L}\frac{v}{r} < -\frac{1}{2} \\ T \cdot r|r|\left[-\frac{1}{32} - \frac{1}{4}\frac{1}{L^3}\left(\frac{v}{r}\right)^3 + \frac{1}{6}\frac{1}{L^4}\left(\frac{v}{r}\right)^4\right], -\frac{1}{2} < -\frac{1}{L}\frac{v}{r} < \frac{1}{2} \\ T \cdot r|r|\left[\frac{1}{6}\frac{1}{L}\left(\frac{v}{r}\right)\right], \frac{1}{2} < -\frac{1}{L}\frac{v}{r} < \infty \end{cases} \tag{12}$$

The proportional coefficient C in Equation (10) is a dimensionless cross-flow coefficient, and its value usually ranges from 0.3 to 0.8.

Equation (9) is a first-order multivariate nonlinear equation about r and δ. By taking δ as a fixed value and conducting cyclic simulations on Equation (9), when the data reach a static state, a corresponding r can be obtained. Let δ take values between −35° and 35°, and then a series of corresponding r can be obtained. By using the least squares method to fit the relationship r → δ, the values of α and β in Equation (8) can be obtained.

The Norrbin nonlinear model [64] can be expressed as

$$\ddot{\psi} = -a_1\dot{\psi} - a_2\dot{\psi}^3 + b\delta \tag{13}$$

The yaw angle is ψ. The rudder angle is δ. The model parameters are $a_1 = \alpha b$, $a_2 = \beta b$, and $b = K/T$. The ship indices are K, T. α, β are nonlinear coefficients. The ship speed and structure determine the values of K, T, α, and β.

According to the NARX model [69], in the absence of considering white noise, Equation (13) is discretized and transformed as shown below.

$$\delta(k+1) = d_1\psi(k) + d_2\psi(k-1) + d_3\psi(k-2) + d_4\left[\psi(k) - \psi(k-1)\right]^3 \tag{14}$$

$d_1 = (1 + a_1 h)/bh^2, d_2 = -(2 + a_1 h)/bh^2, d_3 = 1/bh^2, d_4 = a_2/bh^3$. The sampling moment is k. The sampling period is h. Apparently, fluctuations in parameters K, T, α and β result in changes in parameters $d_i(i = 1, \cdots, 4)$, thereby making the mathematical model uncertain.

3. Ship Course Inverse Model Identification Based on DNFM

The neural fuzzy system is essentially a rule-based fuzzy system. The introduction of neural networks serves only as a structural auxiliary tool. The excessive fuzzy rules will result in redundancy, decrease the flexibility of the controller, and be difficult or even impossible to implement. However, an overly simplistic network structure cannot accurately depict the control model of the controlled object. Due to the complementarity in working mechanisms and the equivalence in functions between fuzzy systems and neural networks, this paper proposes a novel dynamic neural fuzzy model and uses this model to identify the inverse dynamic characteristics of ship motion. The 'dynamic' concept proposed in this paper implies that the DNFM is not preset and adjusted during learning.

3.1. Design of DNFM

The DNFM features a five-layer forward network structure, as depicted in Figure 3. Essentially, this model is a deformed RBF network, and it equals a T-S fuzzy system in function.

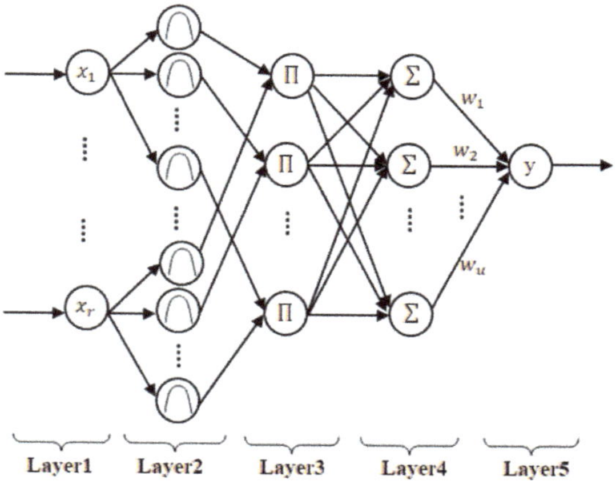

Figure 3. Dynamic neural fuzzy model structure.

In the figure, $x_1, x_2, \cdots, x_r$ represent the input. The output is y. The jth $(j = 1, 2, \cdots, u)$ membership function of the ith $(i = 1, 2, \cdots, r)$ input variable is MF_{ij}. The jth fuzzy rule is R_j. N_j represents the jth normalization node. The result parameter or connection weight of the jth rule is w_j. The total number of rules in the system is u.

The first layer, which is the input layer, has each node representing an input variable, respectively, with the total number of input variables being r. To enhance the generalization ability of the model, typically, the discretized values of a variable at the current step and its previous one or several steps are considered as different input variables, thereby expanding the single-input variable mode to a multi-input variable mode. To improve the online identification ability of the DNFM, in this paper, $\psi(k)$, $\psi(k-1)$, and $\delta(k)$ are jointly taken as the input.

The second layer serves as the membership function layer. Each node represents a Gaussian membership function respectively.

$$\mu_{ij}(x_i) = exp[-\frac{(x_i - c_{ij})^2}{\sigma_j^2}] \tag{15}$$

The jth membership function of x_i is μ_{ij}, and c_{ij} and σ_j are the center and width of the jth Gaussian membership function of x_i.

The third layer serves as the fuzzy rule layer. Each node represents the antecedent, that is, the IF part, of a possible fuzzy rule, respectively. Here, the T-norm product operator is adopted. The Gaussian membership functions of each input variable are multiplied together, and the widths of the membership functions of each variable corresponding to each rule are the same. The output of the jth rule R_j is as follows

$$\varphi_j = exp[-\frac{\|X - C_j\|^2}{\sigma_j^2}] \tag{16}$$

$X = (x_1, x_2, \cdots, x_r) \in \Re^r$, and the center of the jth RBF neural unit is $C_j = (c_{1j}, c_{2j}, \cdots, c_{rj}) \in \Re^r$.

Layer 4 is the normalization layer where each node represents an N-node. Each N-node indicates the proportion of the output of the corresponding rule in the preceding layer to the total output of all rules. The output of the jth node N_j is as follows:

$$\phi_j = \frac{\varphi_j}{\sum_{k=1}^{u} \varphi_k} \tag{17}$$

The fifth layer functions as the output layer. The nodes in this layer represent the output variables of the model, which can be either single-output or multi-output. In this paper, a single output is adopted, and the output quantity is the control rudder angle provided by the DNFM. It is manifested as a linear superposition of all the outputs of the upper layer nodes according to the weights.

$$y(X) = \sum_{k=1}^{u} \omega_k \phi_k \tag{18}$$

The output variable is y, which is the rudder angle. The weight of the kth rule is ω_k. It is specifically manifested as the linear combination of each input variable according to a set of corresponding weight coefficients. The expression is as follows:

$$\omega_k = a_{k0} + a_{k1}x_1 + a_{k2}x_2 + \cdots + a_{kr}x_r \tag{19}$$

Therefore, the output of the model should be obtained:

$$y(X) = \frac{\sum_{i=1}^{u}\left[(a_{i0} + a_{i1}x_1 + \cdots + a_{ir}x_r)exp(-\frac{\|X-C_i\|^2}{\sigma_i^2})\right]}{\sum_{i=1}^{u} exp(-\frac{\|X-C_i\|^2}{\sigma_i^2})} \tag{20}$$

3.2. Design of Learning Algorithm for Dynamic Neural Fuzzy Model

Within the third layer of the model, every node stands for either the IF portion of a rule or an RBF unit. As the quantity of fuzzy rules requiring identification cannot be fixed beforehand, the model's structure cannot be pre-established. Hence, this paper proposes a method that has the ability to automatically figure out the number of rules and simultaneously adjust the parameters and structure. It mainly includes two parts: the determination of fuzzy rules and the determination of weights.

By adopting the hierarchical learning concept, it can determine whether to generate a new rule based on the output error e_i of the model and the coverage range d_{min} of the Gaussian function.

With respect to the ith observation data point (X_i, o_i), wherein X_i serves as the input vector and o_i denotes the expected output, we can obtain the following output error:

$$\|e_i\| = \|o_i - y_i\| \tag{21}$$

y_i is the total output calculated by Formula (20) under the current structure of the model. If $\|e_i\| > k_e$, a new rule is added. The error index is k_e.

The distance $d_i(j)$ between the input value X_i and the center C_j of the existing RBF neural unit can be obtained:

$$d_i(j) = \|X_i - C_j\| \tag{22}$$

The following value should be found out:

$$d_{min} = argmin(d_i(j)) \tag{23}$$

The coverage range is d_{min}. If $d_{min} > k_d$, then the DNFM considers whether to add a new fuzzy rule. Here, k_d represents the effective radius of the coverage range d_{min}.

A new rule is generated only when both $\|e_i\| > k_e$ and $d_{min} > k_d$ are satisfied simultaneously. In the other three possible situations, only the width of the existing RBF unit and the weight of the THEN part need to be adjusted.

If u fuzzy rules are generated by n observation data, then the output of the fourth layer can be obtained:

$$\phi = \begin{bmatrix} \phi_{11} & \cdots & \phi_{1n} \\ \vdots & \vdots & \vdots \\ \phi_{u1} & \cdots & \phi_{un} \end{bmatrix} \tag{24}$$

For any input $X_j(x_{1j}, x_{2j}, \cdots x_{rj})$, the DNFM's output Equation (20) should be rewritten as follows:

$$Y = W\Psi \tag{25}$$

The following equations give W and Ψ.

$$W = (a_{10} \quad \cdots \quad a_{u0} \quad a_{11} \quad \cdots \quad a_{u1} \quad \cdots \quad a_{1r} \quad \cdots \quad a_{ur}) \tag{26}$$

$$\Psi = \begin{bmatrix} \phi_{11} & \cdots & \phi_{1n} \\ \vdots & \vdots & \vdots \\ \phi_{u1} & \cdots & \phi_{un} \\ \phi_{11} \cdot x_{11} & \cdots & \phi_{1n} \cdot x_{1n} \\ \vdots & \vdots & \vdots \\ \phi_{u1} \cdot x_{11} & \cdots & \phi_{un} \cdot x_{1n} \\ \vdots & \vdots & \vdots \\ \phi_{11} \cdot x_{r1} & \cdots & \phi_{1n} \cdot x_{rn} \\ \vdots & \vdots & \vdots \\ \phi_{u1} \cdot x_{r1} & \cdots & \phi_{un} \cdot x_{rn} \end{bmatrix} \quad (27)$$

Suppose $O = (o_1, o_2, \cdots, o_n) \in \Re^n$ is the ideal output. The linear least squares method is adopted to approximate an optimal weight vector $\Psi^* \in \Re^{u\times(r+1)}$ to minimize the error energy. The determination of this optimal weight vector is based on the following equation.

$$W^* = O(\Psi^T\Psi)^{-1}\Psi^T \quad (28)$$

3.3. Inverse Model Identification for Ship Course Control Based on DNFM

Based on the ship course control mathematical model (14), the identification process of the inverse model of ship course control using the DNFM is shown in Figure 4.

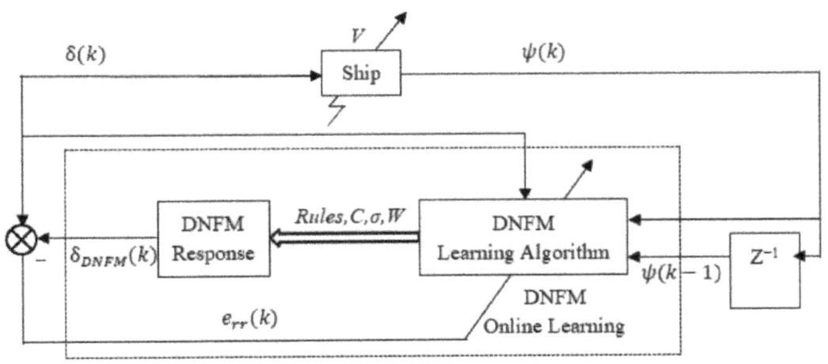

Figure 4. Identification process of inverse model for ship course control.

In the figure, the ship motion model is set such that the modeling parameters K, T, α, and β change with the ship's speed V. Given that online system identification is employed, the sample data (X_i, o_i) utilized in the DNFM learning are real ship sample data rather than the sine wave and sawtooth wave signals used in [54].

The input and output values of the DNFM corresponding to the ith sample data are X_i and o_i. The input for ship course control is the rudder angle $\delta(k)$. The rudder angle $\delta(k)$ can produce an output course $\psi(k)$. In the DNFM structure, $\psi(k)$, $\psi(k-1)$, and $\delta(k)$ are jointly taken as input quantities to determine the DNFM structure, namely the number of fuzzy rules. Additionally, the center C_j, width σ_j of each fuzzy rule (i.e., RBF neural unit), and the weight matrix W of the output layer are calculated. Taking $\psi(k)$ and $\psi(k-1)$ as two input quantities aims to expand the sample sampling range of the model and enhance the generalization ability of the DNFM. The flow of inverse model identification for ship course control based on the DNFM is shown in Figure 5.

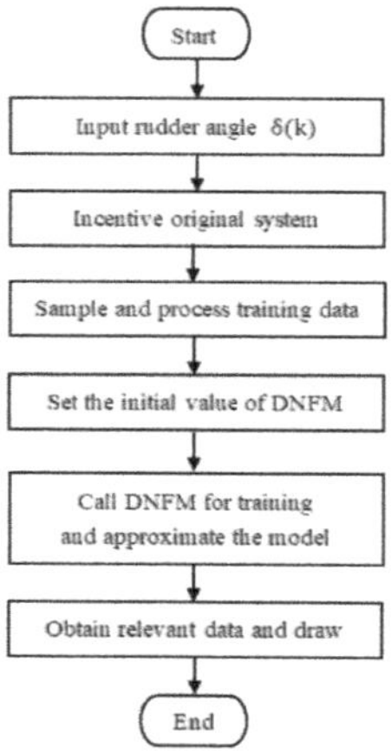

Figure 5. Flow of inverse model identification for ship course control based on DNFM.

Once the DNFM structure is determined, the DNFM response module will generate the corresponding output value $\delta_{\rm DNFM}(\rm k)$. Here, $\rm err(k)$ represents the difference between the input rudder angle $\delta(\rm k)$ of the ship and the output rudder angle $\delta_{\rm DNFM}(\rm k)$ of the DNFM. It reflects the approximation performance and learning efficiency of the DNFM. If $\rm err(k)$ is in $[-\rm err_d, err_d]$ and can be maintained until the end of learning, then the system considers that the DNFM can approximate the inverse dynamic characteristics of $\psi \rightarrow \delta$ in course control and be utilized in the next stage of online ship course control.

4. Fractional-Order $\rm PI^{\lambda}D^{\mu}$ Controller Based on DNFM

4.1. Fractional Calculus and $PI^{\lambda}D^{\mu}$ Controller

Fractional calculus is a further extension of integer-order calculus. Essentially, it is calculus of arbitrary order. Its operation operator is presented as follows.

$$ {}_aD_t^{\alpha} = \begin{cases} \frac{d^{\alpha}}{dt^{\alpha}}, Re(\alpha) > 0 \\ 1, Re(\alpha) = 0 \\ \int_{\alpha}^{t} (d\tau)^{-\alpha}, Re(\alpha) < 0 \end{cases} \tag{29} $$

The upper and lower limits of the operation operator are a and t. The order of calculus is α. And it is a real number. Currently, there are mainly three forms of the definition of fractional calculus.

The calculation formula defined by Caputo:

$$ {}_aD_t^{\alpha} f(t) = \frac{1}{\Gamma(m-\alpha)} \int_a^t \frac{f^m(\tau)}{(t-\tau)^{1-(m-\alpha)}} d\tau \tag{30} $$

Here, $\rm m - 1 < \alpha < m, m \in N$.

The calculation formula defined by Grunwald–Letnikov:

$$ {}_aD_t^{\alpha} f(t) = \lim_{h \to 0} \frac{1}{h^{\alpha}} \sum_{i=0}^{[\frac{t-a}{h}]} (-1)^j \binom{\alpha}{j} f(t - jh) \tag{31} $$

Here, $\binom{\alpha}{j}$ is the binomial coefficient; [x] means x is an integer.

The calculation formula defined by Riemann–Liouville:

$$_aD_t^\alpha f(t) = \frac{1}{\Gamma(m-\alpha)}(\frac{d}{dt})^m \int_a^t \frac{f(\tau)}{(t-\tau)^{1-(m-\alpha)}}d\tau \tag{32}$$

Here, $\Gamma(\bullet)$ is the Gamma function.

To enhance the control accuracy of the PID controller, the theory of fractional calculus is introduced. By extending the integral order λ and differential order μ to the real number domain, the $PI^\lambda D^\mu$ controller is obtained. The expression of its differential equation is presented.

$$u(t) = K_p e(t) + K_i D_t^{-\lambda} e(t) + K_d D_t^\mu e(t) \tag{33}$$

Here, $D_t^\alpha \equiv {}_a^C D_t^\alpha$ is defined by Caputo; $\lambda > 0$ and $\mu > 0$ are arbitrary real orders. Taking the Laplace transform of this fractional calculus gives

$$L\left\{{}_a^C D_t^\alpha f(t)\right\} = s^\alpha F(s) - \sum_{k=0}^{n-1} s^{a-k-1} f^{(k)}(0) \tag{34}$$

From Equations (33) and (34), the transfer function of the $PI^\lambda D^\mu$ controller can be obtained.

$$G_c(s) = K_p + \frac{K_i}{s^\lambda} + K_d s^\mu \tag{35}$$

If $\lambda = 1$ and $\mu = 1$, it becomes a PID controller. If $\lambda = 1$ and $\mu = 0$, it is a PI controller. If $\lambda = 0$ and $\mu = 1$, it is a PD controller. Since λ and μ can be arbitrary positive real numbers, a reasonable selection of them can effectively enhance the control performance.

The $PI^\lambda D^\mu$ controller pertains to a pseudo-polynomial control system. The method of the PID controller cannot be directly applied and requires rationalization and approximation. Presently, the commonly employed approximation methods mainly comprise direct and indirect approximation methods. The effect of an indirect approximation method is evidently superior to that of a direct approximation method and is widely utilized. The Oustaloup approximation method is currently a commonly used indirect approximation method. Transforming s^α into integer-order calculus, with the order taken as N, and the fitting frequency range being (ϖ_b, ϖ_h), it is then converted into an integer-order transfer function.

$$s^\alpha = \left(\frac{d\omega_b}{b}\right)^\alpha \left(\frac{ds^2 + bs\omega_h}{d(1-\alpha)s^2 + bs\omega_h + d\alpha}\right) \left[\frac{1 + \frac{s}{d\omega_b/b}}{1 + \frac{s}{d\omega_h/b}}\right]^\alpha \tag{36}$$

Here, $0 < \alpha < 1$, $b > 0$, $d > 0$, and $s = j\omega$. The fractional-order part of Equation (36), that is $K(s)$, can be expressed in the form of zeros and poles of a rational transfer function:

$$K(s) = \lim_{N\to\infty} K_N(S) = \lim_{N\to\infty} \prod_{k=-N}^{N} \frac{1 + s/\omega'_k}{1 + s/\omega_k} \tag{37}$$

The k-th zero and pole:

$$\omega'_k = \left(\frac{b}{d}\right)^{\frac{2k-\alpha}{2N+1}} \omega_h^{\frac{N+k+\frac{1}{2}(1-\alpha)}{2N+1}} \omega_b^{\frac{N-k+\frac{1}{2}(1+\alpha)}{2N+1}} \tag{38}$$

$$\omega_k = \left(\frac{b}{d}\right)^{\frac{2k+\alpha}{2N+1}} \omega_h^{\frac{N+k+\frac{1}{2}(1+\alpha)}{2N+1}} \omega_b^{\frac{N-k+\frac{1}{2}(1-\alpha)}{2N+1}} \tag{39}$$

Then

$$G(s) = K\left(\frac{ds^2 + b\omega_h s}{d(1-\alpha)s^2 + b\omega_h s + d\alpha}\right) \prod_{k=-N}^{N} \frac{1 + s/\omega'_k}{1 + s/\omega_k} \tag{40}$$

where $K = (\omega_b \omega_h)^{\alpha}$.

4.2. Course Controller Design

The inverse model of ship course control trained in Section 3.3 is connected in parallel with the $PI^{\lambda}D^{\mu}$ controller to construct a DNFM-FOPID controller for course tracking. During the control process, the weights of the output layer of the DNFM are adjusted online to compensate for modeling errors. The structure of the DNFM-FOPID controller is shown in Figure 6.

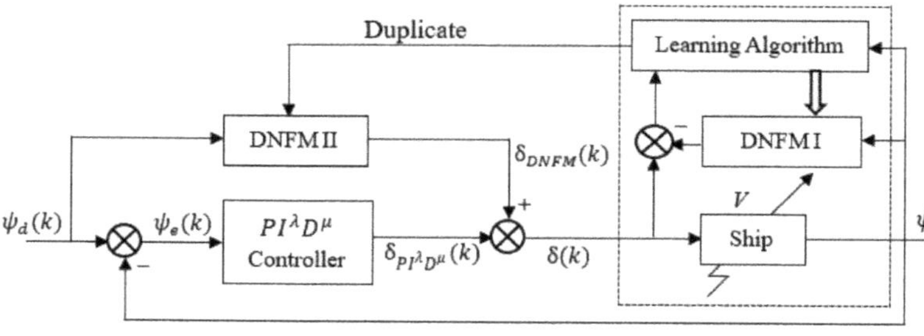

Figure 6. Ship course control system.

As indicated by the dotted box in Figure 6, after the completion of learning ship inverse dynamics, the trained DNFMI is obtained. This model is duplicated as the DNFMII model and connected in parallel with the $PI^{\lambda}D^{\mu}$ controller to generate the control input rudder angle $\delta(k)$.

$$\delta(k) = \delta_{PI^{\lambda}D^{\mu}}(k) + \delta_{DNFM}(k) \tag{41}$$

The DNFMII model is employed to generate a compensation control signal and acts as the main controller. The addition of the $PI^{\lambda}D^{\mu}$ controller is aimed at achieving faster and more accurate course tracking. Although the DNFMII model is a copy of the DNFMI model, its weights will be further adjusted to compensate for modeling errors.

The square of the error:

$$E(k) = \frac{1}{2}\left[\delta(k) - \delta_{DNFM}(k)\right]^2 = \frac{1}{2}\delta^2{}_{PI^{\lambda}D^{\mu}}(k) \tag{42}$$

Using the gradient descent method, the law of online adjustment of weights is shown.

$$W(k+1) = W(k) - \eta\frac{\partial E(k)}{\partial W(k)} \tag{43}$$

Here, η is the learning rate and greater than zero.

From Equations (43) and (25), we can obtain the following:

$$\frac{\partial E(k)}{\partial W(k)} = -[\delta(k) - \delta_{DNFM}(k)] \cdot \Psi(k) \tag{44}$$

Equation (45) is an adaptive law.

$$W(k+1) = W(k) + \eta\delta_{PI^{\lambda}D^{\mu}}(k) \cdot \Psi(k) \tag{45}$$

This controller, by adjusting the output layer weight matrix of the DNFM that meets the training requirements, jointly realizes course tracking control with the fractional-order $PI^{\lambda}D^{\mu}$ controller.

5. Simulations

Taking the 5446TEU large container ship of the COSCO Group as an example, when the modeling parameters are uncertain due to time-varying ship speed, the proposed controller is employed for course tracking control simulation.

The rated speed of this ship is V = 24.5 knots = 12.6028 m/s. The block coefficient $C_b = 0.67$. The mass of the ship m is 3.5453×10^4 tons, and the fully loaded mass can reach 6.5531×10^4 tons. The specific parameters of the ship are shown in Table 1.

Table 1. Parameters of 5446TEU container ship.

Ship Length LOA	Ship Width B	Rudder Area A_δ	No Load Mass m	Design Draft T
280 m	39.8	61 m^2	3.5453×10^4 tons	12.532 m
Two Column Length L	**Gravity Center Distance X_C**	**Square Coefficient C_b**	**Fully Load Mass m**	**Fully Load Draft**
267 m	2.64 m	0.67	6.5531×10^4 tons	14.023 m

From the basic parameters of the ship in Table 1, all dimensionless hydrodynamic derivatives can be obtained. When the ship is sailing at the rated speed V = 24.5 knots = 12.6028 m/s, by using the hydrodynamic derivative formulas introduced in [68], each hydrodynamic derivative at the rated speed can be calculated, as shown in Table 2. Through calculation, it can be obtained that K = 0.2419 and T = 206.7958.

Table 2. Dynamic coefficients under rated speed.

$Y_{\dot{v}}$	$N_{\dot{v}}$	Y_v	N_v	Y_δ
−0.01045	−0.00041371	−0.016338	−0.0050478	0.002567
$Y_{\dot{r}}$	$N_{\dot{r}}$	Y_r	N_r	N_δ
−0.00063623	−0.00058912	0.0038469	−0.0025973	−0.0012835

Since the ship's speed varies during the voyage, K and T also change accordingly. Generally, K_0 and T_0 at the rated speed can be first calculated, and then the corresponding K and T at different speeds can be obtained according to the proportional relationship. The coefficients of the nonlinear forces are calculated from the various parameters of the hydrodynamic derivatives and the parameters of the ship itself. As mentioned earlier, the nonlinear parameters α and β can be obtained through the method of curve fitting the relationship. In this example, the parameters at a speed of 24.5 knots are adopted, and it is obtained that α = 11.6049 and β = 10.1966. The corresponding modeling parameters under different speeds can be calculated, as shown in Table 3.

Table 3. Corresponding modeling parameters under different speeds.

v/knot	K	T	α	β
27.1	0.2676	186.9556	10.4915	7.5343
26.5	0.2617	191.1885	10.7291	8.0578
25.0	0.2468	202.6599	11.3728	9.5969
24.5	0.2419	206.7958	11.6049	10.1966
22.5	0.2222	225.1776	12.6366	13.1644
19.8	0.1955	255.8837	14.3601	19.3172
17.8	0.1758	284.6346	15.9748	26.5858
16.3	0.1609	310.8280	17.4475	34.6171

Although the nonlinear parameters α and β are obtained by fitting the relationship of $\delta \to \dot{\psi}$, the nonlinear parameters are the ones that characterize the maneuvering characteristics of the ship. Therefore, their magnitudes are related to the ship's sailing speed. When the sailing speed is high, the maneuvering characteristics are good and the nonlinearity of the ship is small, that is, α and β are small; conversely, α and β are large.

The ship data of the 5446TEU large container ship [68] can be obtained. The simulation time is set at 600 s. Regarding the ship's working conditions, in the first 120 s and the last 120 s of the simulation, the ship speed is maintained at 16.3 knots ~27.1 knots. In the middle 360 s, the ship is in a stable acceleration stage. The ship speed change is shown in Figure 7.

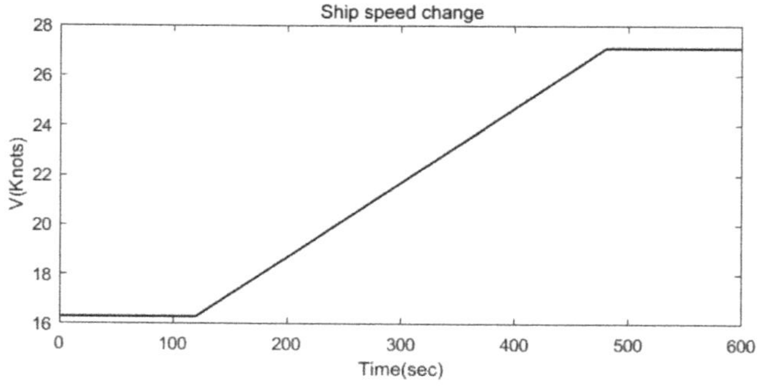

Figure 7. Change in ship speed V.

The above process describes the accelerated sailing of the ship within 600 s. Under the requirements of general modeling accuracy, it is in line with the actual dynamic process of conventional surface ships. On this basis, the established mathematical model of ship motion basically coincides with the sea trial data of the actual ship, and this model can be used for the design of ship motion controllers.

Since the change in modeling parameters is closely related to the ship speed, and K, T, α, and β corresponding with ship speed are given, the following assumptions can be made about their change rules: In the first 120 s and the last 120 s of learning, each parameter, respectively, satisfies the corresponding constant value when the ship speed is 16.3 knots and 27.1 knots. In the middle 360 s, each parameter changes linearly, as shown in Figure 8. That is, under the general modeling accuracy requirement, it conforms to the actual dynamic process of the ship accelerating from 16.3 knots to 27.1 knots within 600 s.

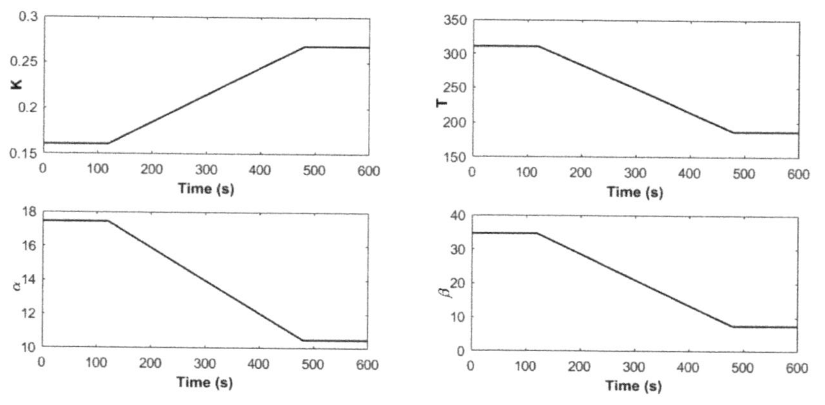

Figure 8. Change in ship model parameters K, T, α, and β.

Set the preset parameters of the DNFM as $[e_{max}, e_{min}, \lambda, d_{max}, d_{min}, \gamma, \sigma_0, k_s, k_w, k_{err}] = [10, 0.1, 0.96, 23.372, 0.2, 0.9621, 10, 1.05, 1.1, 0.1]$, and set the learning rate $\eta = 0.5$. Set the parameters of the $PI^{\lambda}D^{\mu}$ controller as $K_p = 0.18$, $K_i = 0.05$, $K_d = 2.39$, $\lambda = 0.01$, and $\mu = 0.877$, and set the fitting frequency range (ω_b, ω_h) of the fractional-order calculus operator to (0.001,1000).

Experiment 1: Course tracking control. Adjust the weight of the DNFMII in Figure 6 online. The left rudder is positive. The initial course angle is 0°. Set the change rule of the expected course angle as follows.

$$\psi_d(k) = \begin{cases} 45^\circ, \quad 0 \le k < 120 \\ 30^\circ, 120 \le k < 360 \\ 60^\circ, 360 \le k \le 600 \end{cases} \tag{46}$$

The results are shown in Figures 9–14. In Figure 9, we can see that the number of fuzzy rules generated starts from one and reaches a maximum of eight rules, finally stabilizing at seven rules. This means, in the entire course tracking control, that the DNFM only requires seven RBF neural units to describe the inverse dynamic characteristics of $\psi \to \delta$. Figure 10 shows the identification results of the course angle and rudder angle by the DNFM. It can be seen from the results that the identification results of the course angle and rudder angle by the DNFM basically coincide with the actual results of the course angle and rudder angle, with only very few points showing deviations, indicating that during this stage, the controller is in the process of oscillatory adjustment. The root mean square error and Err are maintained in a very small neighborhood near the origin in Figures 11 and 12. As can be seen from Figure 13, under the action of the dynamic neural fuzzy controller, the ship can quickly track the desired course, featuring fast rudder response and small steady-state error. Figure 14 shows the output rudder angle curve. The actual output rudder angle is $\delta(k)$, and its value is obtained by summing the outputs of the DNFM and the fractional-order PID controller. The output rudder angle first turns to the full rudder position, and then quickly adjusts to a very small rudder value. Moreover, after the ship's sailing becomes stable, the rudder angle returns to zero. Until the next time the course changes, the autopilot will repeat the operation under the control of the proposed controller.

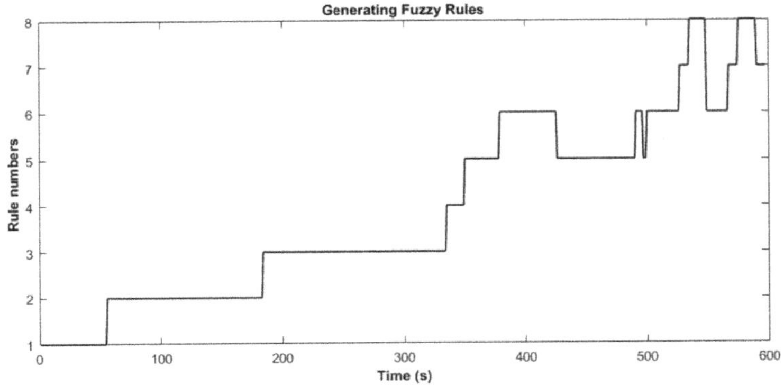

Figure 9. DNFM generating fuzzy rules.

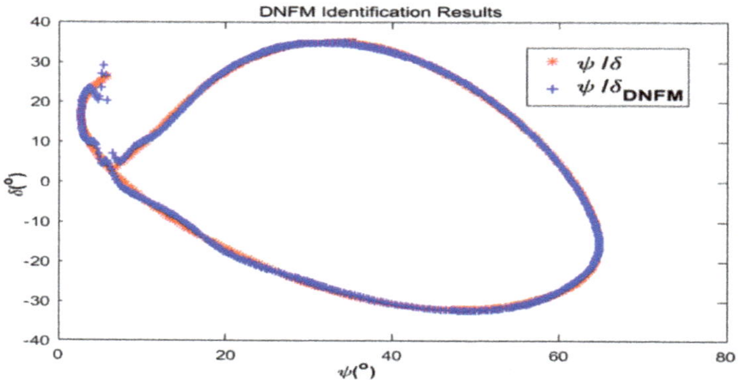

Figure 10. DNFM identification results.

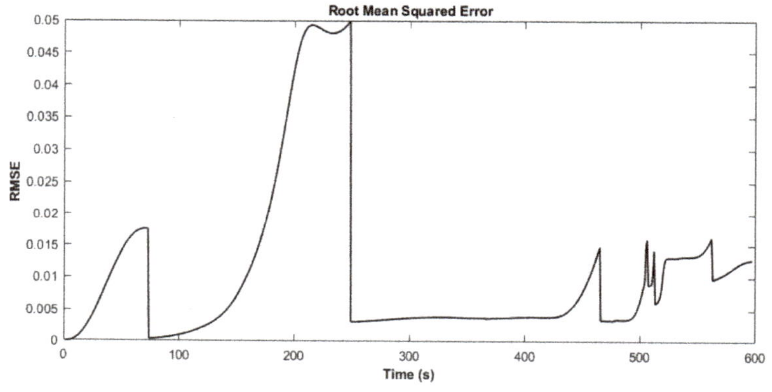

Figure 11. Root mean squared error in learning.

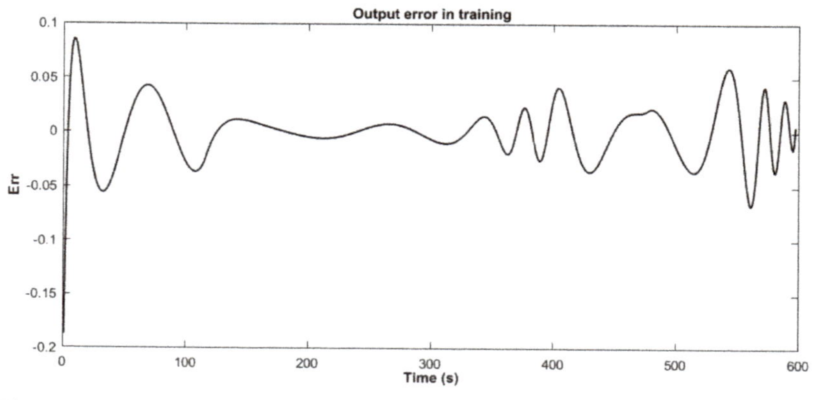

Figure 12. DNFM identification error.

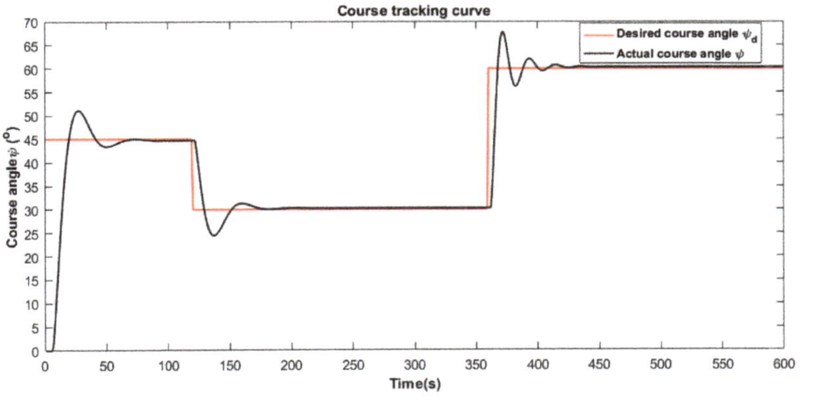

Figure 13. Ship course tracking.

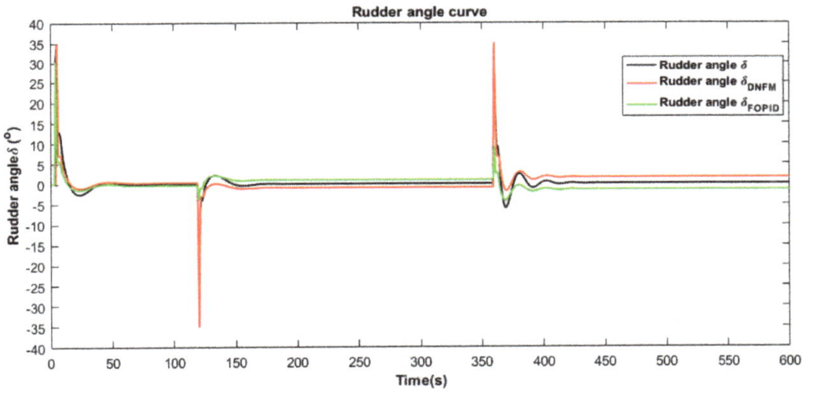

Figure 14. Rudder control for ship course.

When the expected course angle changes, there is a certain degree of oscillation in the control effect. However, the ship's course can gradually approach the expected course and ultimately maintain a stable tracking effect.

Experiment 2: Course control under the interference of a wind and wave environment. For ships sailing at sea, due to their inherent randomness, the uncertain factors caused by external environmental disturbances such as wind, waves, and currents in ship motion control problems are more difficult to handle than the uncertain factors of regular time-varying modeling parameters. Although the presence of a constant uniform current also generates uncertainties, it only leads to a change in the ship's position and has minimal impact on the course control problem. Hence, it is not considered for the time being.

On the basis of the time-varying ship speed condition in the previous section, the influence of wind and wave environmental disturbances is introduced to construct uncertain conditions. The impacts of wind and wave disturbances on ship course control are, respectively, quantified as equivalent control input rudder angles as follows.

$$\delta_w = \delta_{WIND} + \delta_{WAVE} \tag{47}$$

Then, when taking into account the wind and wave environmental disturbances, the Norrbin nonlinear mathematical model Equation (13) for ship course control can be written as follows.

$$\ddot{\psi} = -a_1\dot{\psi} - a_2\dot{\psi}^3 + b(\delta + \delta_w) \tag{48}$$

The wind force is at level 6 and the wind direction is 10°. It is equivalent to the input control rudder angle, as shown in Figures 15 and 16. It can be seen from Figure 16 that the waves cause relatively large disturbances to the ships. The maximum disturbance can reach 7.6 degrees. The simulation time is $N_k = 600$s, the simulation step size is h = 1s, the expected course is $\psi_d = 30°$, and the initial course is $\psi_0 = 0°$.

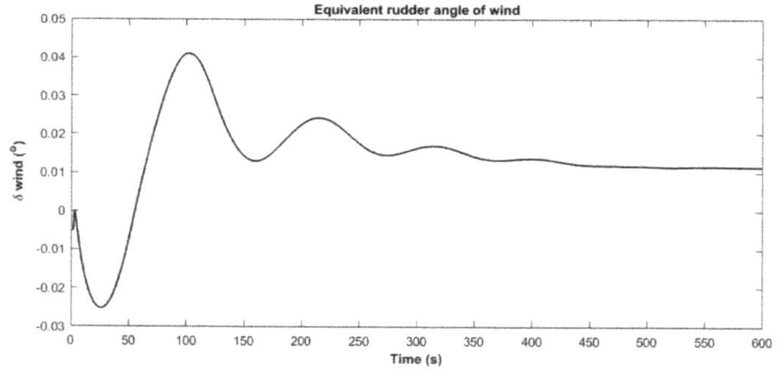

Figure 15. Equivalent rudder angle of wind.

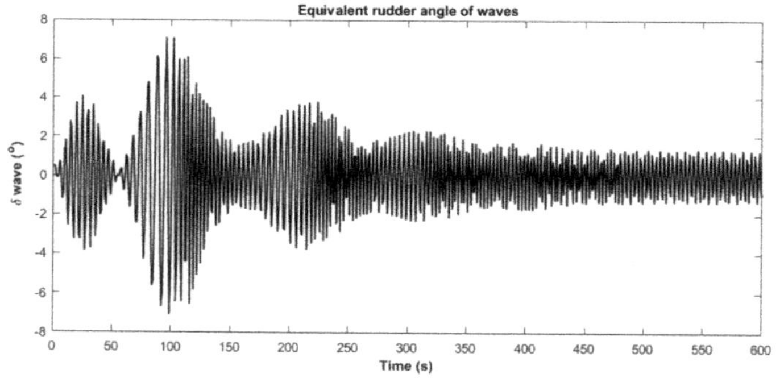

Figure 16. Equivalent rudder angle of waves.

The simulation results are presented in Figures 17–21. In Figure 17, we can see that under the interference of wind and waves, the number of fuzzy rules generated by the controller starts from 1 and a maximum of 18 rules are generated. After 50 s, the number of rules returns to two and finally stabilizes at three. That is to say, in the entire course control, the DNFM only requires three RBF neural units to describe the inverse dynamic characteristics of $\psi \rightarrow \delta$. In Figure 18, only very few points show deviations, indicating that during this stage, the controller is processing the oscillatory adjustment and interference of wind and waves. The root mean square error and Err are maintained in a very small neighborhood near the origin in Figures 19 and 20. In Figure 21, under the interference of wind and waves, the ship can also quickly reach the expected course angle. And during the course adjustment process, the overshoot is very small. The output rudder angle first turns to the full rudder position, and then quickly adjusts to a very small rudder value. Due to the continuous interference of wind and waves, the autopilot will keep adjusting around zero.

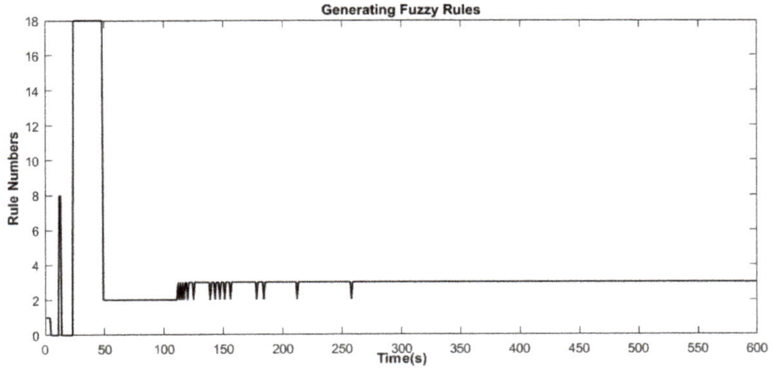

Figure 17. DNFM generating fuzzy rules under wind and wave disturbances.

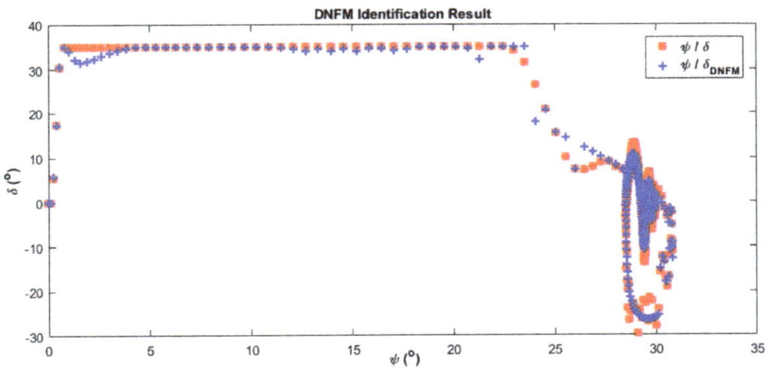

Figure 18. DNFM identification results under wind and wave disturbances.

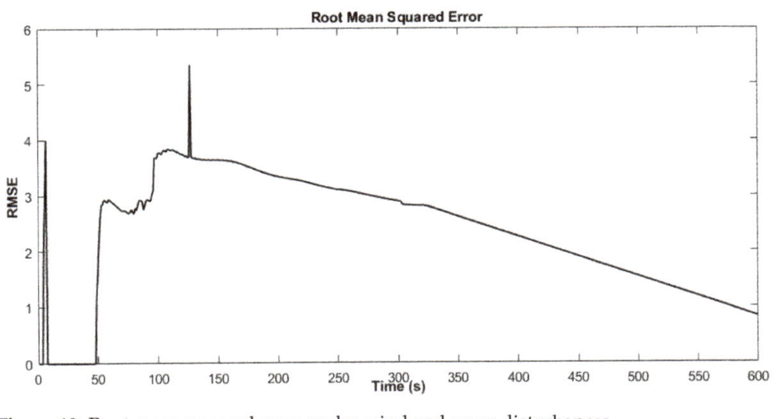

Figure 19. Root mean squared error under wind and wave disturbances.

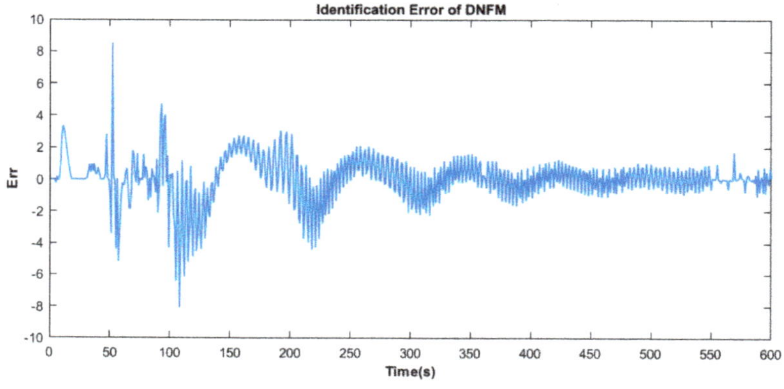

Figure 20. Identification error of DNFM.

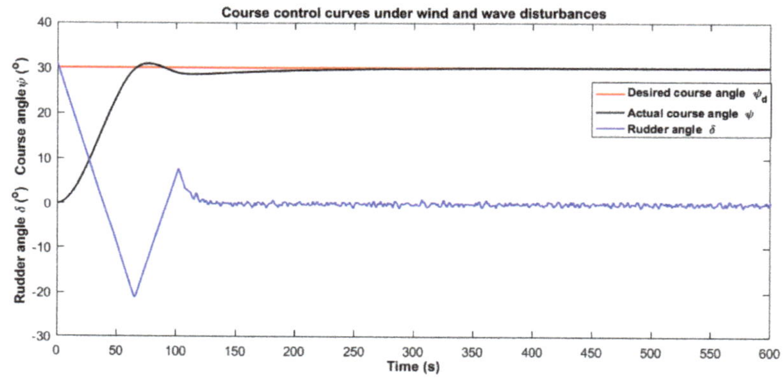

Figure 21. Course control and rudder angle curves under wind and wave disturbances.

As can be seen from Figures 17–21, based on the proposed controller in this paper, it can still reach stability quickly even in the presence of wind and wave interference. The controller has stronger robustness and better adjustment quality. This indicates that the proposed controller is capable of effectively handling the uncertainties introduced by wind and wave disturbances, ensuring stable ship course control and demonstrating its superiority in challenging marine environments.

Experiment 3: Comparison of course control of different algorithms. This paper represents a further improvement of the algorithm proposed in [54], converting the identification of the ship motion inverse dynamics model from an offline state to an online state and utilizing it for ship course control. To further verify the effectiveness, the proposed controller is simultaneously compared with the ANFM-FOPID controller [54], PID controller [64], PSO-PID controller [65], and EA-FOPID controller [66] in a comparative test. The simulation time is $N_k = 600s$, the simulation step size is $h = 1s$, the expected course is $\psi_d = 30°$, and the initial course is $\psi_0 = 0°$. The simulation results are shown in Figures 22 and 23 and Table 4.

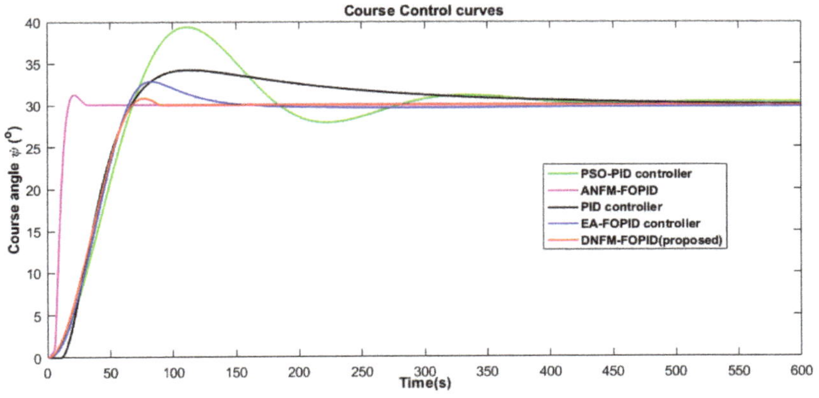

Figure 22. Comparison of five different controllers.

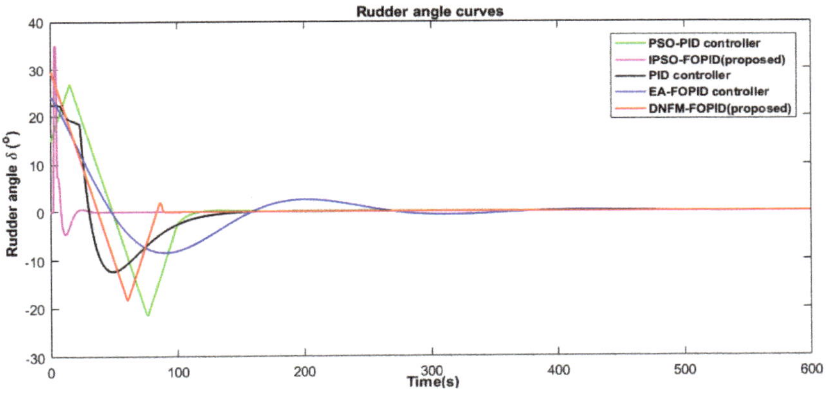

Figure 23. Rudder angle using five different controllers.

Table 4. Performance of five different controllers.

Controller Type	Settling Time t_s	Rise Time t_r	Overshoot M	Steady-State Error ESS
DNFM-FOPID (proposed)	92	68	2.83%	0.007
ANFM-FOPID	32	18	4.27%	0.009
PSO-PID	416	67	31.16%	0.024
PID	519	71	14.25%	0.007
EA-FOPID	163	65	9.53%	0.086

The proposed controller is superior to the ANFM–FOPID controller in overshoot and steady-state error, except that its adjustment time and rise time are lower than those of the ANFM–FOPID controller. The reason for this is that the ANFM–FOPID controller is an offline controller. Compared with the other three controllers, the controller is clearly superior in terms of control performance indicators. This shows that the controller has better control performance and can effectively handle ship course control tasks.

6. Conclusions

This paper represents a further exploration based on the algorithm proposed in [54]. The ship motion inverse dynamics model identification method in the offline state is transformed into model identification in the online state. The trained DNFM is employed as an inverse controller and connected in parallel with the $PI^{\lambda}D^{\mu}$ controller to construct an online DNFM-FOPID controller for ship course control. In the online system identification, both the structure and parameters can be adjusted simultaneously. The input space can be automatically divided, and the number of membership functions and fuzzy rules can be determined. Additionally, the number of fuzzy rules can be automatically adjusted according to pruning rules, that is, neural arithmetic. The dynamic neural fuzzy model features a five-layer forward network structure. Except for the input and output layers, the structures and parameters of the middle three layers are adjusted dynamically. It can be seen from the experimental results that the number of rules output by the DNFM reaches a maximum of 18, that is, the matrix dimension is 3×18. The computational burden is quite small, it does not require the occupation of a large amount of CPU resources, and the computational speed is fast and the response time is short. To verify the effectiveness of the algorithm, three types of experiments were conducted: a course tracking control experiment, course control experiment under wind and wave interference, and comparison experiment of course control using different controllers. The results indicate that the proposed controller can effectively overcome the uncertainty of modeling parameters and the interference of wind and waves in course control. It ensures a high prediction accuracy and control effect, quickly and effectively tracking the expected course.

In Experiment 1, on course tracking control, when the expected course changes, there is a certain oscillation in the control effect. The analysis indicates that the reason for this lies in the fact that the five parameters of the $PI^{\lambda}D^{\mu}$ controller are not in an online tuning mode, resulting in a lack of flexibility for the controller. To address this issue, the focus of the next step will be to introduce an intelligent optimization algorithm based on the controller to tune the parameters of the $PI^{\lambda}D^{\mu}$ controller online and further enhance the control effect.

Author Contributions: Conceptualization, G.L. and M.L.; methodology, G.L; software, G.L. and M.L.; validation, G.L. and Y.L.; data curation, G.L. and X.L.; writing—original draft preparation, G.L.; writing—review and editing, Y.L. and X.L.; project administration, Y.L. and X.Z.; funding acquisition, G.L. and H.J. All authors have read and agreed to the published version of the manuscript.

Funding: This research was funded by the National Natural Science Foundation of China under Grant 62276042, the Basic Scientific Research Project of Liaoning Province Education Department under Grant LJKMZ20220838.

Data Availability Statement: Data are contained within the article.

Conflicts of Interest: The authors declare no conflicts of interest.

References

1. Ariza, R.W.; Leong, Z.Q.; Nguyen, H.; Gamini, J.S. Non-parametric dynamic system identification of ships using multi-output Gaussian Processes. *Ocean Eng.* **2018**, *166*, 26–36. [CrossRef]
2. Sun, Q.; Chen, J.; Zhou, L.; Ding, S.; Han, S. A study on ice resistance prediction based on deep learning data generation method. *Ocean Eng.* **2024**, *301*, 117467. [CrossRef]
3. Wang, Z.H.; Xu, H.T.; Xia, L.; Zou, Z.J.; Soares, C.G. Kernel-based support vector regression for nonparametric modeling of ship maneuvering motion. *Ocean Eng.* **2020**, *216*, 107994.
4. Peng, B.; Gao, D.R.; Wang, M.Q.; Zhang, Y.Q. 3D-STCNN: Spatiotemporal convolutional neural network based on EEG 3D features for detecting driving fatigue. *J. Data Sci. Intell. Syst.* **2024**, *2*, 1–13. [CrossRef]
5. Li, M.; Lv, Z.; Cao, Q.; Gao, J.; Hu, B. Automatic assessment method and device for depression symptom severity based on emotional facial expression and pupil-wave. *IEEE Trans. Instrum. Meas.* **2024**, *73*, 2531215. [CrossRef]
6. Aumtab, C.; Wanichanon, T. Stability and Tracking Control of Nonlinear Rigid-Body Ship Motions. *J. Mar. Sci. Eng.* **2022**, *10*, 153. [CrossRef]
7. Li, F.; Chen, J.; Zhou, L.; Kujala, P. Investigation of ice wedge bearing capacity based on an anisotropic beam analogy. *Ocean Eng.* **2024**, *302*, 117611. [CrossRef]
8. Li, G.; Li, Y.; Chen, H.; Deng, W. Fractional-Order Controller for Course-Keeping of Underactuated Surface Vessels Based on Frequency Domain Specification and Improved Particle Swarm Optimization Algorithm. *Appl. Sci.* **2022**, *12*, 3139. [CrossRef]
9. Jiao, L.C.; Yang, S.Y.; Liu, F.; Wang, S.G.; Feng, Z.X. Seventy years beyond neural networks: Retrospect and prospect. *Chin. J. Comput.* **2016**, *39*, 1697–1716.

10. Kim, H.; Kim, S.H.; Jeon, M.; Kim, J.; Song, S.; Paik, K.J. A study on path optimization method of an unmanned surface vehicle under environmental loads using genetic algorithm. *Ocean Eng.* **2017**, *142*, 616–624. [CrossRef]
11. Song, Y.J.; Han, L.H.; Zhang, B.; Deng, W. A dual-time dual-population multi-objective evolutionary algorithm with application to the portfolio optimization problem. *Eng. Appl. Artif. Intell.* **2024**, *133*, 108638. [CrossRef]
12. Zhu, T.; Ren, R.; Li, Y.; Liu, W. A Model-Based Reinforcement Learning Method with Conditional Variational Auto-Encoder. *J. Data Sci. Intell. Syst.* 2024. [CrossRef]
13. Wang, Q.; Sun, C.; Chen, Y. Adaptive neural network control for course-keeping of ships with input constraints. *Trans. Inst. Meas. Control* **2019**, *41*, 1010–1018. [CrossRef]
14. Zhou, X.B.; Ma, H.J.; Gu, J.G.; Chen, H.L.; Deng, W. Parameter adaptation-based ant colony optimization with dynamic hybrid mechanism. *Eng. Appl. Artif. Intell.* **2022**, *114*, 105139. [CrossRef]
15. Simani, S.; Lam, Y.P.; Farsoni, S.; Castaldi, P. Dynamic Neural Network Architecture Design for Predicting Remaining Useful Life of Dynamic Processes. *J. Data Sci. Intell. Syst.* **2024**, *2*, 141–152. [CrossRef]
16. Deng, W.; Zhang, L.R.; Zhou, X.B.; Zhou, Y.Q.; Sun, Y.Z.; Zhu, W.H.; Chen, H.Y.; Deng, W.Q.; Chen, H.L.; Zhao, H.M. Multi-strategy particle swarm and ant colony hybrid optimization for airport taxiway planning problem. *Inf. Sci.* **2022**, *612*, 576–593. [CrossRef]
17. Lin, Y.H.; Chang, W.J.; Pen, C.L. Fuzzy Steering Control for T-S Fuzzy Model-Based Multiple Ship Systems Subject to Formation and Containment. *Int. J. Fuzzy Syst.* **2023**, *25*, 1782–1794. [CrossRef]
18. Chen, H.; Ru, J.; Long, H.; He, J.; Chen, T.; Deng, W. Semi-supervised adaptive pseudo-label feature learning for hyperspectral image classification in internet of things. *IEEE Internet Things J.* **2024**, *11*, 30754–30768. [CrossRef]
19. Qian, Y.; Hu, D.; Chen, Y.; Fang, Y.; Hu, Y. Adaptive Neural Network-Based Tracking Control of Underactuated Offshore Ship-to-Ship Crane Systems Subject to Unknown Wave Motions Disturbances. *IEEE Trans. Syst. Man Cybern. Syst.* **2022**, *52*, 3626–3637. [CrossRef]
20. Huang, C.; Zhou, X.B.; Ran, X.J.; Wang, J.M.; Chen, H.Y.; Deng, W. Adaptive cylinder vector particle swarm optimization with differential evolution for UAV path planning. *Eng. Appl. Artif. Intell.* **2023**, *121*, 105942. [CrossRef]
21. Narendra, K.S.; Parthasarathy, K. Identification and control of dynamical systems using neural networks. *IEEE Trans. Neural Netw.* **1990**, *1*, 4–27. [CrossRef] [PubMed]
22. Lin, C.-T.; Lee, C.S.G. Neural-network-based fuzzy logic control and decision system. *IEEE Trans. Comput.* **1991**, *40*, 1320–1336. [CrossRef]
23. Sastry, P.S.; Unnikrishnan, K.P. Memory neuron networks for identification and control of dynamical systems. *IEEE Trans. Neural Netw.* **1994**, *5*, 306–319. [CrossRef] [PubMed]
24. Barada, S.; Singh, H. Generating optimal adaptive fuzzy-neural models of dynamical systems with applications to control. *IEEE Trans. Syst. Man Cybern. Part C (Appl. Rev.)* **1998**, *28*, 371–391. [CrossRef]
25. Lee, C.H.; Teng, C.C. Identification and control of dynamic systems using recurrent fuzzy neural networks. *IEEE Trans. Fuzzy Syst.* **2000**, *8*, 349–366.
26. Abiyev, R.H.; Kaynak, O. Type 2 Fuzzy Neural Structure for Identification and Control of Time-Varying Plants. *IEEE Trans. Ind. Electron.* **2010**, *57*, 4147–4159. [CrossRef]
27. Feng, S.; Chen, C.L.P. Fuzzy Broad Learning System: A Novel Neuro-Fuzzy Model for Regression and Classification. *IEEE Trans. Cybern.* **2020**, *50*, 414–424. [CrossRef]
28. Zhao, H.; Wu, Y.; Deng, W. Fuzzy broad neuroevolution networks via multi-objective evolutionary algorithms: Balancing structural simplification and performance. *IEEE Trans. Instrum. Meas.* **2024**. [CrossRef]
29. Yang, G.X.; Guo, C. Simulation study of ship control based on hybrid intelligent technology. *J. Syst. Simul.* **2002**, *14*, 637–640.
30. Deng, W.; Li, K.; Zhao, H. A flight arrival time prediction method based on cluster clustering-based modular with deep neural network. *IEEE Trans. Intell. Transp. Syst.* **2024**, *25*, 6238–6247. [CrossRef]
31. Li, X.; Zhao, H.; Xu, J.; Zhu, G.; Deng, W. APDPFL: Anti-poisoning attack decentralized privacy enhanced federated learning scheme for flight operation data sharing. *IEEE Trans. Wirel. Commun.* **2024**. [CrossRef]
32. Li, M.; Wang, Y.; Yang, C.; Lu, Z.; Chen, J. Automatic diagnosis of depression based on facial expression information and deep convolutional neural network. *IEEE Trans. Comput. Soc. Syst.* **2024**, *11*, 5728–5739. [CrossRef]
33. Huang, C.; Wu, D.Q.; Zhou, X.B.; Song, Y.J.; Chen, H.L.; Deng, W. Competitive swarm optimizer with dynamic multi-competitions and convergence accelerator for large-scale optimization problems. *Appl. Soft Comput.* **2024**, *167*, 112252. [CrossRef]
34. Zhang, X.K.; Lu, X.F. Robust neural network control for sip's course-keeping. *Ship Mech.* **2006**, *10*, 54–58.
35. Ran, X.J.; Suyaroj, N.; Tepsan, W.; Ma, J.H.; Zhou, X.B.; Deng, W. A hybrid genetic-fuzzy ant colony optimization algorithm for automatic K-means clustering in urban global positioning system. *Eng. Appl. Artif. Intell.* **2024**, *137*, 109237. [CrossRef]
36. Xu, J.; Li, T.; Zhang, D.; Wu, J. Ensemble clustering via fusing global and local structure information. *Expert Syst. Appl.* **2024**, *237*, 121557. [CrossRef]
37. Long, H.Y.; Chen, T.; Chen, H.Y.; Zhou, X.B.; Deng, W. Principal space approximation ensemble discriminative marginalized least-squares regression for hyperspectral image classification. *Eng. Appl. Artif. Intell.* **2024**, *133*, 108031. [CrossRef]
38. Xie, P.; Deng, L.; Ma, Y.; Deng, W.Q. EV-Call 120: A new-generation emergency medical service system in China. *J. Transl. Intern. Med.* **2024**, *12*, 209–212. [CrossRef] [PubMed]
39. Zhou, B.; Cai, F.; Zhang, Y.S. The design and simulation of a fuzzy controller for course keeping. *Ship Sci. Technol.* **2004**, *26*, 26–27.

40. Huang, C.; Zhou, X.B.; Ran, X.J.; Liu, Y.; Deng, W.Q.; Deng, W. Co-evolutionary competitive swarm optimizer with three-phase for large-scale complex optimization problem. *Inf. Sci.* **2023**, *619*, 2–18. [CrossRef]
41. Zhang, R.; Chen, Y.; Sun, Z. Path control of a surface ship in restricted waters using sliding mode. *IEEE Trans. Control Syst. Technol.* **2000**, *8*, 722–732. [CrossRef] [PubMed]
42. Deng, W.; Cai, X.; Wu, D.; Song, Y.; Chen, H.; Ran, X.; Zhou, X.; Zhao, H. MOQEA/D: Multi-objective QEA with decomposition mechanism and excellent global search and its application. *IEEE Trans. Intell. Transp. Syst.* **2024**, *25*, 12517–12527. [CrossRef]
43. He, Z.J. Fuzzy controller for ship's course steering. *Ship Sci. Technol.* **2008**, *30*, 58–61, 65.
44. Li, T.Y.; Shu, X.Y.; Wu, J.; Zheng, Q.X.; Lv, X.; Xu, J.X. Adaptive weighted ensemble clustering via kernel learning and local information preservation. *Knowl.-Based Syst.* **2024**, *294*, 111793. [CrossRef]
45. Zhao, H.; Wang, L.; Zhao, Z.; Deng, W. A new fault diagnosis approach using parameterized time-reassigned multisynchrosqueezing transform for rolling bearings. *IEEE Trans. Reliab.* **2024**. [CrossRef]
46. Guo, D.; Wu, E.Q.; Wu, Y.; Zhang, J.; Law, R.; Lin, Y. FlightBERT: Binary Encoding Representation for Flight Trajectory Prediction. *IEEE Trans. Intell. Transp. Syst.* **2023**, *24*, 1828–1842. [CrossRef]
47. Zhou, X.B.; Cai, X.; Zhang, H.; Zhang, Z.H.; Jin, T.; Chen, H.Y.; Deng, W. Multi-strategy competitive-cooperative co-evolutionary algorithm and its application. *Inf. Sci.* **2023**, *635*, 328–344. [CrossRef]
48. Meng, F.C.; Wang, N. Indirect Fuzzy Adaptive Heading Control of Surface Ships. In Proceedings of the 2014 33rd CHINESE CONTROL CONFERENCE (CCC), Nanjing, China, 28–30 July 2014; pp. 4576–4580.
49. Zhang, Z.; Zhang, X.; Zhang, G. ANFIS-based course-keeping control for ships using nonlinear feedback technique. *J. Mar. Sci. Technol.* **2019**, *24*, 1326–1333. [CrossRef]
50. Zhao, H.; Gao, Y.; Deng, W. Defect detection using shuffle Net-CA-SSD lightweight network for turbine blades in IoT. *IEEE Internet Things J.* **2024**, *11*, 32804–32812. [CrossRef]
51. Fang, M.C.; Li, Z.Y. Application of neuro-fuzzy algorithm to portable dynamic positioning control system for ships. *Int. J. Nav. Archit. Ocean Eng.* **2016**, *8*, 38–52. [CrossRef]
52. Deng, W.; Li, X.; Xu, J.; Li, W.; Zhu, G.; Zhao, H. BFKD: Blockchain-Based Federated Knowledge Distillation for Aviation Internet of Things. *IEEE Trans. Reliab.* **2024**. [CrossRef]
53. Hatami, E.; Salarieh, H. Adaptive critic-based neuro-fuzzy controller for dynamic position of ships. *Sci. Iran.* **2015**, *22*, 272–280.
54. Li, G.; Chen, B.; Chen, H.; Deng, W. Fractional-Order $PI^{\lambda}D^{\mu}$ Controller Using Adaptive Neural Fuzzy Model for Course Control of Underactuated Ships. *Appl. Sci.* **2022**, *12*, 5604. [CrossRef]
55. Deng, W.; Chen, X.; Li, X.; Zhao, H. Adaptive federated learning with negative inner product aggregation. *IEEE Internet Things J.* **2023**, *11*, 6570–6581. [CrossRef]
56. Zhang, C.; Wan, L.; Liu, Y. Ship heading control based on fuzzy PID control. In Proceedings of the 2019 34rd Youth Academic Annual Conference of Chinese Association of Automation (YAC), Jinzhou, China, 6–8 June 2019; pp. 607–612.
57. Podlubny, I. Fractional-order System and Fractional-order Control. *IEEE Trans. Autom. Control* **1999**, *44*, 208–214. [CrossRef]
58. Kanagaraj, N. An adaptive neuro-fuzzy inference system to improve fractional order controller performance. *Intell. Autom. Soft Comput.* **2023**, *35*, 3213–3226. [CrossRef]
59. Shalaby, R.; El-Hossainy, M.; Abo-Zalam, B.; Mahmoud, T.A. Optimal fractional-order PID controller based on fractional-order actor-critic algorithm. *Neural Comput. Applic.* **2023**, *35*, 2347–2380. [CrossRef]
60. Li, X.; Zhao, H.; Deng, W. IOFL: Intelligent-optimization-based federated learning for Non-IID data. *IEEE Internet Things J.* **2024**, *11*, 16693–16699. [CrossRef]
61. Zhao, C.; Jiang, M.; Huang, Y. Formal Verification of Fractional-Order PID Control Systems Using Higher-Order Logic. *Fractal Fract.* **2022**, *6*, 485. [CrossRef]
62. Shi, X.; Chen, P.; Chen, L. An Integrated Method for Ship Heading Control Using Motion Model Prediction and Fractional Order Proportion Integration Differentiation Controller. *J. Mar. Sci. Eng.* **2023**, *11*, 2294. [CrossRef]
63. Lin, Y.; Guo, D.; Wu, Y.; Li, L.; Wu, E.Q.; Ge, W. Fuel consumption prediction for pre-departure flights using attention-based multi-modal fusion. *Inf. Fusion* **2024**, *101*, 101983. [CrossRef]
64. Fossen, T.I. *Guidance and Control of Ocean Vehicle*; John Wiley & Sons Ltd: Hoboken, NJ, USA, 1994.
65. Dlabač, T.; Alasan, M. PSO-BASED PID controller design for ship course-keeping autopilot. *Brodogradnja* **2019**, *70*, 1–15. [CrossRef]
66. Zafer, B.I.; Karahan, O. Fractional PID controllers tuned by evolutionary algorithms for robot trajectory control. *Turk. J. Electr. Eng. Comput. Sci.* **2012**, *20*, 1123–1136.
67. Meng, W.; Guo, C.; Liu, Y. Robust Adaptive Path Following for Underactuated Surface Vessels with Uncertain Dynamics. *Mar. Sci. Appl.* **2012**, *11*, 244–250. [CrossRef]
68. Shen, Z. Study on General Fuzzified CMAC Based Ship Motion Intelligent Control and Its Distributed Simulation. Ph.D Thesis, Dalian Maritime University, Dalian, China, 2005.
69. Leontaritis, I.J.; Billings, S.A. Input-output parametric models for non-linear systems part I: Deterministic non-linear systems. *Int. J. Control* **1985**, *41*, 303–328. [CrossRef]

fractal and fractional

MDPI

Article

Hover Flight Improvement of a Quadrotor Unmanned Aerial Vehicle Using PID Controllers with an Integral Effect Based on the Riemann–Liouville Fractional-Order Operator: A Deterministic Approach

Gustavo Delgado-Reyes [1], Jorge Salvador Valdez-Martínez [2,*], Pedro Guevara-López [3,*] and Miguel Angel Hernández-Pérez [1]

1 Institute of Engineering, Universidad Veracruzana, Juan Pablo II, Boca del Río 94294, Veracruz, Mexico; gusdelgado@uv.mx (G.D.-R.); miguelhernandez06@uv.mx (M.A.H.-P.)
2 Academic Division of Industrial Mechanics, Universidad Tecnológica Emiliano Zapata del Estado de Morelos, Av. Universidad Tecnológica No. 1, Emiliano Zapata 62765, Morelos, Mexico
3 Superior School of Mechanical and Electrical Engineering "Culhuacán", Instituto Politécnico Nacional, Av. Santa Ana No. 1000, San Francisco Culhuacan 04440, Mexico City, Mexico
* Correspondence: jorgevaldez@utez.edu.mx (J.S.V.-M.); pguevara@ipn.mx (P.G.-L.)

Abstract: The hovering flight of a quadrotor Unmanned Aerial Vehicle (UAV) refers to maintaining the aircraft in a fixed position in the air, without lateral, vertical, or rotational movements, using only the vehicle's control systems to maintain proper balance in all spatial dimensions. Algorithms and control systems have been developed to continuously adjust motor speeds to counteract deviations from the desired position and achieve effective hovering flight. This paper proposes a set of PID controllers with an integral effect based on the Riemann–Liouville fractional-order approach to improve the hovering flight of a quadrotor UAV. This research innovates by introducing a set of fractional-order PID controllers for UAV hover stability, which offer better adaptability to non-linear dynamics and robustness than traditional PID controllers. Also presented is the development of new performance metrics (MSE, BQC-LR), which allow for more comprehensive control system evaluations. A thorough comparative analysis with conventional control methods demonstrates the superior performance of fractional-order control in real-world simulations. The numerical simulation results show the effectiveness of the proposed Fractional Integral Action PID Controller in the control of UAV hovering flight, while comparative analyses against a classical controller emphasize the benefits of the fractional-order approach in terms of control accuracy.

Keywords: quadrotor UAV; PID fractional-order controller; Riemann–Liouville; hovering flight

1. Introduction

The hovering flight of a quadrotor (also known as a quadcopter) UAV refers to maintaining the aircraft in a stationary position in the air, without lateral, vertical, or rotational movement, solely using the vehicle's control systems to maintain proper balance in all spatial dimensions [1]. This challenge is fundamental in robotics and drone engineering. It has become a significant technical issue due to their practical application in various industries such as aerial photography, surveillance, package delivery, and infrastructure inspection, among others [2]. Achieving hovering flight involves using accelerometers, gyroscopes, and positioning systems like GPS or computer vision [3]. Hovering can be more challenging under certain environmental conditions, such as strong winds, turbulence, or abrupt changes in the drone's payload. Factors like sensor accuracy, feedback latency, and system processing capability also influence a quadcopter's ability to maintain a stable position in the air [4].

The design of controllers that allow for improving the hovering flight of a quadrotor UAV has multiple benefits in areas such as technological development and practical applications; for example, precision agriculture, infrastructure inspection, surveillance, security, operational efficiency, and safety [2,5–7].

At present, efforts to develop algorithms and control systems that continuously adjust motor speeds to counteract any deviation from the desired position and achieve hovering flight effectively have intensified [8]. Some of the most common algorithms used to address this problem include control strategies such as proportional–integral–derivative (PID) control [9,10], linear quadratic regulator (LQR) control [11–13], Model Predictive Control (MPC) [14,15], sliding mode control (SMC) [16,17], Backstepping Control [17,18], Mixed-Integer Linear Programming [19], deep reinforcement learning [20], and more advanced approaches such as optimal control and neural network-based control, among others. Recently, some studies have been carried out that allow us to visualize the importance and relevance of addressing the stability of the hovering flight of a quadrotor UAV. Such are the cases presented in [11,12], where LQR is utilized for different types of UAVs (flying wing and quadcopter), focusing on stability and precision under various flight conditions. These studies are geared towards UAVs with applications in monitoring, search and rescue, and aerial photography. In [13], a comparison is presented between Linear Quadratic Integral (LQI) and LQR control; the study is applied to ground vehicles (SUVs), highlighting the advantages of LQI in critical situations. It addresses the safety of ground vehicles, which is crucial for critical manoeuvres and stability systems. Ref. [14] introduces multitrajectory MPC (mt-MPC), excelling in the safe navigation of drones in unknown environments. It concentrates on autonomous navigation in unfamiliar settings, relevant for complex and exploratory missions. This paper stands out for its focus on uncertainty and robust optimization formulation, whereas others concentrate more on the practical implementation of controllers.

Some aspects can be investigated to improve the hovering flight of a quadcopter UAV. A possible solution that could be used is a fractional-order PID controller (FOPID) [21]. This area of research includes controller design and optimization, tuning techniques, and control structures tailored to this specific application [22,23]. Nevertheless, studying system stability under different flight conditions and workloads and evaluating the controller's robustness and resistance to external disturbances are essential, as is mentioned in [24]. Another critical aspect is investigating how the controller contributes to the greater energy efficiency of the system, resulting in increased flight autonomy [25]. Finally, experimental validation in controlled and real environments is crucial to verify the effectiveness and performance of the fractional-order PID controller in improving the hovering flight of quadcopter UAVs by gathering and comparing empirical data with those of theoretical models [26].

In terms of a control method using fractional calculus, it is possible to find numerous examples, research studies, and academic papers that explore its application across various engineering fields. PID, FOPID, and PID Plus Second-Order Derivative (PIDD^2) were shown in [10], Fractional-order sliding mode control (FOSMC) was presented in [27]. Both approaches show advancements in fractional controllers, but FOSMC offers greater robustness against external disturbances. The fractional-order controllers discussed in [23,26,28], versus Fractional Order Internal Model Control (FO-IMC) [29], apply similar concepts but in different contexts (UAVs versus DC-DC converters). On the other hand, the experimental developments presented in [26,29], in comparison to the simulations demonstrated in [23,27,28], indicate that experimental validation provides a more practical evaluation of these controllers, although simulation studies allow for more exhaustive testing in controlled environments. Regarding application and robustness, distributed and cyber–physical systems [28] versus GPS-free visual control [23] both address robustness issues in specific contexts, showing the adaptability of fractional controllers to different scenarios. In terms of methodological innovation, and the bio-inspired optimization in [10,28] versus the adaptive and robust control in [26,27], it was shown that the combination of optimization techniques and adaptive/robust control stands out as a move towards more efficient and precise solutions.

A fractional-order sliding mode control (FO-SMC) approach was presented for a quadcopter with six degrees of freedom (6-DOF) to manage its attitude and altitude control in [30]. It was demonstrated that the FO-SMC eliminates tracking errors in both aspects, even in the presence of external disturbances. To validate the robustness of the proposed controller, simulations were conducted, comparing rise time, settling time, and overshoot between a Proportional Derivative (PD) controller and FO-SMC. Although FO-SMC showed a slight increase in rise time, it significantly improved the overshoot and settling time, facilitating trajectory planning and formation. Furthermore, a simulation under wind disturbances revealed that FO-SMC completely rejected them, whereas the PD controller failed. The FO-SMC was implemented in hardware under wind disturbances, showing positive results. Ref. [31] shows the results of a Particle Swarm Optimization (PSO) algorithm based on fractional-order (FO) operators, which was proposed to enhance and optimize the PID control system in a quadcopter, aiming to reduce computation time and improve its efficiency in finding solutions. Additionally, the system's settling time and peak response were improved by using a fractional-order PID (FOPID) controller. The simulation results indicated that FOPID offers more advantages than traditional PID. In the proposed algorithm, the UAV model was simulated multiple times, evaluating its response parameters based on a cost metric, and after several iterations, the FOPID coefficients were adjusted to achieve the system's best response. PSO was used to tune the FOPID parameters, resulting in improved accuracy compared to the traditional PID.

In the research in [32], a fractional-order PID (FOPID) control strategy was proposed to achieve high-precision control in a UAV system, leveraging the multi-degree-of-freedom characteristic of FOPID. The PSO algorithm was used in combination with the simulated annealing (SA) algorithm to optimize FOPID parameters, enhancing the quadcopter's performance. It is important to mention that PSO is a bio-inspired optimization algorithm that allows faster and more accurate searches compared to other methods such as genetic algorithms (GAs). The simulation results show that the PSO-SA combination significantly reduces convergence time and improves accuracy, enhancing response speed and reducing overshoot, making it suitable for UAV control. In [33], a robust controller based on the combination of linear active disturbance rejection control (LADRC) and fractional-order PID (FOPID), named FOLADRC, was developed. This controller, designed for quadcopter UAVs, demonstrates greater accuracy and stability against uncertainties and external disturbances compared to traditional PID and LADRC controllers. The Sparrow Search Algorithm (SSA) was used to optimize the controller's parameters, enhancing its ability to avoid suboptimal solutions. The system was tested on realistic UAV models, improving their accuracy, speed, and robustness in experimental tests, with shorter settling times and a greater resistance to disturbances.

A surrogate-based analysis and optimization (SBAO) approach for relatively high-order FOPID controllers was developed in [34]. This approach was validated in two case studies: a simulated quadcopter model for waypoint navigation and a real-time two-rotor helicopter system. This work expands the research on UAV PID control, focusing on multivariable, coupled, dynamic FOPID controllers and improving the design efficiency of these controllers through surrogate models. This enhances their control performance in complex and dynamic systems, allowing for precise tuning for specialized applications, particularly in fractional-order systems. A fractional-order PID (FOPID) was developed by [35] for controlling the lateral channel of a small fixed-wing UAV using time-domain optimization criteria. The FOPID exceeded other reference methods, such as the Skogestad method and an optimized integer PID, providing smoother and faster responses. The results indicated that system identification via ARX plays a crucial role in performance and that approximating fractional control to an integer-based control can degrade controller performance. The system's behavior also showed some non-linearity, affecting controller performance, which was validated using the FlightGear simulator and the SenecaII aircraft model. The paper [36] presented an adaptive fault-tolerant cooperative control (FTCC) scheme for networked UAVs. This approach employs fractional-order calculus and recurrent neural networks (RNNs) to learn the non-linear and strongly coupled terms generated by actuator faults and wind disturbances. It was the first work to address fault-induced terms during networked UAV formation flight using neural networks with self-feedback loops. The scheme is based on a PID controller with error transformation and fractional-order sliding surfaces to enhance control accuracy. Furthermore, a stability analysis was included, demonstrating the system's robustness and suitability for real-time practical applications. To validate its functionality, a hardware-in-the-loop experimental platform was implemented using the open-source Pixhawk autopilot, which is widely used in aviation, terrestrial, and marine communities.

The issue of a variable load in quadcopter trajectory tracking under wind disturbances was addressed in [37]. The variable load generated too much weight and altered the quadcopter's moment of inertia. To mitigate these effects, a sliding mode controller with a back-stepping approach was proposed, allowing the ideal trajectory to be followed and reducing wind disturbances when no load was present. To compensate for the overweight caused by the load, a method employing a correction coefficient was introduced. Additionally, the controller's sliding surfaces were upgraded to fractional-order surfaces, which reduced undesirable fluctuations in rotor position, orientation, stability, and speed. This adaptive approach enabled variable load estimation, achieving a controller that was robust against mass uncertainties in altitude control. In [38], a control strategy was proposed to improve tracking accuracy and resistance to interference in quadcopters. This strategy employed a controller based on quaternion error, combining fractional-order PID control with S-plane control. Trajectory tracking experiments shown that compared to fractional-order PID, this method has greater resistance to wind disturbances and better tracking accuracy. Simulations conducted in Matlab SIMULINK confirm that the use of quaternions avoided singularity problems and simplified the calculation of orientation and stability angles. Furthermore, the fractional-order S-plane controller obtained the advantages of conventional PID controllers, allowing for finer adjustments to the system's order, resulting in smoother control. The results demonstrated that, compared to fractional-order PID, the S-plane controller provided greater precision and robustness to the control system.

These papers demonstrate that fractional-order controllers, whether adaptive, sliding mode, or visual, offer significant improvements in the performance, robustness, and disturbance rejection of UAVs and distributed systems. Additionally, the use of bio-inspired optimization algorithms, such as Particle Swarm Optimization (PSO) and the Salp Swarm Algorithm (SSA), are crucial for the precise tuning of parameters in fractional-order controllers, enhancing their performance in practical applications. Concerning the experimental validation of controllers in physical prototypes and energy converters, it was observed that this is essential to demonstrate the effectiveness of the proposed techniques, providing tangible results that support theoretical conclusions and simulations. Thus, it can be deduced that each work presents a specific and advanced approach to the control and stability of different types of vehicles, adapting to their respective environments and mission needs. Their comparison shows how different control methodologies can be applied and optimized to achieve various technological applications, as well as precision, stability, and safety goals.

Finally, one of the most relevant contributions of this research is studying how including a fractional-order PID into a drone's hovering flight allows for the possibility of improving the stability and precision of its flight compared to a traditional controller. This implies the extent to which the fractional-order controller parameters influence flight performance and the drone's ability to adapt to environmental changes and external disturbances. This could lead to more robust and safe flight under various conditions. Therefore, the goal of this paper is to improve the hovering flight of a quadcopter UAV using a fractional-order PID controller. This could help obtain greater precision and stability in various critical applications, such as aerial photography, surveillance, and infrastructure inspection, which would enhance the quality and usefulness of the collected data. Additionally, achieving more stable and precise flight could lead to greater operational efficiency, reducing costs and increasing productivity in sectors such as precision agriculture. An improvement in UAV operational safety is also expected, especially in critical applications such as border surveillance or disaster monitoring, where flight reliability and stability are crucial. Technologically, this research could contribute to the advancement and evolution of UAV technology, opening up new opportunities for more advanced and sophisticated applications across various industries and sectors. Finally, improving hovering flight could reduce UAV operational and maintenance costs, making this technology more accessible and cost-effective for many users and applications.

To achieve the objective of this paper, we first aim to describe the mathematical modeling of UAV kinematics, defined by the combination of the translational and rotational dynamics of the quadrotor. This results in a non-linear mathematical model expressed in state-space format. To simplify our design analysis, this model is converted into a linear format, as linear models are mathematically more manageable than their non-linear counterparts. This conversion facilitates the examination of the system's stability, controllability, and observability. Moreover, it enables the use of classical control methods, such as PID control, linear quadratic regulator (LQR), and Model Predictive Control (MPC). It is also possible to develop robust controllers that can be optimized to enhance quadrotor performance across various operating conditions. In terms of computational complexity, linear models demand fewer resources for their simulation and control, which is advantageous for real-time hardware and software applications, thus making the system more practical and accessible. Once the linear model is obtained, a classical control system for the "hovering" of a quadrotor is designed, as well as a set of PID controllers with an integral effect based on a fractional-order operator. This is carried out to compare the hovering flight of both control systems and analyze the results using the mean squared error (MSE) and the best performance in the closed-loop response (BQC-LR) between them.

This paper is structured as follows: The first section presents the introduction. The second section addresses the problem statement. The third section covers the topics necessary to understand the general context of this research. The fourth section presents the obtained results. The fifth section provides an analysis and discussion of the results. Finally, the sixth section presents the conclusions.

2. Problem Statement

The current state of the art for improving the hovering flight of a quadcopter UAV using fractional-order PI or PID controllers involves a combination of theoretical and practical advancements in UAV control and dynamic system control theory. Some relevant aspects include the following:

1. The development of fractional-order controllers: Considerable research has been conducted on designing and implementing fractional-order controllers in various areas, including UAV systems. These controllers offer advantages over conventional controllers in terms of their adaptability, robustness, and performance in the presence of non-linearities and disturbances [28,39].

2. Stability and performance studies: Studies have been conducted to analyze the stability and performance of fractional-order control systems applied to UAVs, including quadcopters. These studies provide crucial information about stability limits and optimal operating conditions to improve hovering flight [40,41].

3. Experimentation and validation: Experiments and tests have been conducted under real conditions to validate the effectiveness of fractional-order controllers in improving the hovering flight of quadcopter UAVs [26]. These studies provide empirical data supporting the effectiveness and feasibility of this technology in practical applications.

4. Ongoing research: Research in this field focuses on optimizing control algorithms, reducing costs, and improving the energy efficiency of UAV systems. Additionally, innova-

tive alternative approaches have been explored to address specific challenges related to stability and hovering flight performance [27].

Based on the above, there is a problem in that some dynamic system simulation software does not have functions required for implementing fractional-order controllers, thus complicating their implementation and the evaluation of their performance. In this sense, in this work a classical order control system (COCSH) based on PID controllers is implemented for the hover flight control of a UAV. Subsequently, said control system is adapted to the fractional approach, obtaining a FOCSH and, thus, a set of PID controllers whose integral action is implemented according to the Riemann–Liouville operator through self-developed programming functions. To evaluate the improvement in the performance of the FOCSH system compared to the COCSH, an objective and quantitative description based on the analysis of mean squared errors (MSEs) is used and derived from this analysis, and then a closed-loop response improvement index (BQC-LR) is proposed which identifies the system with the best quality in its closed-loop response. Finally, as one of the contributions of this work, two criteria are proposed for the selection of the order of the fractional controllers that constitute the FOCSH, which show the robustness of the system even when the mathematical model of the UAV tends to behave in a non-linear manner.

3. Mathematical Modeling of an Unmanned Aerial Vehicle (UAV)

3.1. UAV Kinematics

The mathematical representation of a quadrotor-type Unmanned Aerial Vehicle (UAV) model is presented in this section. This UAV is equipped with four actuators, and by controlling these, its possible movements are determined: movement along the x axis, y axis, and z axis. To achieve movements along the x, y, and z axes, the quadrotor's orientation must be adjusted by altering its pitch angle θ (also known as pitch movement), roll angle ϕ (also known as roll movement), and yaw angle ψ (also known as yaw movement), which are defined with respect to the UAV's orientation framework. As shown in Figure 1, the quadrotor operates under two frames of reference, C and I. The C frame represents the non-inertial frame of reference, that is, it is a coordinate axis whose center is the quadrotor's center of mass; meanwhile, the I frame represents the inertial frame of reference, which is fixed to the land's surface (the ground or platform on which the drone initiates its flight).

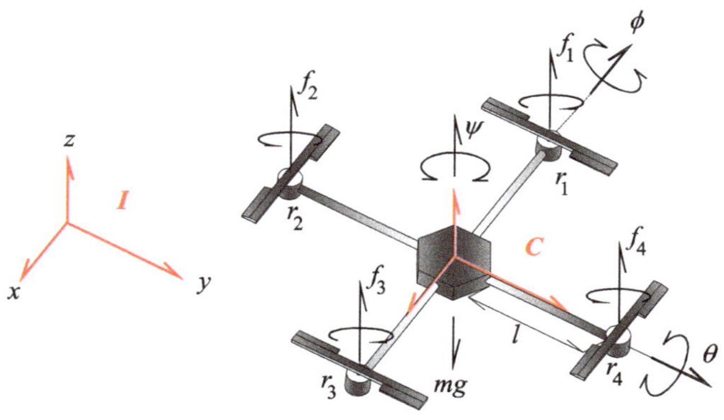

Figure 1. Kinematics and frames of reference of UAV quadrotor.

The movements that a quadrotor can perform, as shown in Figure 2, are as follows:

(a) Pitch movement allows the UAV to move along the x axis. To achieve this, the quadrotor needs to increase the angular velocity of rotor 1 (referred to as r_1 in Figure 1), reduce the angular velocity of rotor 3 (r_3), and maintain a medium velocity in rotors 2 (r_2) and 4 (r_4), generating the torque τ_θ. When these actions are performed, the UAV tilts, resulting in the generation of the angle θ (see Figure 2).

(b) Roll movement allows the drone to move along the y axis. For this movement, the quadrotor needs to increase the angular velocity of rotor 2 (r_2), reduce the angular velocity of rotor 4 (r_4), and maintain an intermediate velocity in rotors 1 (r_1) and 3 (r_3), generating the torque τ_ϕ. When these actions are performed, the UAV tilts, resulting in the generation of the angle ϕ.

(c) Yaw movement occurs in the drone's vertical plane. To produce this movement, the angular velocity of rotors 1 (r_1) and 3 (r_3) must be increased or decreased by the same amount, while the angular velocity of rotors 2 (r_2) and 4 (r_4) must be adjusted in the opposite direction, creating the torque τ_ψ. Consequently, the quadrotor tends to rotate around its center of mass, generating the angle ψ (see Figure 2).

(d) By generating thrust, the drone can move along the z axis. This movement is achieved by maintaining the same angular speed (but not the same direction of rotation) in the four rotors of the quadrotor so that the thrust is greater than the gravitational force exerted on the aircraft.

By generating roll ϕ, yaw ψ and pitch θ angles, it is possible to modify the quadrotor's spatial position. The previously described movements must be controlled simultaneously for the UAV to hover or follow specific trajectories assigned through a flight plan. It is important to note that even slight inclinations due to small changes in the UAV's angles can complicate altitude control, thereby compromising the system's stability.

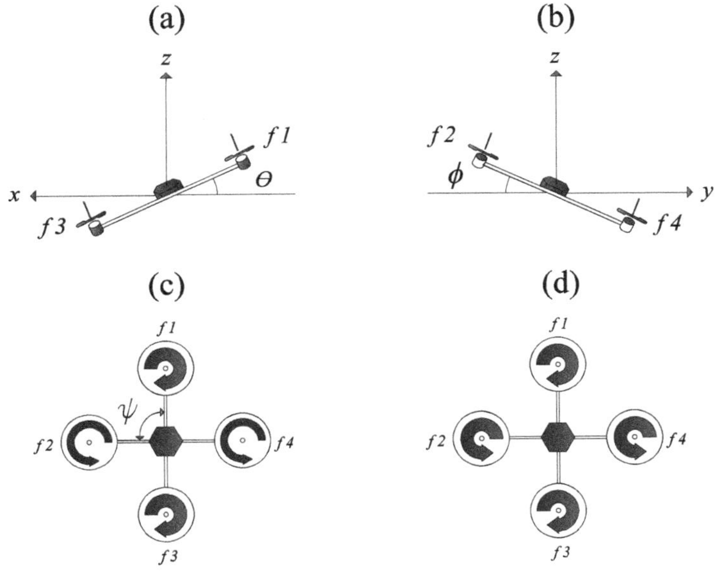

Figure 2. A quadrotor's possible movements: (**a**) pitch movement, (**b**) roll movement, (**c**) yaw movement, and (**d**) movement along the z axis (thrust).

Considering the above, it is helpful to define the mathematical representation of the drone's orientation. The UAV's non-linear dynamic model is extracted from [24,42], where it is completely detailed. This specifies that the UAV's dynamics are defined by a combination of translational and simplified rotational dynamics, leading to the mathematical expressions presented below.

$$\begin{aligned}
\ddot{\psi} &= \frac{\tau_\psi}{I_{zz}} + \frac{(I_{xx} - I_{yy})\dot{\phi}\dot{\theta}}{I_{zz}} + \dot{\phi}\dot{\theta}, \\
\ddot{\theta} &= \frac{\tau_\theta}{I_{yy}} + \frac{(I_{zz} - I_{xx})\dot{\psi}\dot{\phi}}{I_{yy}} - \dot{\psi}\dot{\phi}, \\
\ddot{\phi} &= \frac{\tau_\phi}{I_{xx}} + \frac{(I_{yy} - I_{zz})\dot{\psi}\dot{\theta}}{I_{xx}} + \dot{\psi}\dot{\theta}, \\
\ddot{x} &= (S\psi S\phi + C\psi S\theta C\phi)\frac{u}{m}, \\
\ddot{y} &= (-C\psi S\phi + S\psi S\theta C\phi)\frac{u}{m}, \\
\ddot{z} &= (C\theta C\phi)\frac{u}{m} - g,
\end{aligned} \tag{1}$$

The inertial parameters are represented by I_{xx}, I_{yy}, and I_{zz}, and m signifies the mass of the quadrotor. The model's inputs are defined according to (2).

$$\begin{aligned}
u &= f_1 + f_2 + f_3 + f_4, \\
\tau_\psi &= (f_1 + f_3)l - (f_2 + f_4)l, \\
\tau_\theta &= (f_1 - f_3)l, \\
\tau_\phi &= (f_2 - f_4)l,
\end{aligned} \tag{2}$$

with l denoting the length of the UAV's arms, g representing gravitational acceleration, and u representing the thrust that constitutes the input to translational dynamics, while the torques τ_θ, τ_ϕ, and τ_ψ denote the inputs to rotational dynamics [24]. Lastly, the forces f_i generated by each rotor that are incorporated into the quadrotor are defined as follows:

$$f_i = k_i \omega_i^2, \quad i = 1,\ 2,\ 3,\ 4. \tag{3}$$

From (3), it can be inferred that ω_i signifies the angular velocity of each rotor and k_i is a constant of proportionality.

3.2. UAV State-Space Model

The UAV quadrotor model mentioned earlier in (1) and (2) can be expressed in a state-space format as follows:

$$\dot{\mathbf{X}} = \mathbf{f}(\mathbf{X}, \mathbf{U}). \tag{4}$$

Subsequently, the state vector $\mathbf{X} = (x_1, x_2, x_3, \ldots, x_{12})$ is defined as illustrated in Table 1.

Table 1. Definition of state-space variables.

State-Space Variables	State-Space Variables
$x_1 = \phi$	$x_7 = x$
$x_2 = \dot{\phi} = \dot{x}_1$	$x_8 = \dot{x} = \dot{x}_7$
$x_3 = \theta$	$x_9 = y$
$x_4 = \dot{\theta} = \dot{x}_3$	$x_{10} = \dot{y} = \dot{x}_9$
$x_5 = \psi$	$x_{11} = z$
$x_6 = \dot{\psi} = \dot{x}_5$	$x_{12} = \dot{z} = \dot{x}_{11}$

Based on the previously defined state vector, the quadrotor's non-linear state-space model can be represented as follows:

$$\dot{\mathbf{X}} = \mathbf{f}(\mathbf{X}, \mathbf{U}) = \begin{bmatrix} x_2 \\ x_4x_6a_1 + x_4x_6 + b_1u_2 \\ x_4 \\ x_2x_6a_2 - x_2x_6 + b_2u_3 \\ x_6 \\ x_2x_4a_3 + x_2x_4 + b_3u_4 \\ x_8 \\ u_xu_1/m \\ x_{10} \\ u_yu_1/m \\ x_{12} \\ (u_zu_1/m) - g \end{bmatrix}, \tag{5}$$

where

$$\begin{aligned} u_1 &= u, \\ u_2 &= \tau_\phi, \\ u_3 &= \tau_\theta, \\ u_4 &= \tau_\psi, \\ a_1 &= (I_{yy} - I_{zz})I_{xx}, \\ a_2 &= (I_{zz} - I_{xx})I_{yy}, \\ a_3 &= (I_{xx} - I_{yy})I_{zz}, \\ b_1 &= 1/I_{xx}, \\ b_2 &= 1/I_{yy}, \\ b_3 &= 1/I_{zz}, \\ u_x &= (S\psi S\phi + C\psi S\theta C\phi) = (Sx_5Sx_1 + Cx_5Sx_3Cx_1), \\ u_Y &= (-C\psi S\phi + S\psi S\theta C\phi) = (-Cx_5Sx_1 + Sx_5Sx_3Cx_1), \\ u_Z &= (C\theta C\phi) = (Cx_3Cx_1). \end{aligned} \tag{6}$$

The values presented in Table 2 are utilized for the numerical simulation of the UAV.

Table 2. UAV model parameters.

Description	Parameter Value
UAV mass	$m = 1.4$ kg
UAV rotor mass	0.088 kg
UAV arm length	$l = 0.28$ m
x axis inertia	$I_{xx} = 0.0116\,\text{kgm}^2$
y axis inertia	$I_{yy} = 0.0116\,\text{kgm}^2$
z axis inertia	$I_{zz} = 0.0232\,\text{kgm}^2$
conversion constant f_u to $f_{\tau\psi}$	T_q

3.3. Linearization of the UAV Mathematical Model

The non-linear quadrotor model defined by Equations (5) and (6) has been linearized around the equilibrium point $P_e = (\phi, \theta, \psi, \ddot{z}) = (0, 0, 0, 0)$. Consequently, the linear version of the non-linear model is derived and presented in the general state-space format defined in Equation (7).

$$\begin{aligned} \dot{\mathbf{X}}(t) &= \mathbf{A}\mathbf{X}(t) + \mathbf{B}\mathbf{U}(t), \\ \mathbf{Y}(t) &= \mathbf{C}\mathbf{X}(t) + \mathbf{D}\mathbf{U}(t), \end{aligned} \tag{7}$$

where $\mathbf{U}(t) \in \mathbb{R}^m$ is the input vector, $\mathbf{Y}(t) \in \mathbb{R}^p$ is the output vector, and $\mathbf{X}(t) \in \mathbb{R}^n$ is the state vector. $\mathbf{A} \in \mathbb{R}^{n\times n}$ is the system matrix, $\mathbf{B} \in \mathbb{R}^{n\times m}$ is the input matrix, $\mathbf{C} \in \mathbb{R}^{p\times n}$ is the output matrix, and $\mathbf{D} \in \mathbb{R}^{n\times q}$ is the input distribution matrix. The coefficients of the matrices **A**, **B**, **C**, and **D** are obtained from the linearized state-space model expressed in the continuous-time domain, as shown in (9).

An alternative way of expressing the linear dynamic model of the UAV is through the transfer function matrix $\mathbf{G_{UAV}}(s)$ obtained by performing the matrix operations defined by (8). This version of the linear model could be useful for the design of control systems for the hovering flight of a drone.

$$\mathbf{G_{UAV}}(s) = \mathbf{C}(s\mathbf{I} - \mathbf{A})^{-1}\mathbf{B}. \tag{8}$$

$$\mathbf{A} = \left.\frac{\partial \mathbf{f}}{\partial \mathbf{X}}\right|_{P_e} = \begin{bmatrix} 0 & 1 & 0 & 0 & 0 & 0 & 0 & 0 & 0 & 0 & 0 & 0 \\ 0 & 0 & 0 & 0 & 0 & 0 & 0 & 0 & 0 & 0 & 0 & 0 \\ 0 & 0 & 0 & 1 & 0 & 0 & 0 & 0 & 0 & 0 & 0 & 0 \\ 0 & 0 & 0 & 0 & 0 & 0 & 0 & 0 & 0 & 0 & 0 & 0 \\ 0 & 0 & 0 & 0 & 0 & 1 & 0 & 0 & 0 & 0 & 0 & 0 \\ 0 & 0 & 0 & 0 & 0 & 0 & 0 & 0 & 0 & 0 & 0 & 0 \\ 0 & 0 & 0 & 0 & 0 & 0 & 0 & 1 & 0 & 0 & 0 & 0 \\ 0 & 0 & g & 0 & 0 & 0 & 0 & 0 & 0 & 0 & 0 & 0 \\ 0 & 0 & 0 & 0 & 0 & 0 & 0 & 0 & 0 & 1 & 0 & 0 \\ -g & 0 & 0 & 0 & 0 & 0 & 0 & 0 & 0 & 0 & 0 & 0 \\ 0 & 0 & 0 & 0 & 0 & 0 & 0 & 0 & 0 & 0 & 0 & 1 \\ 0 & 0 & 0 & 0 & 0 & 0 & 0 & 0 & 0 & 0 & 0 & 0 \end{bmatrix},$$

$$\mathbf{B} = \left.\frac{\partial \mathbf{f}}{\partial \mathbf{U}}\right|_{P_e} = \begin{bmatrix} 0 & 0 & 0 & 0 \\ 0 & l/I_{xx} & 0 & 0 \\ 0 & 0 & 0 & 0 \\ 0 & 0 & l/I_{yy} & 0 \\ 0 & 0 & 0 & 0 \\ 0 & 0 & 0 & l/I_{zz} \\ 0 & 0 & 0 & 0 \\ 0 & 0 & 0 & 0 \\ 0 & 0 & 0 & 0 \\ 0 & 0 & 0 & 0 \\ 0 & 0 & 0 & 0 \\ 1/m & 0 & 0 & 0 \end{bmatrix}, \tag{9}$$

$$\mathbf{C} = \mathbf{I}_{12\times 12},$$
$$\mathbf{D} = \mathbf{0}_{12\times 4}.$$

4. Main Results

This section presents the classical order control system for quadrotor hovering (COCSH), which has been developed using parameters obtained from the dynamic matrices of the quadrotor's linear model, as defined in the previous section. Next, we adapt and modify the COCSH to include a fractional approach, leading to the creation of a Fractional Order Control System (FOCSH). This development serves as the main contribution of this research.

4.1. Classical Order Control System Design

The classical order control system for hovering drone flight (COCSH) has been designed as detailed in Figure 3, taking into consideration the reduced dynamics model expressed by (1) and (2), or alternatively, by (5) and (6). The main purpose of this system is to generate the appropriate control actions for the quadrotor to reach and maintain hovering at a desired position in three-dimensional space with the coordinates (x_{ref}, y_{ref}, z_{ref}), and with a desired orientation ψ_{ref}, thus ensuring the asymptotic convergence to zero of the errors e_y, e_x, e_ϕ, e_θ, e_ψ, and e_z. As shown in this figure, the COCSH consists of six proportional–integral–derivative (PID) controllers, whose control actions are described by the expressions denoted in (10). These controllers are coupled and organized in such a way that they form two feedback loops: there are inner-loop controllers and outer-loop controllers. Based on the latter expression, it is important to highlight that kp_y, kp_x, kp_ϕ, kp_θ, kp_ψ, kp_z, ki_y, ki_x, ki_ϕ, ki_θ, ki_ψ, ki_z, kd_y, kd_x, kd_ϕ, kd_θ, kd_ψ, and $kd_z \in \mathbb{R}$ are constant gains that have been calculated by tuning the PID controllers that constitute the COCSH.

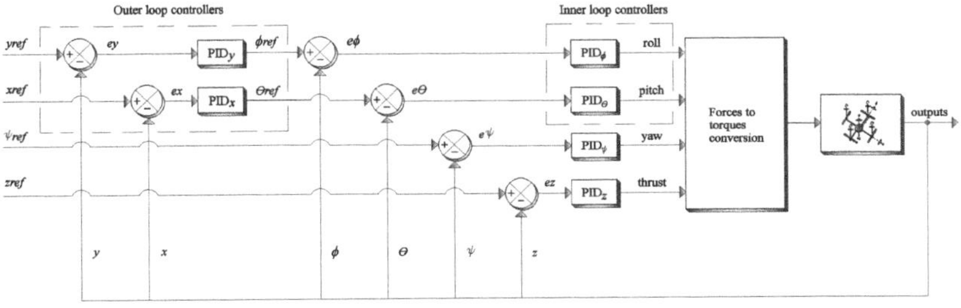

Figure 3. Classical order control system for quadrotor hovering.

$$
\begin{aligned}
\phi ref(t) &= kp_y ey(t) + ki_y \int_0^t ey(t)dt + kd_y \frac{dey(t)}{dt}, \\
\theta ref(t) &= kp_x ex(t) + ki_x \int_0^t ex(t)dt + kd_x \frac{dex(t)}{dt}, \\
roll(t) &= kp_\phi e\phi(t) + ki_\phi \int_0^t e\phi(t)dt + kd_\phi \frac{de\phi(t)}{dt}, \\
pitch(t) &= kp_\theta e\theta(t) + ki_\theta \int_0^t e\theta(t)dt + kd_\theta \frac{de\theta(t)}{dt}, \\
yaw(t) &= kp_\psi e\psi(t) + ki_\psi \int_0^t e\psi(t)dt + kd_\psi \frac{de\psi(t)}{dt}, \\
thrust(t) &= kp_z ez(t) + ki_z \int_0^t ez(t)dt + kd_z \frac{dez(t)}{dt}.
\end{aligned}
\tag{10}
$$

It is worth noting that, alongside the translation and rotation dynamics of the aircraft, the dynamics of its actuators have also been taken into account for the COCSH design. In this context, these rotor-type motors (r_1, r_2, r_3, r_4), which are integrated into the drone, have been modeled as direct current (DC) motors. To simplify the control system's design, the dynamics of each rotor of the drone have been modeled as those of a DC motor and approximated to a first-order dynamic system following the methodology outlined in [43]. This results in the transfer function $R_i(s)$, defined by (11), where the angular velocity ω_i is considered the output and the excitation voltage V_a the input.

$$
R_i(s) = \frac{\omega_i(s)}{V_a(s)} = \frac{0.975}{0.25s + 1}. \tag{11}
$$

Considering the UAV model parameters described in Table 2 and the equilibrium point $P_e = (\phi, \theta, \psi, \ddot{z}) = (0, 0, 0, 0)$, and taking the Laplace transform, with initial conditions set to zero, of the expressions that define the UAV's reduced rotational dynamics implicit in (1), the following second-order transfer functions implicit in (12)–(15) are obtained. These are used to tune all PID controllers that are not part to the outer loop (see Figure 3). These transfer functions are not suitable for describing abrupt dynamics in the UAV, as they have been derived using the small-angle approximation denoted by the equilibrium point P_e. It is worth noting that in this transfer function, the variables ϕ, θ, ψ, and z are considered as outputs, while the variables u_2, u_3, u_4, and u_1 (as denoted by expression (6)), are considered as inputs, respectively. In this context, for the design of the PID controllers that are not part of the outer loop of the UAV hover control system (COCSH), it is presumed that the approximate orientation and elevation dynamics are interconnected with the actuator dynamics in a cascaded manner; in other words, it is assumed that the overall dynamics, which take into account the effects of the actuators, can be roughly represented as the product of the transfer functions displayed below:

$$G_{UAV\phi}(s) = R_i(s)G_{\phi approx}(s) = \frac{0.975}{0.25s+1}\frac{0.28}{0.0116s^2}, \tag{12}$$

$$G_{UAV\theta}(s) = R_i(s)G_{\theta approx}(s) = \frac{0.975}{0.25s+1}\frac{0.28}{0.0116s^2}, \tag{13}$$

$$G_{UAV\psi}(s) = R_i(s)G_{\psi approx}(s) = \frac{0.975}{0.25s+1}\frac{0.28T_q}{0.0232s^2}, \tag{14}$$

$$G_{UAVz}(s) = R_i(s)G_{zapprox}(s) = \frac{0.975}{0.25s+1}\frac{1}{1.4s^2}. \tag{15}$$

To carry out the design of the PID controllers that do not belong to the outer loop, the transfer functions defined by (12)–(15) are used. These are considered individually as a part that requires control and as single-input single-output (SISO) systems. Thus, a PID controller is tuned for each of these expressions, resulting in four closed-loop systems with negative feedback, where the controllers represented by the transfer functions indicated in (18)–(21) are employed. It is important to mention that expressions (16)–(21) correspond to the Laplace transforms of the expressions denoted by (10), considering initial conditions equal to zero.

$$PID_y(s) = \frac{\Phi_{ref}(s)}{Ey(s)} = \frac{kd_y\left(s^2 + \frac{kp_y}{kd_y}s + \frac{ki_y}{kd_y}\right)}{s} = \frac{k_y\left[s^2 + (a_y+b_y)s + a_yb_y\right]}{s}, \tag{16}$$

$$PID_x(s) = \frac{\Theta_{ref}(s)}{Ex(s)} = \frac{kd_x\left(s^2 + \frac{kp_x}{kd_x}s + \frac{ki_x}{kd_x}\right)}{s} = \frac{k_x\left[s^2 + (a_x+b_x)s + a_xb_x\right]}{s}, \tag{17}$$

$$PID_\phi(s) = \frac{Roll(s)}{E\phi(s)} = \frac{kd_\phi\left(s^2 + \frac{kp_\phi}{kd_\phi}s + \frac{ki_\phi}{kd_\phi}\right)}{s} = \frac{k_\phi\left[s^2 + (a_\phi+b_\phi)s + a_\phi b_\phi\right]}{s}, \tag{18}$$

$$PID_\theta(s) = \frac{Pitch(s)}{E\theta(s)} = \frac{kd_\theta\left(s^2 + \frac{kp_\theta}{kd_\theta}s + \frac{ki_\theta}{kd_\theta}\right)}{s} = \frac{k_\theta\left[s^2 + (a_\theta+b_\theta)s + a_\theta b_\theta\right]}{s}, \tag{19}$$

$$PID_{\psi}(s) = \frac{Yaw(s)}{E\psi(s)} = \frac{kd_{\psi}\left(s^2 + \frac{kp_{\psi}}{kd_{\psi}}s + \frac{ki_{\psi}}{kd_{\psi}}\right)}{s} = \frac{k_{\psi}\left[s^2 + (a_{\psi} + b_{\psi})s + a_{\psi}b_{\psi}\right]}{s}, \tag{20}$$

$$PID_z(s) = \frac{Thrust(s)}{Ez(s)} = \frac{kd_z\left(s^2 + \frac{kp_z}{kd_z}s + \frac{ki_z}{kd_z}\right)}{s} = \frac{k_z\left[s^2 + (a_z + b_z)s + a_z b_z\right]}{s}. \tag{21}$$

To carry out the design of the PID controllers in the outer loop, the transfer functions defined by (22) and (23) are used. These expressions represent to the closed-loop transfer functions that describe two of the four feedback control systems previously discussed. When designing these outer-loop controllers, each of these closed-loop transfer functions are considered individually as the part requiring control and treated as single-input single-output (SISO) systems. Thus, a PID controller is adjusted for each of these expressions, creating two closed-loop systems with negative feedback, utilizing the controllers defined by the transfer functions represented in (16) and (17).

$$G_{PID_{\phi}-G_{UAV\phi}}(s) = \frac{\Phi(s)}{\Phi_{ref}(s)} = \frac{\frac{k_{\phi}(s+a_{\phi})(s+b_{\phi})}{s}\,\frac{0.273}{0.0029s^3+0.0116s^2}}{1+\frac{k_{\phi}(s+a_{\phi})(s+b_{\phi})}{s}\,\frac{0.273}{0.0029s^3+0.0116s^2}}, \tag{22}$$

$$G_{PID_{\theta}-G_{UAV\theta}}(s) = \frac{\Theta(s)}{\Theta_{ref}(s)} = \frac{\frac{k_{\theta}(s+a_{\theta})(s+b_{\theta})}{s}\,\frac{0.273}{0.0029s^3+0.0116s^2}}{1+\frac{k_{\theta}(s+a_{\theta})(s+b_{\theta})}{s}\,\frac{0.273}{0.0029s^3+0.0116s^2}}. \tag{23}$$

It is important to mention that to tune all the PID controllers that make up the control system for the drone's hovering flight (COCSH), the root locus technique in the continuous-time domain has been used. To accomplish this, unit step-type reference signals have been utilized for the sake of simplicity.

4.2. Fractional-Order Control System Design

In this section, the classical order control system for hovering drone flight (COCSH) is modified by adapting it to the Riemann–Liouville fractional calculus approach. This modification allows for the development of a new control system called the Fractional-Order Control System for hovering drone flight (FOCSH). This reinforces the advantages that can be observed and identifies potential enhancements in the transient response of the FOCSH in comparison to the COCSH. To validate this modification, the next section provides numerical simulations applied to a quadrotor UAV model, with the results assessed using performance indices. In order to adapt the COCSH to the proposed FOCSH, the first step involves considering the following definitions of fractional calculus:

Definition 1. *The iterated Cauchy integral of order $n_c \in \mathbb{N}$ is defined by the following expression [42]:*

$$\begin{gathered} {}^{C}_{a}I^{n_c}_{x}\{f(x)\} = \frac{1}{(n_c-1)!}\int_a^x (x-t)^{n_c-1}f(t)dt, \\ {}^{C}_{a}I^{0}_{x}\{f(x)\} = f(x). \end{gathered} \tag{24}$$

Definition 2. *Euler's Gamma function is one that generalizes the factorial concept and extends it to the real-number set. This function is described by the following expression:*

$$\Gamma(n) = \int_a^{\infty} t^{n-1}e^{-t}dt,\ n \in \mathbb{R}. \tag{25}$$

The following expressions define some of the fundamental properties of Euler's Gamma function [42]:

$$\begin{aligned}\Gamma(n+1) &= n\Gamma(n),\\ \Gamma(n) &= (n-1)!,\\ \Gamma(1) &= \Gamma(2) = 1,\\ \Gamma\left(\frac{1}{2}\right) &= \sqrt{\pi}.\end{aligned} \tag{26}$$

When taking into account the definition and properties of Euler's Gamma function within the context of the Cauchy iterated integral, the fractional-order integral operator outlined in Definition 3 is established.

Definition 3. *The Riemann–Liouville integral operator of order $\alpha \geq 0$, when $a \geq 0$, is defined as*

$$\begin{aligned}{}_{a}^{RL}I_{x}^{\alpha}\{f(x)\} &= \frac{1}{\Gamma(\alpha)}\int_{a}^{x}(x-t)^{\alpha-1}f(t)dt,\\ {}_{a}^{RL}I_{x}^{0}\{f(x)\} &= f(x),\end{aligned} \tag{27}$$

where $\Gamma(\alpha)$ is Euler's Gamma function and $\alpha \in \mathbb{R}$ [42].

In the next step, taking into account the earlier definitions and the characteristics of the Gamma function, all PID controllers forming the Classical Order Control System (COCSH) for hovering drone flight are adjusted to modify their integral effect, as outlined in (27), within the FOCSH, as illustrated in (28).

$$\begin{aligned}\phi ref(t) &= kp_{y}ey(t) + ki_{y}\,{}_{0}^{RL}I_{t}^{\alpha}\{ey(t)\} + kd_{y}\frac{dey(t)}{dt},\\ \theta ref(t) &= kp_{x}ex(t) + ki_{x}\,{}_{0}^{RL}I_{t}^{\alpha}\{ex(t)\} + kd_{x}\frac{dex(t)}{dt},\\ roll(t) &= kp_{\phi}e\phi(t) + ki_{\phi}\,{}_{0}^{RL}I_{t}^{\alpha}\{e\phi(t)\} + kd_{\phi}\frac{de\phi(t)}{dt},\\ pitch(t) &= kp_{\theta}e\theta(t) + ki_{\theta}\,{}_{0}^{RL}I_{t}^{\alpha}\{e\theta(t)\} + kd_{\theta}\frac{de\theta(t)}{dt},\\ yaw(t) &= kp_{\psi}e\psi(t) + ki_{\psi}\,{}_{0}^{RL}I_{t}^{\alpha}\{e\psi(t)\} + kd_{\psi}\frac{de\psi(t)}{dt},\\ thrust(t) &= kp_{z}ez(t) + ki_{z}\,{}_{0}^{RL}I_{t}^{\alpha}\{ez(t)\} + kd_{z}\frac{dez(t)}{dt}.\end{aligned} \tag{28}$$

Finally, for the implementation of the FOCSH, the classical order controllers of the control scheme shown in Figure 3 are replaced by fractional-order PID controllers, whose control actions are described by (28). It is important to highlight that the gains of the controllers that make up the FOCSH are the same as those obtained by tuning the COCSH controllers.

5. Validation and Discussion of the Results

The results of the numerical simulation carried out in MATLAB R2022b® software are presented in this section, comparing the COCSH and the FOCSH applied to the non-linear dynamic model of a UAV given by (5) and (6), as described in Section 3. For this purpose, the UAV model parameters listed in Table 2 are considered. These simulations are designed to demonstrate the performance and possible enhancements in the transient behavior of the closed-loop response when the proposed COCSH approach is utilized, and then compare it with the FOCSH performance, with both approaches involving the UAV performing hovering maneuvers. Therefore, to validate the results, a performance index such as the mean squared error (MSE) is proposed.

5.1. Closed-Loop Response Comparison for UAV Hover Flight Using COCSH vs. FOCSH

To evaluate the advantages of the closed-loop response of the FOCSH compared to the COCSH when the quadrotor performs hover movements, we set out to conduct a set of numerical simulations in which a simulation time of 90 s is established, considering an iteration step of $\Delta t = 25$ miliseconds, with the aim of having convergence with the continuous-time controllers' design technique. It is important to highlight that in these numerical simulations, both control systems are coupled with the non-linear dynamic model of the UAV, with the aim of enabling the quadrotor to reach its desired orientation, ψ_{ref}, and maintain hovering in two desired positions in three-dimensional space (P_{ref1} and P_{ref2}), whose coordinates, expressed in meters, are described by the following expression:

$$\begin{aligned} P_{ref1} &= \left(x_{ref1}, y_{ref1}, z_{ref1}\right) = (11, -11, 1.3), \\ P_{ref2} &= \left(x_{ref2}, y_{ref2}, z_{ref2}\right) = (1, -1, 13.3). \end{aligned} \tag{29}$$

The COCSH and FOCSH control systems have been designed to enable the drone to reach points P_{ref1} and P_{ref2} in three-dimensional space, despite disturbances arising from various sources, such as atmospheric turbulence, sudden changes in payload, and gusts of wind.

The closed-loop response of the COCSH, when the drone hovers with an orientation of $\psi = 0$ radians at points P_{ref1} and P_{ref2} in three-dimensional space, is shown in Figures 4 and 5, respectively.

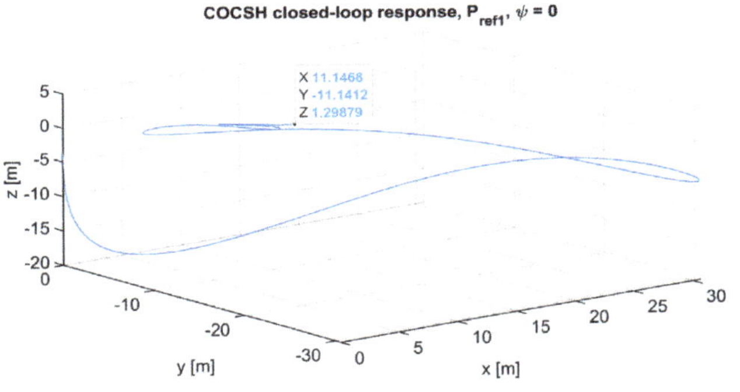

Figure 4. COCSH closed-loop response when the drone hovers at point P_{ref1} with an orientation of $\psi = 0$ radians.

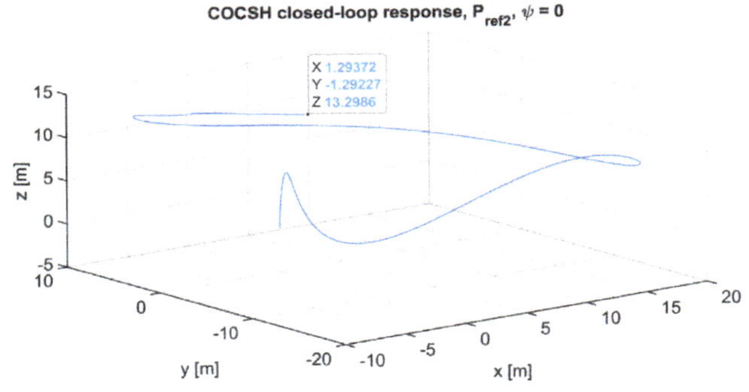

Figure 5. COCSH closed-loop response when the drone hovers at point P_{ref2} with an orientation of $\psi = 0$ radians.

The closed-loop response of the COCSH when the drone hovers with an orientation of $\psi = 0.4$ radians at points P_{ref1} and P_{ref2} in three-dimensional space can be seen in Figures 6 and 7, respectively.

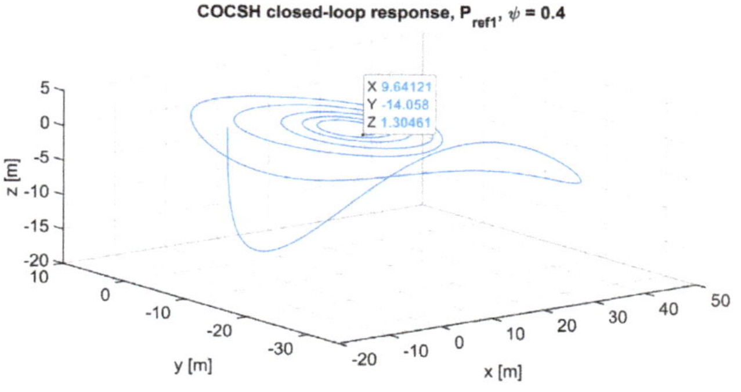

Figure 6. COCSH closed-loop response when the drone hovers at point P_{ref1} with an orientation of $\psi = 0.4$ radians.

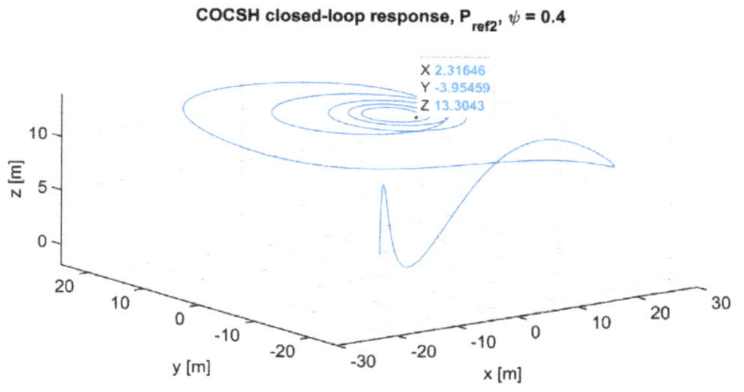

Figure 7. COCSH closed-loop response when the drone hovers at point P_{ref2} with an orientation of $\psi = 0.4$ radians.

The closed-loop response of the FOCSH when the drone hovers with an orientation of $\psi = 0$ radians at points P_{ref1} and P_{ref2} in three-dimensional space can be seen in Figures 8 and 9, respectively. For both points in three-dimensional space, $\alpha = 0.99$ has been used.

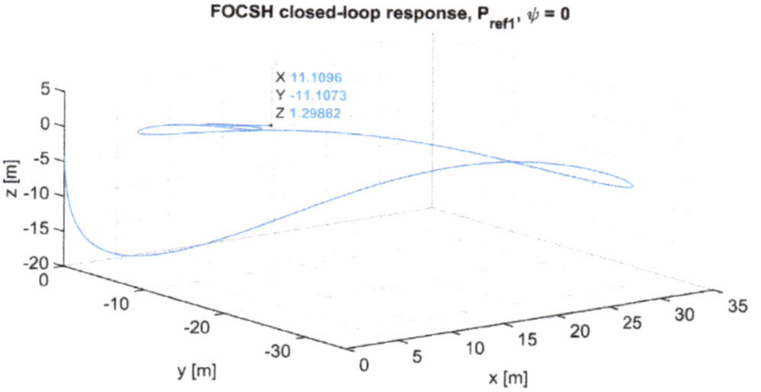

Figure 8. FOCSH closed-loop response when the drone hovers at point P_{ref1} with $\alpha = 0.99$ and an orientation of $\psi = 0$ radians.

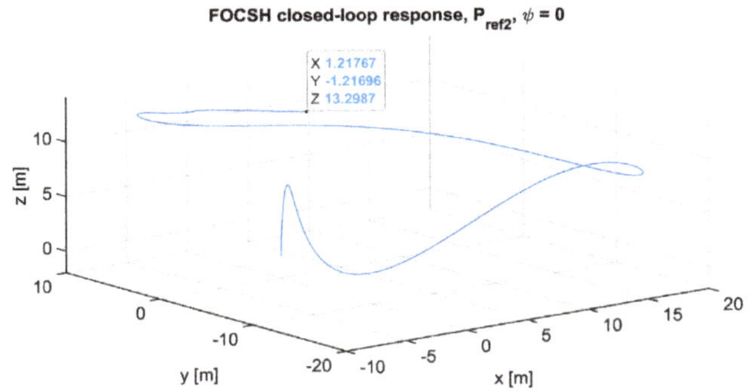

Figure 9. FOCSH closed-loop response when the drone hovers at point P_{ref2} with $\alpha = 0.99$ and an orientation of $\psi = 0$ radians.

The closed-loop response of the FOCSH when the drone hovers with an orientation of $\psi = 0.4$ radians at points P_{ref1} and P_{ref2} in three-dimensional space can be seen in Figures 10 and 11, respectively. It is important to mention that for point P_{ref1}, $\alpha = 1.098$ was used, while for point P_{ref2} $\alpha = 1.0044$ was used.

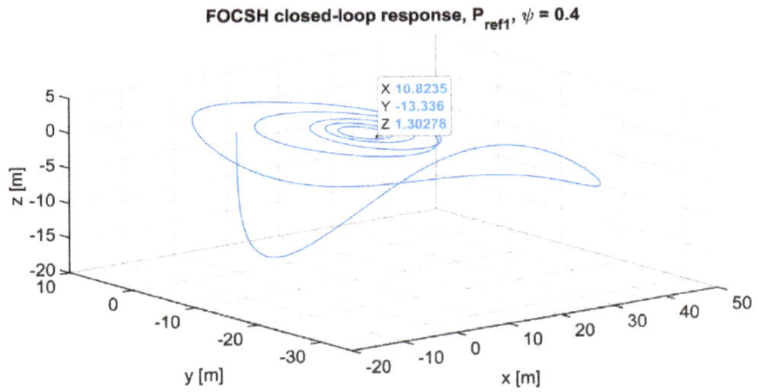

Figure 10. FOCSH closed-loop response when the drone hovers at point P_{ref1} with $\alpha = 1.098$ and an orientation of $\psi = 0.4$ radians.

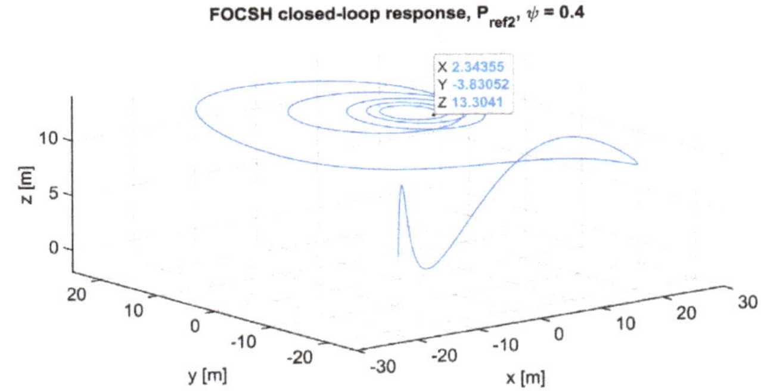

Figure 11. FOCSH closed-loop response when the drone hovers at point P_{ref2} with $\alpha = 1.0044$ and an orientation of $\psi = 0.4$ radians.

5.2. Discussion

In order to highlight one of the contributions of this paper, Figures 12–19 present the superimposed results of the closed-loop response for the hovering flight of the drone at points P_{ref2} and P_{ref2} in three-dimensional space, conducted using the COCSH and the FOCSH. These figures demonstrate that both closed-loop responses, depending on the α parameter (used only for the FOCSH implementation), exhibit similar dynamics and convergence times. However, in most cases, the FOCSH enables the drone to reach points P_{ref1} and P_{ref2} more quickly (see Figures 4–11). Therefore, it is necessary to use performance indices that describe the efficiency of the classical and fractional approaches. In this context, we propose quantitatively and qualitatively assessing the drone hovering performance of the control systems through the mean squared error (MSE). In this sense, a comprehensive analysis of the evolution of the mean squared error of the drone's ability to reach the final position in a three-dimensional space during stationary flight (hovering) was carried out. The MSE was calculated with the aim of evaluating the accuracy and stability of both control systems, thus allowing us to denote the improvements in the closed-loop responses. This analysis allows for quantifying the drone's deviations from its target position, providing a comprehensive measure of the controller's performance in terms of positional stability under stationary conditions.

To carry out this analysis, a recursive MSE calculation scheme was designed in order to obtain a temporary mean squared error throughout the drone's stabilization process, as well as a global measure of the system's behavior. The accuracy of the final hovering position (rather than the trajectory) allows the evaluation of the controller's performance in its ability to precisely reach and maintain the desired position under various operational conditions and external disturbances. In this way, the calculation of the MSE becomes a fundamental tool in validating the control system's effectiveness in real-world environments and to optimize its performance for future applications.

Figures 20–33 show the dynamics of the MSE performance indices of the closed-loop responses when the drone hovers at the three-dimensional-space points P_{ref1} and P_{ref2} using the COCSH and FOCSH control systems. These indices serve as a quantitative test of the drone's convergence quality, where a small MSE magnitude indicates a minor error in the translational positions when the drone hovers at points P_{ref1} and P_{ref2}. Additionally, to indicate the best performance in the closed-loop response between the COCSH and the FOCSH, an indicator of the Best Quality of the Closed-Loop Response (BQC-LR) as described in Table 3 is proposed. This table specifies that the MSE is lower in all cases for the proposed FOCSH approach.

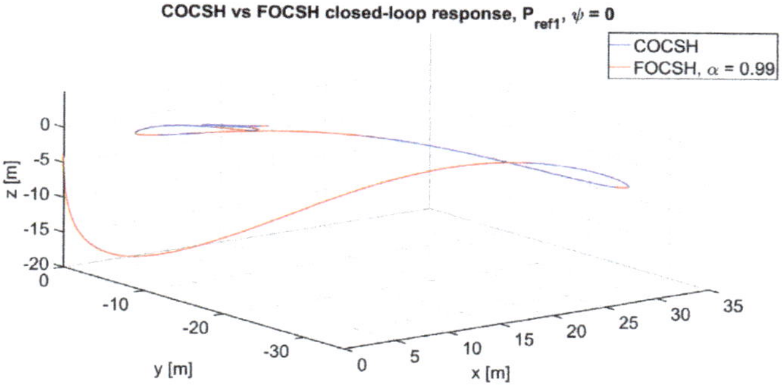

Figure 12. COCSH and FOCSH (with $\alpha = 0.99$) closed-loop response comparison when the drone hovers at point P_{ref1} with an orientation of $\psi = 0$ radians.

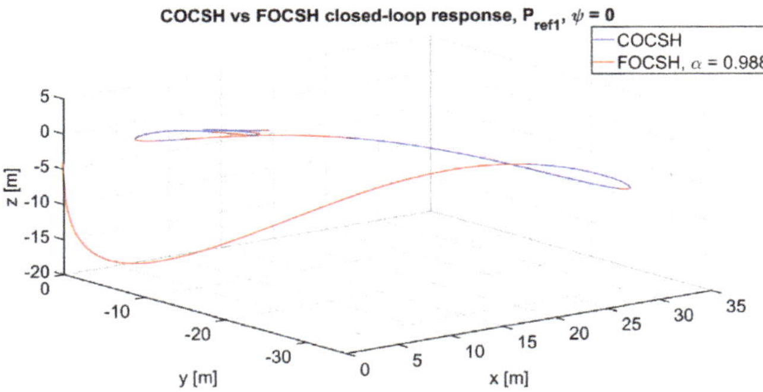

Figure 13. COCSH and FOCSH (with $\alpha = 0.9882$) closed-loop response comparison when the drone hovers at point P_{ref1} with an orientation of $\psi = 0$ radians.

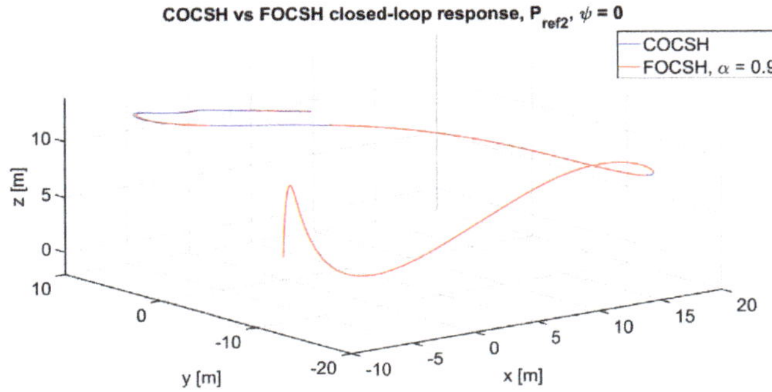

Figure 14. COCSH and FOCSH (with $\alpha = 0.99$) closed-loop response comparison when the drone hovers at point P_{ref2} with an orientation of $\psi = 0$ radians.

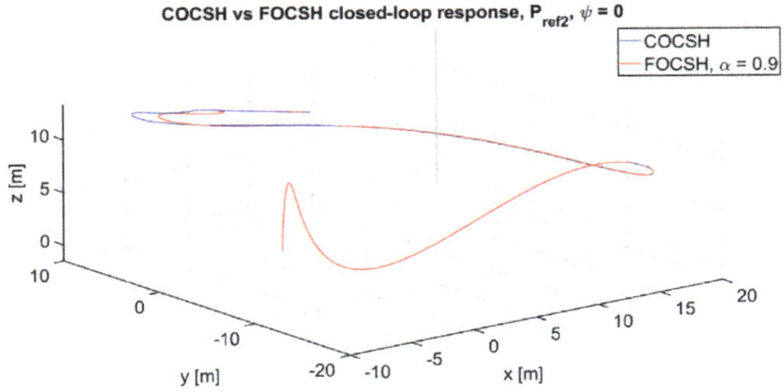

Figure 15. COCSH and FOCSH (with $\alpha = 0.9$) closed-loop response comparison when the drone hovers at point P_{ref2} with an orientation of $\psi = 0$ radians.

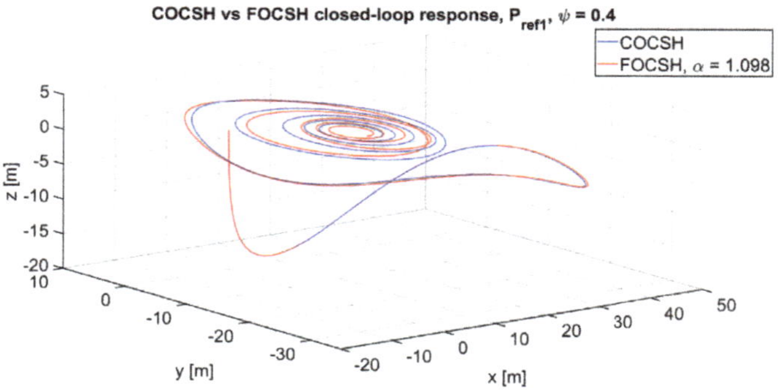

Figure 16. COCSH and FOCSH (with $\alpha = 1.098$) closed-loop response comparison when the drone hovers at point P_{ref1} with an orientation of $\psi = 0.4$ radians.

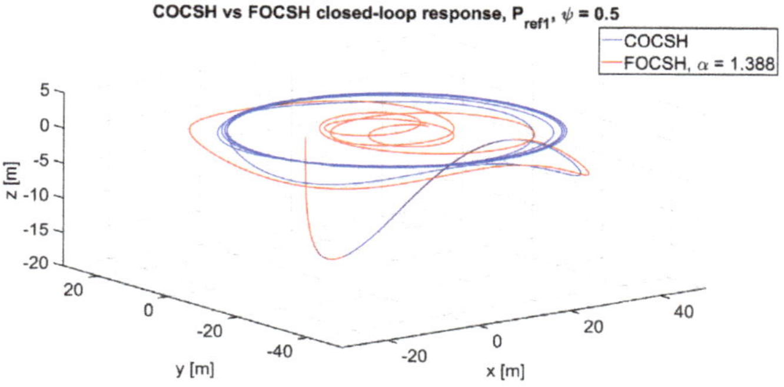

Figure 17. COCSH and FOCSH (with $\alpha = 1.388$) closed-loop response comparison when the drone hovers at point P_{ref1} with an orientation of $\psi = 0.5$ radians.

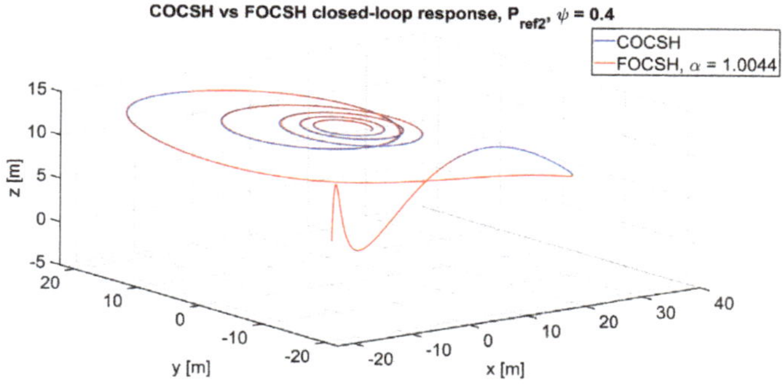

Figure 18. COCSH and FOCSH (with $\alpha = 1.0044$) closed-loop response comparison when the drone hovers at point P_{ref2} with an orientation of $\psi = 0.4$ radians.

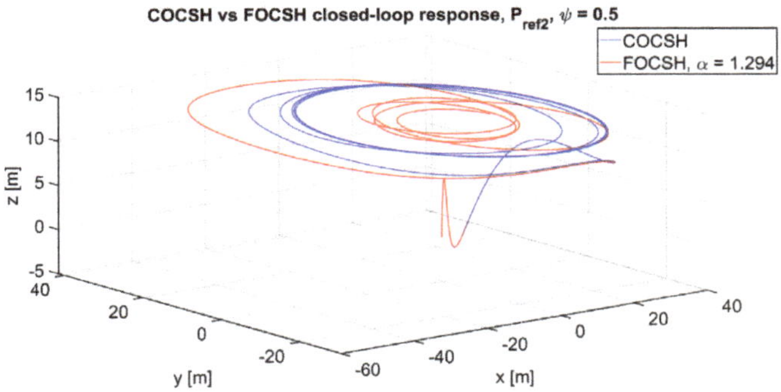

Figure 19. COCSH and FOCSH (with $\alpha = 1.294$) closed-loop response comparison when the drone hovers at point P_{ref2} with an orientation of $\psi = 0.5$ radians.

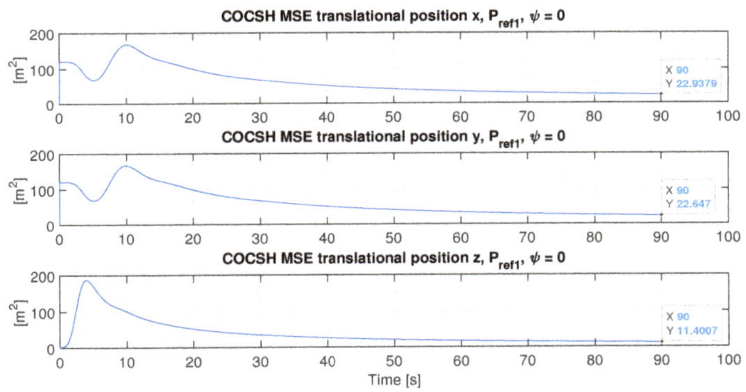

Figure 20. COCSH MSE translational positions x, y, and z when the drone hovers at point P_{ref1} with an orientation of $\psi = 0$ radians.

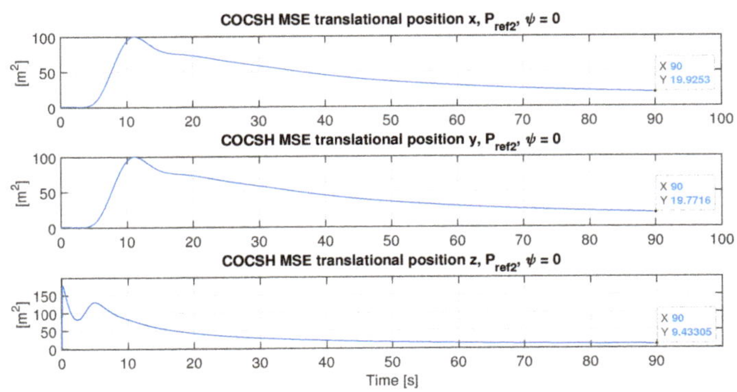

Figure 21. COCSH MSE translational positions x, y, and z when the drone hovers at point P_{ref2} with an orientation of $\psi = 0$ radians.

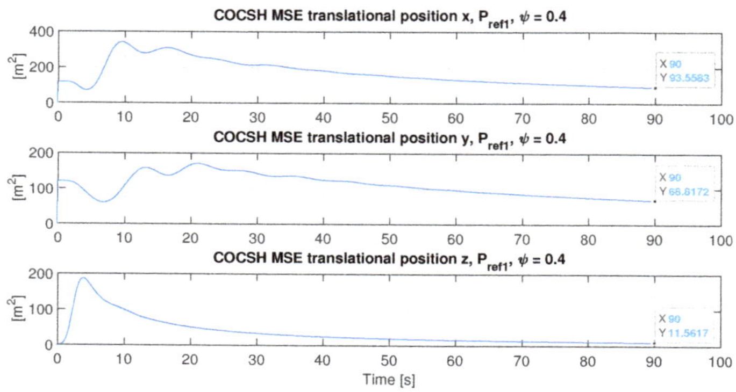

Figure 22. COCSH MSE translational positions x, y, and z when the drone hovers at point P_{ref1} with an orientation of $\psi = 0.4$ radians.

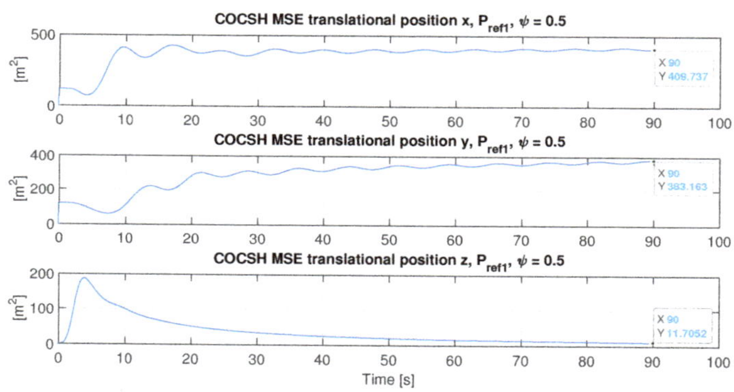

Figure 23. COCSH MSE translational positions x, y, and z when the drone hovers at point P_{ref1} with an orientation of $\psi = 0.5$ radians.

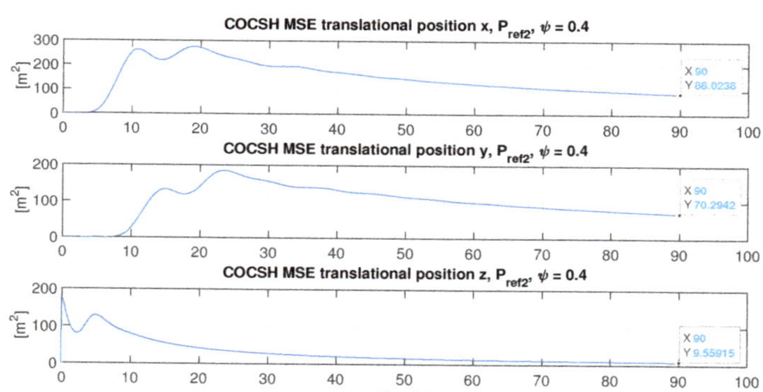

Figure 24. COCSH MSE translational positions x, y, and z when the drone hovers at point P_{ref2} with an orientation of $\psi = 0.4$ radians.

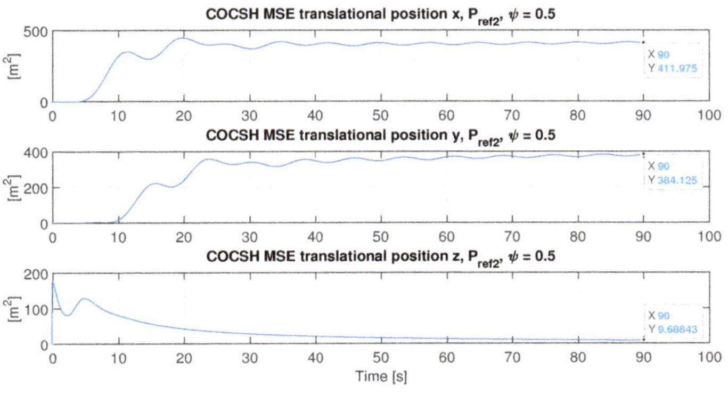

Figure 25. COCSH MSE translational positions x, y, and z when the drone hovers at point P_{ref2} with an orientation of $\psi = 0.5$ radians.

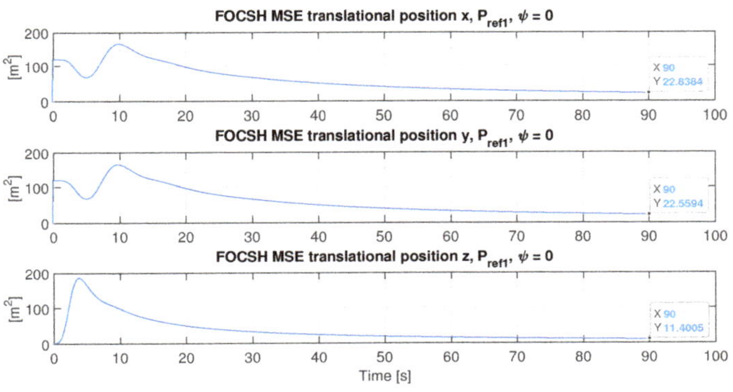

Figure 26. FOCSH MSE translational positions x, y, and z when the drone hovers at point P_{ref1} with $\alpha = 0.99$ and an orientation of $\psi = 0$ radians.

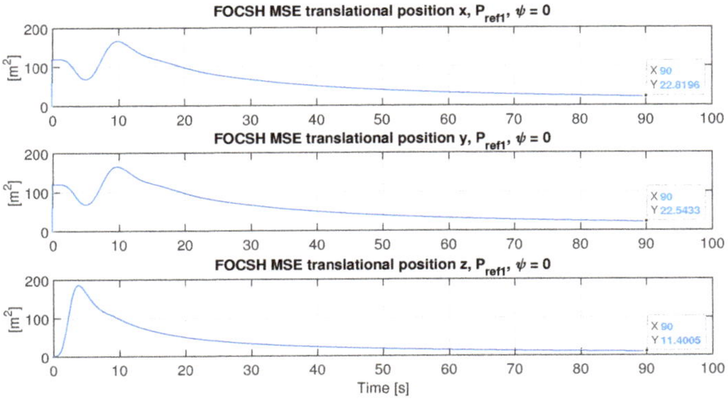

Figure 27. FOCSH MSE translational positions x, y, and z when the drone hovers at point P_{ref1} with $\alpha = 0.9882$ and an orientation of $\psi = 0$ radians.

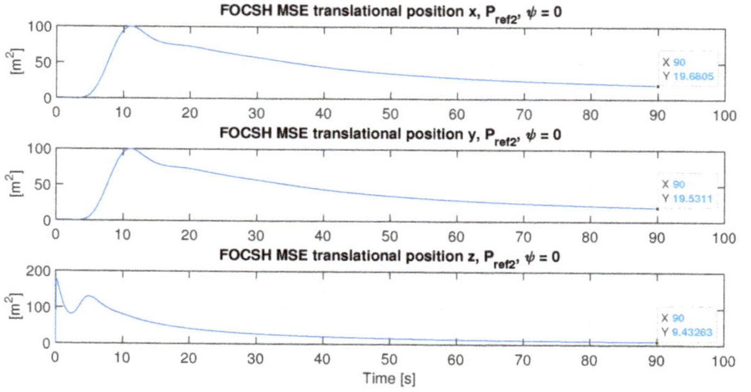

Figure 28. FOCSH MSE translational positions x, y, and z when the drone hovers at point P_{ref2} with $\alpha = 0.99$ and an orientation of $\psi = 0$ radians.

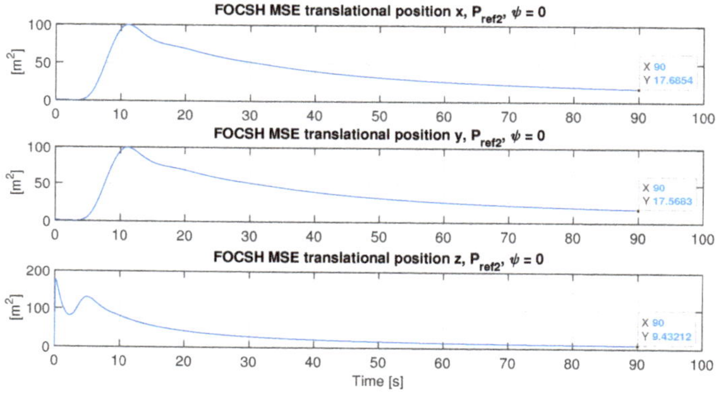

Figure 29. FOCSH MSE translational positions x, y, and z when the drone hovers at point P_{ref2} with $\alpha = 0.9$ and an orientation of $\psi = 0$ radians.

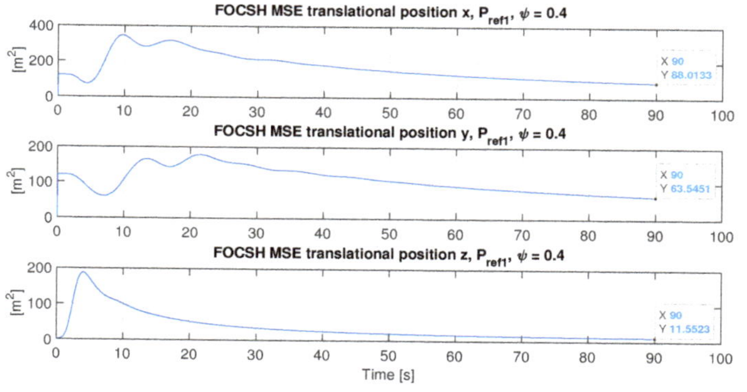

Figure 30. FOCSH MSE translational positions x, y, and z when the drone hovers at point P_{ref1} with $\alpha = 1.098$ and an orientation of $\psi = 0.4$ radians.

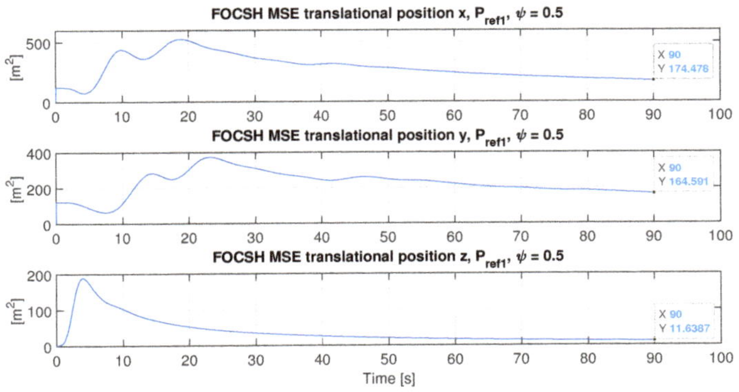

Figure 31. FOCSH MSE translational positions x, y, and z when the drone hovers at point P_{ref1} with $\alpha = 1.388$ and an orientation of $\psi = 0.5$ radians.

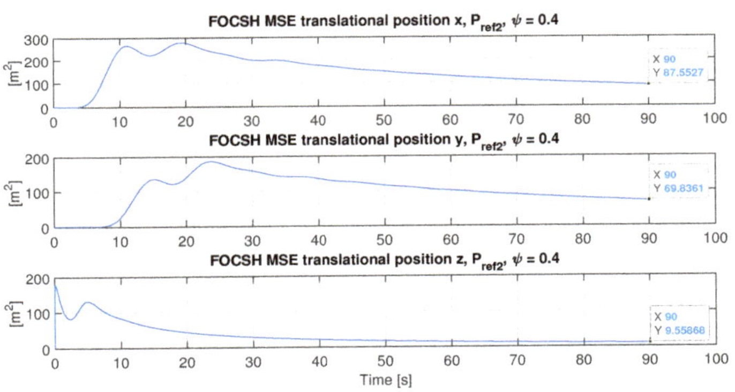

Figure 32. FOCSH MSE translational positions x, y, and z when the drone hovers at point P_{ref2} with $\alpha = 1.0044$ and an orientation of $\psi = 0.4$ radians.

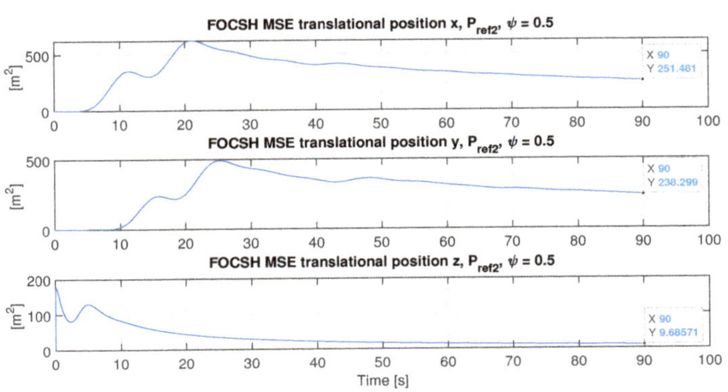

Figure 33. FOCSH MSE translational positions x, y, and z when the drone hovers at point P_{ref2} with $\alpha = 1.294$ and an orientation of $\psi = 0.5$ radians.

Based on Table 3, it can be inferred that in every case, the mean squared error (MSE) is higher for the COCSH approach than for the proposed FOCSH approach. Furthermore, considering the BQC-LR indicator, which denotes the best control system with a special character (*), it is possible to conclude that the FOCSH performs better in hover flight across all cases, since its MSE performance index is lower than that of the COCSH.

Finally, it is important to mention that through an analysis based on the simulation results shown in Figures 4–19 and the MSE performance indices shown in Figures 20–33, which have been obtained by coupling the COCSH and FOCSH control systems to the non-linear model of the UAV described by (1), (2), (5) and (6), it can be established that from the improvement in the closed-loop response of the FOCSH compared to the COCSH, the selection of the alpha parameter could be made considering the following two proposed criteria:

The selection of the α parameter for FOCSH implementation should be made according to expression (30), when the model is described by (1), (2), (5) and (6), due to small values of ψ tending to behave in a linear manner, like the model described by (7)–(9), thus enhancing the controller's robustness.

$$\alpha = \lim_{\Delta\alpha \to 0} \{1 - \Delta\alpha\}. \tag{30}$$

The selection of the α parameter for FOCSH implementation should be made according to expression (31), when the model is described by (1), (2), (5) and (6), due to large values of ψ tending to behave in a non-linear manner, thus enhancing the controller's robustness.

$$\alpha = \lim_{\Delta\alpha \to 0} \{1 + \Delta\alpha\}. \tag{31}$$

To denote the ability of the FOCSH to handle non-linearities and the improvement in its closed-loop response, Figures 34 and 35 are presented, in which a simulation time of 300 s is considered. Figure 34 shows a comparison of the closed-loop response between COCSH and FOCSH when the drone reaches a hover at point P_{ref1}, with a desired orientation ψ = 0.5 radians (rotation around its vertical axis). When analyzing the drone's trend over a simulation time of 300 s, it can be observed that the elevation control along the z axis works for both COCSH and FOCSH. However, in the case of the COCSH, the drone follows an orbital flight around the reference point P_{ref1}, which it is unable to reach as the simulation progresses. In contrast, for the FOCSH, with a value of α = 1.388, this orbital flight converges around the reference point P_{ref1}.

A similar behavior is shown in Figure 35 when the drone reaches a hover at point P_{ref2} with a desired orientation ψ = 0.5 radians. In the course of a simulation time of 300 s, it can be seen that in the case of the COCSH, the drone once again follows an orbital flight around the reference point P_{ref2} without reaching it as the simulation progresses. In contrast, for the FOCSH, with α = 1.294, this orbital flight converges towards the reference point P_{ref2}.

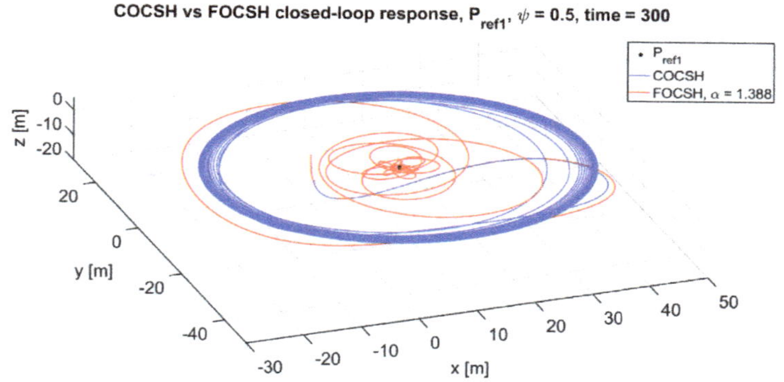

Figure 34. COCSH and FOCSH closed-loop response comparison when the drone hovers at point P_{ref1} with an orientation of $\psi = 0.5$ radians and considering a simulation time of 300 s.

Table 3. SMCO and SMFO performance indices and BQC-LR indicator. * denotes the best control system.

	MSE Translational Position		BQC-LR	
Hover at Point	**COCSH**	**FOCSH**	**COCSH**	**FOCSH**
		with $\alpha = 0.99$:		
$P_{ref1}, \psi = 0$	$x = 22.9379$	$x = 22.8384$		
$P_{ref1}, \psi = 0$	$y = 22.6470$	$y = 22.5594$		
$P_{ref1}, \psi = 0$	$z = 11.4007$	$z = 11.4005$		
		with $\alpha = 0.9882$:		
$P_{ref1}, \psi = 0$	$x = 22.9379$	$x = 22.8196$		*
$P_{ref1}, \psi = 0$	$y = 22.6470$	$y = 22.5433$		*
$P_{ref1}, \psi = 0$	$z = 11.4007$	$z = 11.4005$		*
		with $\alpha = 0.99$:		
$P_{ref2}, \psi = 0$	$x = 19.9253$	$x = 19.6805$		
$P_{ref2}, \psi = 0$	$y = 19.7716$	$y = 19.5311$		
$P_{ref2}, \psi = 0$	$z = 9.43305$	$z = 9.43283$		
		with $\alpha = 0.9$:		
$P_{ref2}, \psi = 0$	$x = 19.9253$	$x = 17.6854$		*
$P_{ref2}, \psi = 0$	$y = 19.7716$	$y = 17.5683$		*
$P_{ref2}, \psi = 0$	$z = 9.43305$	$z = 9.43212$		*
		with $\alpha = 1.098$:		
$P_{ref1}, \psi = 0.4$	$x = 93.5583$	$x = 88.0133$		*
$P_{ref1}, \psi = 0.4$	$y = 68.8172$	$y = 63.5451$		*
$P_{ref1}, \psi = 0.4$	$z = 11.5617$	$z = 11.5523$		*
		with $\alpha = 1.388$:		
$P_{ref1}, \psi = 0.5$	$x = 409.737$	$x = 174.478$		*
$P_{ref1}, \psi = 0.5$	$y = 383.163$	$y = 164.591$		*
$P_{ref1}, \psi = 0.5$	$z = 11.7052$	$z = 11.6387$		*
		with $\alpha = 1.0044$:		
$P_{ref2}, \psi = 0.4$	$x = 88.0238$	$x = 87.5527$		*
$P_{ref2}, \psi = 0.4$	$y = 70.2942$	$y = 69.8361$		*
$P_{ref2}, \psi = 0.4$	$z = 9.55915$	$z = 9.55868$		*
		with $\alpha = 1.294$:		
$P_{ref2}, \psi = 0.5$	$x = 411.975$	$x = 251.481$		*
$P_{ref2}, \psi = 0.5$	$y = 384.125$	$y = 238.299$		*
$P_{ref2}, \psi = 0.5$	$z = 9.68843$	$z = 9.68571$		*

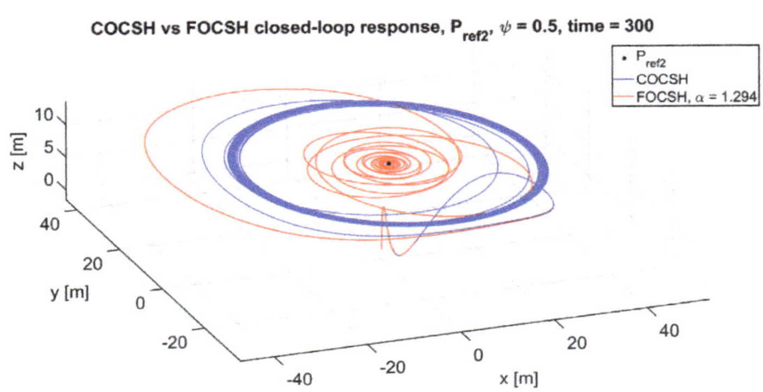

Figure 35. COCSH and FOCSH closed-loop response comparison when the drone hovers at point P_{ref2} with an orientation of $\psi = 0.5$ radians and considering a simulation time of 300 s.

It is important to emphasize that the expressions defining the controllers that constitute the COCSH are described by Equations (10)–(23). Furthermore, it is emphasized that these have been tuned using the root locus technique in the continuous-time domain for practical and straightforward purposes. Consequently, these have been adapted to the fractional approach using the same gains to conduct a fair comparison with classical controllers. It is also mentioned that self-developed functions have been used for the implementation of the proposed fractional-order controllers. In this sense, we proposed conducting a set of numerical simulations in which a simulation time of 90 s was established, with an iteration step of $\Delta t = 25$ ms, to create convergence with the continuous-time controller's design technique. Therefore, the contribution of this work lies in something other than the stability analysis of the COCSH (something already explored in depth by other authors). Instead, it lies in highlighting the improvement in the closed-loop response of the FOCSH when the system (UAV) tends to behave in a non-linear manner, demonstrating the robustness of the FOCSH. Finally, in control systems, performance improvements are often significant, even if they are very small, since a small variation in the closed-loop response could have consequences such as system instability or the saturation of some electronic signal conditioning stages. From the point of view of energy savings in UAVs, small variations translate into unnecessary energy costs that deteriorate the helpful life of the battery in use. Therefore, in this research work, performance indices are proposed that quantitatively demonstrate the performance improvements between fractional PID control (which also presents an alternative solution to classic controllers) and classic control.

6. Conclusions

In this paper, we present a thorough comparative analysis of two control systems designed to enhance the hovering flight of a quadrotor UAV: the Classical Order Control System for Hovering (COCSH) and the Fractional Order Control System for Hovering (FOCSH). Through numerical simulations and performance index analyses, the following detailed conclusions were drawn:

Performance of the FOCSH Controller: The FOCSH demonstrated superior performance compared to the COCSH in all studied scenarios. This was evidenced by the lower mean squared error (MSE) indices for the FOCSH, indicating its greater accuracy in maintaining the UAV's hover position and the better quality of its closed-loop response. The improved accuracy resulted in more precise translational positions, contributing to the overall stability and effectiveness of the UAV's hovering capabilities.

Robustness and Stability: The FOCSH not only enhanced the precision of the UAV's hovering but also proved to be more robust against external disturbances, such as atmospheric turbulence and sudden changes in payload. This robustness is crucial for maintaining flight stability under adverse environmental conditions, which are common challenges in UAV operations. The FOCSH's ability to handle such disturbances effectively ensures a more reliable and secure flight, reducing the risk of deviations and instability.

Application of Fractional Calculus: The implementation of fractional calculus in the FOCSH's PID controllers allowed for a more precise adaptation of the integral effect, thereby improving the system's ability to manage non-linearities and disturbances. The fractional approach offers an advantage over traditional controllers by providing greater flexibility and a more accurate tuning of control parameters. This adaptability is key in navigating complex and dynamic environments where traditional control methods may fall short.

Impact and Potential Applications: The findings have significant implications for various industrial applications, including precision agriculture, infrastructure inspection, and surveillance. The improvement in hovering stability and precision not only enhances the quality of the data collected but also improves the operational safety of UAVs. Additionally, this increased efficiency leads to reduced operational costs, as it minimizes the need for additional flights and potential repairs due to unstable flights.

In summary, this study demonstrates that fractional-order controllers, specifically the FOCSH, significantly enhance the ability of UAVs to maintain precise and stable hovering. This provides an effective solution to the increasing demands seen in UAV applications across different sectors. The results underscore the importance of further researching and developing fractional-order controllers to continue advancing UAVs' system performance and capabilities.

Author Contributions: Conceptualization, G.D.-R., J.S.V.-M., P.G.-L. and M.A.H.-P.; methodology, G.D.-R. and J.S.V.-M.; software, G.D.-R. and M.A.H.-P.; validation, G.D.-R., J.S.V.-M., P.G.-L. and M.A.H.-P.; formal analysis, G.D.-R. and P.G.-L.; investigation, G.D.-R. and J.S.V.-M.; resources, G.D.-R., J.S.V.-M., P.G.-L. and M.A.H.-P.; data curation, G.D.-R. and P.G.-L.; writing—original draft preparation, G.D.-R., J.S.V.-M. and P.G.-L.; writing—review and editing, G.D.-R., J.S.V.-M. and P.G.-L.; visualization, G.D.-R. and J.S.V.-M.; supervision, G.D.-R., J.S.V.-M., P.G.-L. and M.A.H.-P.; project administration, G.D.-R. and J.S.V.-M.; funding acquisition, G.D.-R., J.S.V.-M., P.G.-L. and M.A.H.-P. All authors have read and agreed to the published version of the manuscript.

Funding: This research received no external funding.

Data Availability Statement: Data is contained within the article.

Conflicts of Interest: The authors declare no conflicts of interest.

References

1. Frigioescu, T.-F.; Condruz, M.R.; Badea, T.A.; Paraschiv, A. A Preliminary Study on the Development of a New UAV Concept and the Associated Flight Method. *Drones* **2023**, *7*, 166. [CrossRef]
2. Telli, K.; Kraa, O.; Himeur, Y.; Ouamane, A.; Boumehraz, M.; Atalla, S.; Mansoor, W. A Comprehensive Review of Recent Research Trends on Unmanned Aerial Vehicles (UAVs). *Systems* **2023**, *11*, 400. [CrossRef]
3. Arafat, M.Y.; Alam, M.M.; Moh, S. Vision-Based Navigation Techniques for Unmanned Aerial Vehicles: Review and Challenges. *Drones* **2023**, *7*, 89. [CrossRef]
4. Saunders, J.; Saeedi, S.; Li, W. Autonomous aerial robotics for package delivery: A technical review. *J. Field Robot.* **2023**, *41*, 1. [CrossRef]
5. Delavarpour, N.; Koparan, C.; Nowatzki, J.; Bajwa, S.; Sun, X. A Technical Study on UAV Characteristics for Precision Agriculture Applications and Associated Practical Challenges. *Remote Sens.* **2021**, *13*, 1204. [CrossRef]
6. Liang, H.; Lee, S.-C.; Bae, W.; Kim, J.; Seo, S. Towards UAVs in Construction: Advancements, Challenges, and Future Directions for Monitoring and Inspection. *Drones* **2023**, *7*, 202. [CrossRef]
7. Balestrieri, E.; Daponte, P.; De Vito, L.; Picariello, F.; Tudosa, I. Sensors and Measurements for UAV Safety: An Overview. *Sensors* **2021**, *21*, 8253. [CrossRef]
8. Hoffmann, G.; Huang, H.; Waslander, S.; Tomlin, C. Quadrotor helicopter flight dynamics and control: Theory and experiment. In Proceedings of the AIAA Guidance, Navigation and Control Conference and Exhibit, Hilton Head, SC, USA, 20–23 August 2007. [CrossRef]
9. Pan, F.; Liu, L.; Xue, D. Optimal PID controller design with Kalman filter for Qball-X4 quad-rotor unmanned aerial vehicle. *Trans. Inst. Meas. Control.* **2017**, *39*, 1785–1797. [CrossRef]
10. Abdulridha, N.M.; Mary, A.H.; Jasim, H.H. Optimized PID, FOPID and PIDD2 for Controlling UAV Based on SSA. *Am. Sci. Res. J. Eng. Technol. Sci.* **2023**, *92*, 77–90.
11. Priyambodo, T.K.; Dhewa, O.A.; Susanto, T. Model of Linear Quadratic Regulator (LQR) Control System in Waypoint Flight Mission of Flying Wing UAV. *J. Telecommun. Electron. Comput. Eng.* **2020**, *12*, 43–49.
12. Dhewa, O.; Dharmawan, A.; Priyambodo, T.K. Model of Linear Quadratic Regulator (LQR) Control Method in Hovering State of Quadrotor. *J. Telecommun. Electron. Comput. Eng.* **2017**, *9*, 135–143.
13. Omar, M.F.; Ghazali, R.; Saadon, I.M.; Aripin, M.K.; Sam, Y.M.; Soon, C.C. Optimal Control for Sport Utility Vehicle System Using Linear Quadratic Integral Control Approach. *J. Telecommun. Electron. Comput. Eng.* **2018**, *10*, 143–149.
14. Saccani, D.; Cecchin, L.; Fagiano, L. Multitrajectory Model Predictive Control for Safe UAV Navigation in an Unknown Environment. *IEEE Trans. Control. Syst. Technol.* **2023**, *31*, 1982–1997. [CrossRef]
15. Singh, B.K.; Kumar, A. Model predictive control using LPV approach for trajectory tracking of quadrotor UAV with external disturbances. *Aircr. Eng. Aerosp. Technol.* **2023**, *95*, 607–618. [CrossRef]
16. Bouadi, H.; Bouchoucha, M.; Tadjine, M. Sliding mode control based on backstepping approach for an UAV type-quadrotor. *Int. J. Mech. Mechatron. Eng.* **2007**, *1*, 39–44.
17. Jiang, X.; Su, C.; Xu, Y.; Liu, K.; Shi, H.; Li, P. An adaptive backstepping sliding mode method for flight attitude of quadrotor UAVs. *J. Cent. South Univ.* **2018**, *25*, 616–631. [CrossRef]
18. Madani, T.; Benallegue, A. Backstepping control for a quadrotor helicopter. In Proceedings of the 2006 IEEE/RSJ International Conference on Intelligent Robots and Systems, Beijing, China, 9–13 October 2006. [CrossRef]
19. Richards, A.; How, J. Aircraft trajectory planning with collision avoidance using mixed integer linear programming. In Proceedings of the 2002 American Control Conference (IEEE Cat. No. CH37301), Anchorage, AK, USA, 8–10 May 2002. [CrossRef]
20. Wang, C.; Wang, J.; Wang, J.; Zhang, X. Deep-reinforcement-learning-based autonomous UAV navigation with sparse rewards. *IEEE Internet Things J.* **2020**, *7*, 6180–6190. [CrossRef]

21. Monje, C.; Vinagre, B.; Feliu, V.; Chen, Y. Tuning and auto-tuning of fractional order controllers for industry applications. *Control. Eng. Pract.* **2008**, *16*, 798–812. [CrossRef]
22. Yoon, J.; Doh, J. Optimal PID control for hovering stabilization of quadcopter using long short term memory. *Adv. Eng. Inform.* **2022**, *53*, 101679. [CrossRef]
23. Shang, B.; Liu, J.; Zhao, T.; Chen, Y. Fractional order robust visual servoing control of a quadrotor UAV with larger sampling period. In Proceedings of the International Conference on Unmanned Aircraft Systems (ICUAS), Arlington, VA, USA, 7–10 June 2016. [CrossRef]
24. Delgado, G.; Valdez, J.S.; Hernández, M.Á.; Pérez, K.R.; García, P.J. Quadrotor Real-Time Simulation: A Temporary Computational Complexity-Based Approach. *Mathematics* **2022**, *10*, 2032. [CrossRef]
25. Chan, Y.Y.; Kam, K.H.; Lee, C.K.M.; Li-Ta, H.; Keung, K.L. Wind dynamic and energy-efficiency path planning for unmanned aerial vehicles in the lower-level airspace and urban air mobility context. *Sustain. Energy Technol. Assess.* **2023**, *57*, 103202. [CrossRef]
26. Timis, D.D.; Muresan, C.I.; Dulf, E.-H. Design and Experimental Results of an Adaptive Fractional-Order Controller for a Quadrotor. *Fractal Fract.* **2022**, *6*, 204. [CrossRef]
27. Al-Dhaifallah, M.; Al-Qahtani, F.M.; Elferik, S.; Saif, A.-W.A. Quadrotor Robust Fractional-Order Sliding Mode Control in Unmanned Aerial Vehicles for Eliminating External Disturbances. *Aerospace* **2023**, *10*, 665. [CrossRef]
28. Dulf, E.H.; Timis, D.; Muresan, C.I. Robust Fractional Order Controllers for Distributed Systems. *Acta Polytech. Hung.* **2017**, *14*, 163–176. [CrossRef]
29. Jain, S.; Hote, Y.; Dehuri, P.; Mittal, D.; Siddhartha, V. Experimental validation of fractional order internal model controller design on buck and boost converter. *Meas. Control.* **2020**, *54*, 748–766. [CrossRef]
30. Haseeb, A.; Ali, A.; Hasan, I. Fractional Order Sliding Mode Controller for Heading and Heave Control of Quadcopter. In Proceedings of the 2nd International Conference on on Engineering, Natural and Social Sciences (ICENSOS), Konya, Turkey, 4–6 April 2023.
31. Sadigh, R. Optimizing PID Controller Coefficients Using Fractional Order Based on Intelligent Optimization Algorithms for Quadcopter. In Proceedings of the 6th RSI International Conference on Robotics and Mechatronics (IcRoM), Tehran, Iran, 23–25 October 2018. [CrossRef]
32. Shao, K.; Liu, J.; Li, M. Fractional order PID control of quadrotor UAV based on SA-PSO algorithm. In Proceedings of the Third International Conference on Computer Science and Communication Technology (ICCSCT 2022), Beijing, China, 30–31 July 2022. [CrossRef]
33. Song, J.; Hu, Y.; Su, J.; Zhao, M.; Ai, S. Fractional-Order Linear Active Disturbance Rejection Control Design and Optimization Based Improved Sparrow Search Algorithm for Quadrotor UAV with System Uncertainties and External Disturbance. *Drones* **2022**, *6*, 229. [CrossRef]
34. Moness, M.; Abdelghany, M.B.; Mohammed, K.M.; Mohamed, M.; Moustafa, A.M. Surrogate Optimal Fractional Control for Constrained Operational Service of UAV Systems. *Drones* **2024**, *8*, 141. [CrossRef]
35. Moreira, E.; Shiroma, P. Design of fractional PID controller in time-domain for a fixed-wing unmanned aerial vehicle. In Proceedings of the 2017 Latin American Robotics Symposium (LARS) and 2017 Brazilian Symposium on Robotics (SBR), Curitiba, Brazil, 8–11 November 2017. [CrossRef]
36. Yu, Z.; Zhang, Y.; Jiang, B.; Su, C.Y.; Fu, J.; Jin, Y.; Chai, T. Fractional order PID-based adaptive fault-tolerant cooperative control of networked unmanned aerial vehicles against actuator faults and wind effects with hardware-in-the-loop experimental validation. *Control. Eng. Pract.* **2021**, *114*, 1–12. [CrossRef]
37. Vahdanipour, M.; Khodabandeh, M. Adaptive fractional order sliding mode control for a quadrotor with a varying load. *Aerosp. Sci. Technol.* **2019**, *86*, 737–747. [CrossRef]
38. Li, J.; Chen, P.; Chang, Z.; Zhang, G.; Guo, L.; Zhao, C. Trajectory Tracking Control of Quadrotor Based on Fractional-Order S-Plane Model. *Machines* **2023**, *11*, 672. [CrossRef]
39. Okoronkwo, J.O.; Mati, A.; Yusuf, J.; Olarinoye, G.; Abubakar, A.S. Development of Fractional Order Controller for Load Frequency Control of Multi Area Interconnected Microgrid Power System. In Proceedings of the 2021 1st International Conference on Multidisciplinary Engineering and Applied Science (ICMEAS), Abuja, Nigeria, 15–16 July 2021. [CrossRef]
40. Huang, Y.; Nitsche, M.; Kanso, E. Stability versus maneuverability in hovering flight. *Phys. Fluids* **2015**, *27*, 061706. [CrossRef]
41. Kim, D.; Lee, Y.; Oh, S.; Park, Y.; Choi, J.; Park, D. Aerodynamic analysis and static stability analysis of Manned/unmanned distributed propulsion aircrafts using actuator methods. *J. Wind. Eng. Ind. Aerodyn.* **2021**, *214*, 104648. [CrossRef]
42. Hernández-Pérez, M.A.; Delgado-Reyes, G.; Borja-Jaimes, V.; Valdez-Martínez, J.S.; Cervantes-Bobadilla, M. An Adaptation of a Sliding Mode Classical Observer to a Fractional-Order Observer for Disturbance Reconstruction of a UAV Model: A Riemann–Liouville Fractional Calculus Approach. *Mathematics* **2023**, *11*, 4876. [CrossRef]
43. Cando, L.; Hernandez, M.A.; Delgado, G.; Galván, R.; Orozco, R. Design and Implementation of a Rotating Hydrodynamic System for Electrochemical Noise Measurement. *ECS Trans.* **2023**, *110*, 79–91. [CrossRef]

fractal and fractional

MDPI

Article

Robust Fractional-Order PI/PD Controllers for a Cascade Control Structure of Servo Systems

Vo Lam Chuong, Ngo Hong Nam, Le Hieu Giang and Truong Nguyen Luan Vu *

Faculty of Mechanical Engineering, Ho Chi Minh City University of Technology and Education, 01 Vo Van Ngan St, Thu Duc City, Ho Chi Minh City 71300, Vietnam; chuongvl@hcmute.edu.vn (V.L.C.); namnh.ncs@hcmute.edu.vn (N.H.N.); gianglh@hcmute.edu.vn (L.H.G.)

* Correspondence: vuluantn@hcmute.edu.vn

Abstract: In this paper, a cascade control structure is suggested to control servo systems that normally include a servo motor in coupling with two kinds of mechanism elements, a translational or rotational movement. These kinds of systems have high demands for performance in terms of fastest response and no overshoot/oscillation to a ramp function input. The fractional-order proportional integral (FOPI) and proportional derivative (FOPD) controllers are addressed to deal with those control problems due to their flexibility in tuning rules and robustness. The tuning rules are designed in the frequency domain based on the concept of the direct synthesis method and also ensure the robust stability of controlled systems by using the maximum sensitivity function. The **M-Δ** structure, using multiplicative output uncertainties for both control loops simultaneously, is addressed to justify the robustness of the controlled systems. Simulation studies are considered for two kinds of plants that prove the effectiveness of the proposed method, with good tracking of the ramp function input under the effects of the disturbances. In addition, the robustness of the controlled system is illustrated by a structured singular value (μ) plot in which its value is less than 1 over the frequency range.

Keywords: fractional-order proportional integral controller (FOPI); fractional-order proportional derivative controller (FOPD); cascade control structure; robust stability; servo systems

1. Introduction

In general, a servo system is an electromechanical drive that includes a servo motor coupled with a ball screw or a rotational load as a mechanism element. Servo systems have been widely used in industrial machines such as computerized numerical controls (CNCs), robots, and so on, with strict requirements for speed and position control, which are essentially fast responses, and no overshoots or oscillations to a ramp function input. The well-known control structure for these systems consists of two control loops in a cascade scheme [1], also called cascade control structure (CCS), which was first introduced in chemical process control [2] and gradually became common in industry applications. This structure consists of two control loops: the inner or secondary loop and the outer or primary loop. The main advantage of this scheme is to reduce the disturbance of the inner loop and improve the servo response of the outer one. This means that if a disturbance affects the inner loop, it will be attenuated before having an impact on the primary output. Therefore, currently, most of the commercial controllers for servo systems utilize this configuration with a proportional integral controller (PI) for the velocity control in the inner loop and a proportional (P) or a proportional derivative (PD) controller for the position control in the outer loop [3–8]. However, it is still difficult to derive analytical tuning rules to meet the increasingly high requirements of motion control and also ensure robustness to system uncertainties [9].

In the literature, for cascade control systems, many studies have been conducted to improve the control performances. Lee et al. [10] proposed a method for tuning the PID controller based on the internal model control (IMC) structure for both secondary and primary loops. Many researchers deploy this approach to solve the tuning problems for a variety of plants such as first-order plus time delay, second-order plus time delay, integrating systems, and unstable systems [11–16]. However, all mentioned methods are only suitable for the processes with sluggish responses and overshoot at the output being accepted. To improve system performance, various works have been reported to use some advanced control techniques such as cascaded sliding mode control [17–19], robust adaptive control [20,21], adaptive fuzzy control [9,22], and so on. These approaches have shown better performances compared to traditional controllers. However, it is hard for them to find their way into industrial applications due to their complexity and the requirement of enhanced modeling parameters. In addition, some nonlinear strategies, especially sliding mode control, normally have an issue in manipulated variables such as discontinuous control signals or ripple effects, which could cause harm to actuators.

Recently, fractional calculus has attracted researchers' attention in the control field in terms of combination with some conventional controllers such as fractional-order PID, fractional-order nonlinear control [18,19], and fractional-order fuzzy control [23]. Due to the most popular PID controller in industrial applications, the generalization of the PID controller, which is called a fractional-order PID controller (FOPID) or $PI^{\lambda}D^{\mu}$ [24], for which, λ and μ are the fractional-order integral and derivative, respectively, deserves to be studied thoroughly for future controllers. The FOPID controller affords more flexibility in tuning rules due to having two extra tuning parameters and more robustness than the integer-order one. In the last two decades, there have been many works that have been proposed to tune the FOPID controller. Normally, they could be categorized into two approaches: the frequency-based method and the time domain-based heuristic algorithm. In the frequency domain, the internal model control (IMC)-based scheme or Bode's ideal transfer function is used to reduce the number of tuning parameters; and finally, some criteria are addressed to ensure robust performance, such as using constraints on phase margin and gain crossover frequency [25–30]. However, most are only available for single-input, single-output (SISO) systems. Therefore, expanding some existing design methods to a complex structure is necessary and deserves to attract more attention from researchers. In the second approach, some evolutionary algorithms such as the genetic algorithm (GA) and particle swarm optimization (PSO) are used to solve the control design problems [31,32]. In [32], the authors proposed the multi-objective PSO to design a robust FOPID controller for first-order plus delay time processes. The objective function considered both system performances in terms of set-point tracking as well as disturbance rejection and system robustness using the maximum sensitivity function. However, this is only verified by SISO systems. Recently, some modern metaheuristic methods [23] such as marine predators algorithm (MPA) [33] or a combination with reinforcement learning for online tuning [34] also have been proposed. From the review, most FOPID controllers are used for SISO or non-linear systems, there are only a few works that use fractional-order controllers for cascade structures in the literature. In [35], the authors proposed the FOPI controller based on internal model control for the inner loop and a PD controller for the outer loop of a parallel cascade control. However, the robust stability is only investigated by simulation studies without a specific systematic method.

In this work, a unified approach to tuning rules for both closed loops in the cascade control system is proposed. In the secondary loop, the FOPI controller is suggested, and its tuning rules are based on the direct synthesis method in the frequency domain. To guarantee the robustness of the controlled system, the maximum sensitivity function is adopted to find out the range of frequency where the M_s value is as close to 1.2 as possible [15,27,36]. Whereas, in the primary loop, the FOPD controller will be adopted and the corresponding tuning rules are similar to those of the inner loop to meet the high requirements of a position servo system, i.e., fast response with no overshoot and oscillation. To justify the robust stability of the whole control system (CCS), the **M-Δ** structure with multiplicative output uncertainties is considered [36]. Moreover, in this study, the parametric uncertainties of both control loops are investigated simultaneously.

Simulation studies will be conducted in this work to illustrate the performance of the controlled system.

This paper is organized as follows. Section 2 briefly introduces fractional calculus and fractional-order controllers including FOPD and FOPI in the frequency domain that are used in this study. In addition, the mechanism systems are also introduced in this section. The proposed method is mentioned in Section 3 including the control structure and also the tuning rules for both control loops. The robust stability is analyzed in Section 4. The simulation studies with two different mechanisms will be discussed in Section 5. Finally, conclusions are given in Section 6.

2. Preliminaries

2.1. Fractional Calculus

Fractional calculus is a generalization of ordinary calculus. It develops a functional operator, D, associated to the order, v ($v \in \Re$), that generalizes usual derivatives (for positive v) and integrals (for negative v). The most commonly used is the Riemann–Liouville definition [37], which is generalized by the following equation:

$$ {}_aD_t^v f(t) = \frac{1}{\Gamma(n-v)} \frac{d^n}{dt^n} \int_a^t \frac{f(\tau)}{(t-\tau)^{v-n+1}} dt,\ n-1 < v < n \quad (1) $$

where $\Gamma(\bullet)$ denotes Euler's gamma function. a and t are the limits.

Note that the Laplace transform of the fractional derivative/integral in Equation (1) follows the rule for zero initial condition for order v ($0 < v < 1$):

$$ L\{{}_aD_t^{\pm v} f(t)\} = s^{\pm v} F(s) \quad (2) $$

2.2. Fractional Linear Model

For a single-input, single-output linear time invariant system where both input ($u(t)$) and output ($y(t)$) signals are relaxed at $t = 0$, the fractional-order differential equation can be expressed as:

$$ \sum_{i=0}^{n} a_i D_0^{\alpha_i} y(t) = \sum_{j=0}^{m} b_j D_0^{\beta_j} u(t) \quad (3) $$

Using Laplace transformation, Equation (3) can derive the following transfer function:

$$ G(s) = \frac{Y(s)}{U(s)} = \frac{b_m s^{\beta_m} + b_{m-1} s^{\beta_{m-1}} + \ldots + b_0 s^{\beta_0}}{a_n s^{\alpha_n} + a_{n-1} s^{\alpha_{n-1}} + \ldots + a_0 s^{\alpha_0}} \quad (4) $$

where α_i and β_i are arbitrary, real, and positive.

2.3. The FOPD/FOPI Controller in Frequency Domain

Based on the fractional-order PID controller ($PI^{\lambda}D^{\mu}$) proposed by Podlubny [24], two fractional-order PD and PI controllers are derived as follows, respectively:

$$ G_{c1}(s) = K_{p1} + K_{d1} s^{\alpha}\ (0 < \alpha < 2) \quad (5) $$

$$ G_{c2}(s) = K_{p2} + \frac{K_{i2}}{s^{\beta}}\ (0 < \beta < 2) \quad (6) $$

By substituting $s = j\omega$ into (5) and (6), the FOPD and FOPI controllers are presented in the frequency domain in Equations (7) and (8), respectively:

$$ G_{c1}(j\omega) = (K_{p1} + K_{d1}\omega^{\alpha}\cos\gamma_d) + jK_{d1}\omega^{\alpha}\sin\gamma_d \quad (7) $$

$$ G_{c2}(j\omega) = \left(K_{p2} + \frac{K_{i2}\cos\gamma_i}{\omega^{\beta}}\right) - j\frac{(K_{i2}\sin\gamma_i)}{\omega^{\beta}} \quad (8) $$

where $\gamma_d = \frac{\pi\alpha}{2}$ and $\gamma_i = \frac{\pi\beta}{2}$.

2.4. System Modeling

In general, a single-axis servo system includes a servo motor that is used to drive a ball screw for translational movement or a rotational mechanism. In this work, a DC servo motor is considered due to its simplicity in modeling and effectiveness in positioning control. Normally, the control system has both the position and velocity control loops where the velocity and position information are estimated based on feedback signals of an encoder and the parameters of the mechanism. The mathematical model of a DC servo motor is available in numerous studies [4–6] and is also described as a block diagram in Figure 1.

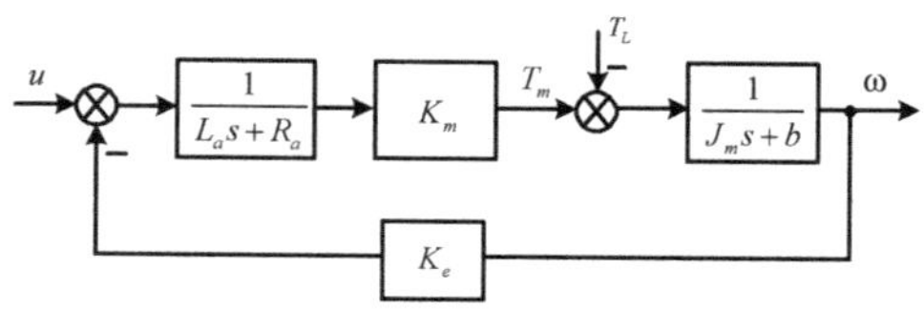

Figure 1. Block diagram of a DC servo motor.

where,
R_a: armature resistance (Ohm); L_a: armature inductance (H)
K_e: back emf constant (V/rad/s); K_m: torque constant (Nm/A)
J_m: inertial moment of the motor shaft (kgm^2)
ω: angular velocity of the motor shaft (rad/s)
u: applied voltage to the motor

From Figure 1, the equivalent transfer function of the DC servo motor is obtained as follows:

$$G_M(s) = \frac{\omega(s)}{U(s)} = \frac{K_m}{L_a J_m s^2 + (L_a b + R_a J_m)s + R_a b + K_m K_e} \quad (9)$$

$$G_M(s) = \frac{K_2}{\tau_2 s^2 + \tau_1 s + 1} \quad (10)$$

where $K_2 = \frac{K_m}{K_m K_e + R_a b}$, $\tau_2 = \frac{L_a J_m}{K_m K_e + R_a b}$, $\tau_1 = \frac{L_a b + R_a J_m}{K_m K_e + R_a b}$.

Case 1: a ball screw system, also called a feed drive system

The block diagram of the mechanical element [6,8] is obtained, as seen in Figure 2:

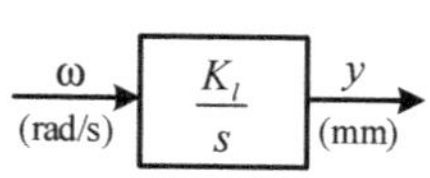

Figure 2. The block diagram of the ball screw.

where ω is the angular velocity (rad/s) of the motor; $K_l = \frac{l}{2\pi}$, and l is the lead of the ball screw (mm); and y is the position of the sliding stable (mm). Therefore, the primary transfer function in this case is obtained:

$$G_l(s) = \frac{K_l}{s} \quad (11)$$

Case 2: a rotational mechanism

The block diagram, in this case, can be simplified as Figure 3 [6]:

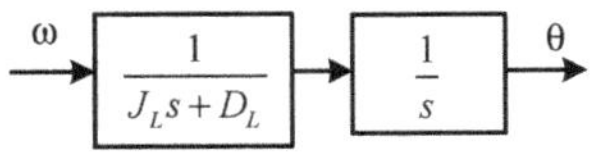

Figure 3. The block diagram of a rotational load.

where J_L is the inertia moment of the rotational load; D_L is viscous damping; and θ is the rotational angle output (rad). From Figure 3, the transfer function of the primary plant is derived as follows:

$$G_L(s) = \frac{K_L}{s(\tau s + 1)} \tag{12}$$

where $K_L = \frac{1}{D_L}; \tau = \frac{J_L}{D_L}$.

3. Analytical Design of FOPI/FOPD Controllers for the Cascade Scheme

3.1. The Control System Structure

The general cascade control system is shown in Figure 4, where $G_{p1}(s)$ and $G_{p2}(s)$ are the transfer functions of the primary and secondary plants, respectively. Additionally, as mentioned above, $G_{p2}(s)$ represents the transfer function of the motor Equation (10) and $G_{p1}(s)$ describes the transfer function of the load in both cases (translational and rotational movement); $G_{c1}(s)$ and $G_{c2}(s)$ are the primary and secondary controllers; and d_1 and d_2 denote load disturbances affecting each control loop, respectively.

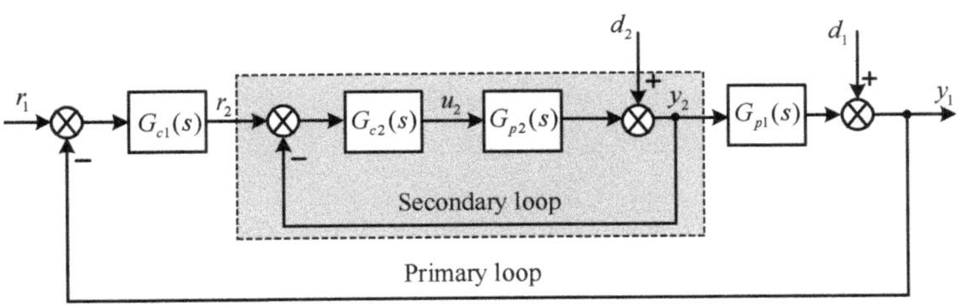

Figure 4. The cascade control system structure.

According to Section 2.4, a dynamic second-order system is used for the secondary model Equation (10).

$$G_{p_2}(s) = G_M(s) = \frac{K_2}{\tau_1 s^2 + \tau_2 s + 1} \tag{13}$$

Whereas, for the primary loop, an integrating system or an integrating first-order stable system is adopted as Equation (11) or Equation (12) for both cases of the load, respectively:

$$G_{p_1}(s) = G_l(s) = \frac{K_l}{s} \tag{14}$$

$$G_{p_1}(s) = G_L(s) = \frac{K_L}{s(\tau s + 1)} \tag{15}$$

3.2. Design of Secondary Controller-Based Direct Synthesis Method

Considering the secondary control loop where its input and output are r_2 and y_2, respectively, the closed-loop transfer function is obtained:

$$\frac{y_2}{r_2} = \frac{G_{c2}G_{p2}}{1 + G_{c2}G_{p2}} \tag{16}$$

From Equation (16), the function of the controller is derived as follows:

$$G_{c2} = \frac{1}{\widetilde{G}_{p2}} \frac{(y_2/r_2)_d}{1-(y_2/r_2)_d} \tag{17}$$

where $\widetilde{G}_{p2}$ is the model of the system in the secondary loop, ideally $\widetilde{G}_{p2} = G_{p2}$; $(y_2/r_2)_d$ represents the desired response of the controlled system. In this paper, a first-order transfer function is chosen for the inner loop; therefore, its transfer function has the following form:

$$\left(\frac{y_2}{r_2}\right)_d = \frac{1}{\tau_{c2}s+1} \tag{18}$$

where τ_{c2} is the desired time response of the controlled loop and τ_{c2} is chosen to ensure the fastest response with no overshoot of the controlled system.

Replacing (13) and (18) into (17), the general form of G_{c2} is obtained:

$$G_{c2} = \frac{\tau_1 s^2 + \tau_2 s + 1}{K_2} \frac{1}{\tau_{c2}s} \tag{19}$$

The complex form in frequency domain of the controller is derived by substituting $s = j\omega$ into Equation (19):

$$G_{c2} = \frac{\tau_2}{K_2\tau_{c2}} - j\frac{1-\tau_1\omega^2}{K_2\tau_{c2}\omega} \tag{20}$$

Comparing Equations (8) and (20), the control parameters of the secondary control loop are derived:

$$K_{i2} = \frac{\omega^\beta(1-\tau_1\omega^2)}{K_2\tau_{c2}\omega \sin\gamma_i} \tag{21}$$

$$K_{p2} = \frac{\tau_2}{K_2\tau_{c2}} - \frac{\omega^\beta(1-\tau_1\omega^2)\cos\gamma_i}{K_2\tau_{c2}\omega^{\beta+1}\sin\gamma_i} \tag{22}$$

3.3. FOPD Controller Design for the Primary Control Loop

The block diagram of the system is reduced in Figure 5 where $(y_2/r_2)_d$ the desired closed-loop transfer function of the inner loop as mentioned above. The closed-loop transfer function of the outer loop (from r_1 to y_1) is described as Equation (23):

$$\frac{Y_1}{R_1} = \frac{G_{c_1}(s)(y_2/r_2)_d G_{p1}(s)}{1+G_{c_1}(s)(y_2/r_2)_d G_{p1}(s)} \tag{23}$$

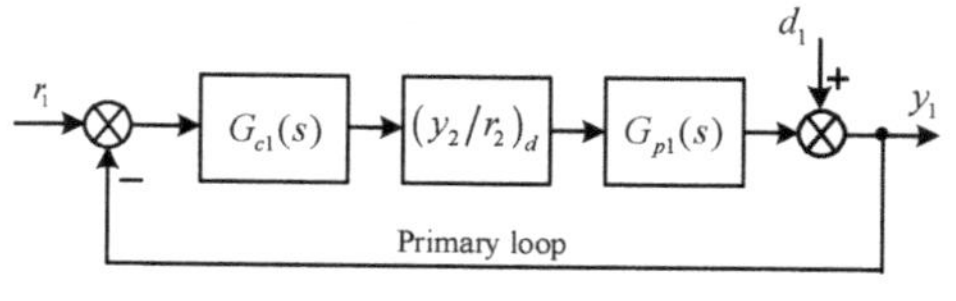

Figure 5. The control structure design of the primary control loop.

Case 1: The primary plant is a feed drive system, so its transfer function is: $G_{p1}(s) = \frac{K_l}{s}$

The equivalent of the primary system is considered as the desired transfer function of the inner loop in series with the primary plant. Therefore, the primary transfer function is obtained as follows:

$$G_p(s) = (y_2/r_2)_d G_{p1}(s) = \frac{1}{\tau_{c2}s+1}\frac{K_l}{s} \tag{24}$$

Similarly to Equation (17), the primary controller is also derived:

$$G_{c1} = \frac{1}{G_p}\frac{(y_1/r_1)}{1-(y_1/r_1)} \simeq \frac{1}{\widetilde{G}_p}\frac{(y_1/r_1)_d}{1-(y_1/r_1)_d} \tag{25}$$

Ideally, $\widetilde{G}_p(s) = G_p(s) = \frac{1}{\tau_{c2}s+1}\frac{K_l}{s}$; and, in this case,

$$\left(\frac{y_1}{r_1}\right)_d = \frac{1}{\tau_{c1}s^{\lambda}+1}\ (1<\lambda<2) \tag{26}$$

where λ is the fractional order; τ_{c1} is the desired response time of the primary output.

Replacing Equation (26) and $\widetilde{G}_p$ into Equation (25), the primary controller is obtained:

$$G_{c1} = \frac{s(\tau_{c2}s+1)}{K_l}\frac{1}{\tau_{c1}s^{\lambda}} = \frac{(\tau_{c2}s+1)}{K_l}\frac{1}{\tau_{c1}s^{\lambda_1}},\ \text{where } \lambda_1 = \lambda - 1\ (0<\lambda_1<1) \tag{27}$$

Substituting $s = j\omega$ to convert G_{c1} into the frequency domain in a complex form:

$$G_{c1} = \frac{\tau_{c2}\omega\sin\gamma_1 + \cos\gamma_1}{\tau_{c1}K_l\omega^{\lambda_1}} + j\frac{\tau_{c2}\omega\cos\gamma_1 - \sin\gamma_1}{\tau_{c1}K_l\omega^{\lambda_1}} \tag{28}$$

Comparing Equations (7) and (28), the control parameters of the primary control loop are obtained:

$$K_{d1} = \frac{\tau_{c2}\omega\cos\gamma_1 - \sin\gamma_1}{\tau_{c1}K_l\omega^{\lambda_1+\alpha}\sin\gamma_d} \tag{29}$$

$$K_{p1} = \frac{\sin(\gamma_1+\gamma_d) - \tau_{c2}\omega\cos(\gamma_1+\gamma_d)}{K_l\tau_{c1}\omega^{\lambda_1}\sin\gamma_d} \tag{30}$$

Case 2: The primary plant is a rotational mechanism, $G_{p1}(s) = \frac{K_L}{s(\tau s+1)}$

In this case, the equivalent transfer function of the primary plant is:

$$G_p(s) = (y_2/r_2)_d\, G_{p1}(s) = \frac{1}{\tau_{c2}s+1}\frac{K_L}{s(\tau s+1)} \tag{31}$$

Therefore, the primary controller could be calculated as follows:

$$G_{c1}(s) = \frac{1}{G_p}\frac{(y_1/r_1)}{1-(y_1/r_1)} = \frac{1}{\widetilde{G}_p}\frac{(y_1/r_1)_d}{1-(y_1/r_1)_d} \tag{32}$$

Ideally, $\widetilde{G}_p(s) = G_p(s) = \frac{1}{\tau_{c2}s+1}\frac{K_L}{s(\tau s+1)}$, and $(y_1/r_1)_d$ is chosen as Equation (26) (case 1); therefore, the primary controller is rewritten as follows:

$$G_{c1}(s) = \frac{s(\tau_{c2}s+1)(\tau s+1)}{K_L}\frac{1}{\tau_{c1}s^{\lambda}} = \frac{(\tau_{c2}s+1)(\tau s+1)}{K_L\tau_{c1}s^{\lambda_1}} \tag{33}$$

where $\lambda_1 = \lambda - 1\ (0<\lambda_1<1)$.

Similarly to case 1, substituting $s = j\omega$ to Equation (33) and rewriting G_{c1} in a complex form:

$$G_{c1} = \frac{(1-\tau\tau_{c2}\omega^2)\cos\gamma_1 + \omega(\tau+\tau_{c2})\sin\gamma_1}{K_L\tau_{c1}\omega^{\lambda_1}} + j\frac{(\tau+\tau_{c2})\omega\cos\gamma_1 - (1-\tau\tau_{c2}\omega^2)\sin\gamma_1}{K_L\tau_{c1}\omega^{\lambda_1}} \tag{34}$$

Comparing Equations (7) and (34), the control parameters of the primary control loop are obtained in this case:

$$K_{d1} = \frac{(\tau+\tau_{c2})\omega\cos\gamma_1 - (1-\tau\tau_{c2}\omega^2)\sin\gamma_1}{K_L\tau_{c1}\omega^{\lambda_1+\alpha}\sin\gamma_d} \tag{35}$$

$$K_{p1} = \frac{(1 - \tau\tau_{c2}\omega^2)\sin(\gamma_1 + \gamma_d) - (\tau + \tau_{c2})\omega\cos(\gamma_1 + \gamma_d)}{K_L \tau_{c1}\omega^{\lambda_1}\sin\gamma_d} \tag{36}$$

3.4. Tuning Procedure Using Maximum Sensitivity Function

From the previous section, it can be seen that the controller parameters are tuned based on some parameters that play an important role in achieving better performance as well as maintaining robustness to model uncertainties. In the proposed method, the maximum sensitivity function is adopted to guarantee the robust stability of each control loop. For a classical feedback control system, the maximum sensitivity function is defined as the following equation:

$$M_s = \max_{\omega}|S(j\omega)| \tag{37}$$

where $S = (1 + L)^{-1}$, and L is an open-loop transfer function of the system, which normally includes the controller and the system model. Normally, $|S|$ is small at low frequencies and reaches 1 at high frequencies. However, at some intermediate frequencies, in practice, a peak value of M_s can be larger than 1, which degrades the system performance. Therefore, the peak value of M_s can be used to measure the robustness of the controlled system. In general, for both robust stability and performance, the M_s value should be close to 1. Therefore, in most of the literature, the typical M_s range is chosen from 1.2 to 2 to ensure the robust stability of control systems [36].

For the cascade control scheme in Figure 1, the maximum sensitivity function is suggested for both control loops (inner and outer) as a tuning criterion and its value is assigned as close as possible to 1.2. In this work, this value is used to choose the appropriate frequency to calculate the control parameters of the inner loop (ω in Equations (21) and (22)), and this frequency is addressed for the outer loop as well. The algorithm to obtain ω is simple and is described as follows (Algorithm 1):

Algorithm 1: The guideline for tuning parameters based on M_s value

```
Initialization
ω = 1; choose τ_c2 and β
while (ω < 1000) do
    Compute K_p2, K_i2 according to Equations (21) and (22) respectively
    Calculate M_s for each set of control parameters
     ω = ω + 1
end while
Choose appropriate ω to get value of M_s being closed to 1.2
end
```

4. Robustness Analysis

In this work, the proposed tuning rules are based on the models of actual systems, which are usually represented by nominal models. The differences between the actual system and its model are considered model mismatch or model uncertainty and may degrade some performance indices in terms of servomechanism and regulator problems. Therefore, it is important to analyze the robust stability of model-based methods for control systems in the presence of uncertainties. The multiple sources of parametric and/or unmodelled dynamics uncertainty are commonly grouped into a single lumped perturbation of a specific structure [36].

In this paper, multiplicative output uncertainty is suggested for each control loop of the cascade structure. Therefore, robust stability analysis is carried out herein by considering the multiplicative output uncertainties of each system parameters simultaneously, as shown in Figure 6.

It can be expressed as follows:

$$\Pi_{O1}: \overline{G}_{p1} = (1 + E_{O1})G_{p1};\ E_{O1} = W_1\Delta_1 \tag{38}$$

$$\Pi_{O2}: \overline{G}_{p2} = (1 + E_{O2})G_{p2};\ E_{O2} = W_2\Delta_2 \tag{39}$$

where Π_{Oi} denotes the set of output perturbed system models; $\overline{G}_{pi}$ is the transfer function of the actual model including its nominal model (G_{p1}, G_{p2}) and lumped uncertainty at the output represented by E_{O1}, E_{O2}. From Equations (38) and (39), Δ_1, Δ_2 are normalized perturbation with $|\Delta_i(j\omega)| \leq 1,\ \forall\omega$.

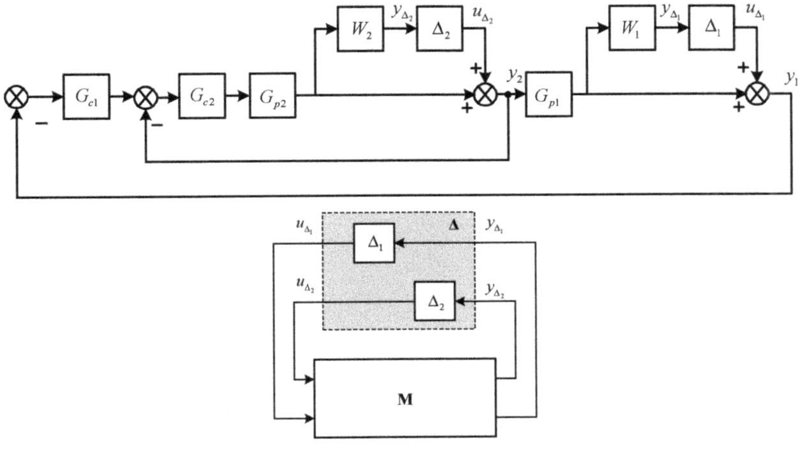

Figure 6. The **M-Δ** structure using multiplicative output uncertainties of a cascade control system.

According to multiplicative relative uncertainties, the weighting scalar W_i captures the variation of uncertainties over frequencies and has to be chosen to ensure [36]:

$$|W_i(j\omega)| \geq l_I(\omega) = \max_{\overline{G}_{pi}\in\Pi_{oi}} \left| \frac{\overline{G}_{pi}(j\omega) - G_{pi}(j\omega)}{G_{pi}(j\omega)} \right|,\ \forall\omega \tag{40}$$

To represent unmodelled dynamics, the common simple form of the multiplicative weight is as follows:

$$W_i = \frac{\tau_w s + r_0}{(\tau_w / r_\infty)s + 1} \tag{41}$$

where r_0 is the relative uncertainties at low frequencies and r_∞ is the magnitude of the weight at high frequencies. It is obvious that W_i is considered as a high-pass filter and, therefore, τ_w is chosen as a cut-off frequency ($1/\ \tau_w$) to ensure that the weight W_i has a larger magnitude than the largest magnitude of the relative uncertainties in Equation (40).

The structure singular value (denoted *SSV*, *mu*, or μ) is suggested to measure the robustness of the control system [36]. In Figure 6, the cascade control scheme with the multiplicative output uncertainty of each loop is rearranged in the **M-Δ** structure where $\mathbf{\Delta} = diag(\Delta_1, \Delta_2)$ and **M** is a 2 × 2 matrix which includes all remaining blocks such as plants, controllers, and weight factors. To derive **M**, the relationships between y_Δ and u_Δ are obtained first:

$$\begin{cases} y_{\Delta_1} = W_1 G_{p1}(1 + G_{p2}G_{c2} + G_{p1}G_{p2}G_{c1}G_{c2})^{-1}(u_{\Delta_2} - G_{p2}G_{c1}G_{c2}u_{\Delta_1}) \\ y_{\Delta_2} = (1 + G_{p2}G_{c2} + G_{p1}G_{p2}G_{c1}G_{c2})^{-1}[-G_{c1}G_{c2}G_{p2}W_2u_{\Delta_1} - (G_{p2}G_{c2} + G_{p1}G_{p2}G_{c1}G_{c2})W_2u_{\Delta_2}] \end{cases} \tag{42}$$

Therefore, each element of **M** can be derived as follows:

$$M_{11} = -W_1 G_{p1} G_{p2} G_{c1} G_{c2} \left(1 + G_{p2} G_{c2} + G_{p1} G_{p2} G_{c1} G_{c2}\right)^{-1} \tag{43}$$

$$M_{12} = W_1 G_{p1} \left(1 + G_{p2} G_{c2} + G_{p1} G_{p2} G_{c1} G_{c2}\right)^{-1} \tag{44}$$

$$M_{21} = -W_2 G_{p2} G_{c1} G_{c2} \left(1 + G_{p2} G_{c2} + G_{p1} G_{p2} G_{c1} G_{c2}\right)^{-1} \tag{45}$$

$$M_{22} = -W_2 G_{p2} \left(G_{c2} + G_{c1} G_{c2} G_{p1}\right) \left(1 + G_{p2} G_{c2} + G_{p1} G_{p2} G_{c1} G_{c2}\right)^{-1} \tag{46}$$

The μ-synthesis states that the control system is stable for all allowed perturbations with $|\Delta_i(j\omega)| \leq 1,\ \forall \omega$ if and only if $\mu(\mathbf{M}(j\omega)) < 1,\ \forall \omega$ [36].

5. Simulation Study

In this paper, the performance indices including the integral absolute error (IAE), the integral time-weighted absolute error (ITAE), and the total variant (TV) are addressed and the smaller values of those indicate better performance. We use the motor and load parameters as seen in Table 1 [38].

Table 1. System parameters.

Symbols	Values	Units
L_a	8×10^{-5}	H
R_a	0.316	Ohm
J_m	1.34×10^{-5}	kgm^2
b	1.82×10^{-5}	Nm/rad/s
K_m	0.03	Nm/A
K_e	0.03	V/rad/s
l	0.01	m
J_L	6.75×10^{-5}	kgm^2
D_L	1.82×10^{-4}	Nm/rad/s

From the parameters in Table 1 and Equation (10), the transfer function of the motor is derived:

$$G_{p2}(s) = \frac{33.1217}{0.00001835 s^2 + 0.0468 s + 1} \tag{47}$$

5.1. *The Secondary Loop Control Design (Velocity Control)*

From Equations (21) and (22), two controller parameters of the inner loop could be obtained. The desired time constant is chosen as $\tau_{c2} = 0.001$ and the fractional order of the integral term $\beta = 1.2$. In this work, to guarantee the robustness of each single loop, the maximum sensitivity function (M_s) is adopted to play a part in tuning control parameters. The proposed algorithm (Section 3.4) is used in this case to obtain values of M_s with respect to ω as in Figure 7. From the figure, it can be seen that the M_s value converges to 1.2 when increasing ω. However, when ω is greater than 230 (rad/s), M_s becomes undetermined. Therefore, in this case, we choose $\omega = 200$ (M_s = 1.232), and as a result of that, the control parameters are obtained as follows:

$$G_{c2}(s) = 1.426 + \frac{24.365}{s^{1.2}} \tag{48}$$

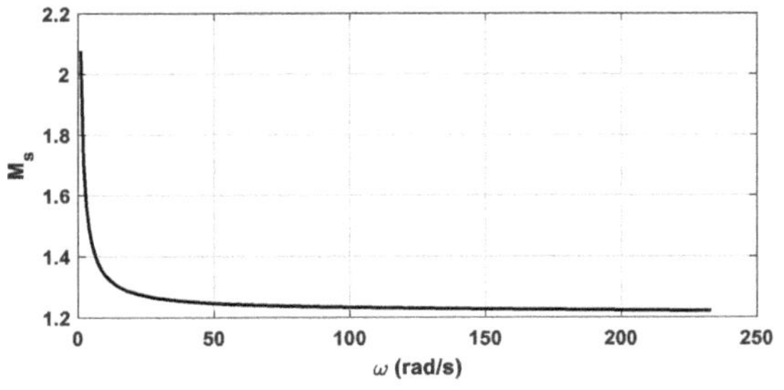

Figure 7. The relationship between maximum sensitivity peak and frequency.

5.2. The Primary Loop Control Design (Position Control)

For the outer loop, in this work, two kinds of plants are considered as mentioned above, a sliding stable or feed drive system and a rotational mechanism. Using the parameters in Table 1, the transfer function of each case is obtained as follows:

$$\text{Case 1}: \; G_{p1}(s) = \frac{0.0016}{s} \tag{49}$$

In this case, $\alpha = 0.6$, $\tau_{c1} = 0.03$, and from Equation (26), $\lambda = 1.1$ is chosen. From Equations (29) and (30), the control parameters are obtained:

$$G_{c1}(s) = 12,196 + 26.0769s^{0.6} \tag{50}$$

To justify the effectiveness of the proposed method, the controller settings of both control loops are simulated by providing a ramp function input. In addition, to investigate the effects of disturbances in terms of d_1 for the primary output and d_2 for the secondary output as in Figure 4, two step changes of these are inserted at times $t = 1$ (s) and $t = 5$ (s), respectively. Figures 8 and 9 illustrate the closed-loop responses at the primary output (position) and secondary output (velocity) in this case. From the figures, it can be seen that the tracking control completely meets the servo performance requirements (fast response and no overshoot or oscillation). The disturbance of the secondary loop (d_2) does not have any effect on the primary output. In contrast, at time $t = 5$, there is a variation in position due to d_1; however, the primary controller maintains the performance in an excellent way. The control signals are shown in Figure 10 to prove that the proposed method has smooth control signals, which is the disadvantage of some advanced control techniques.

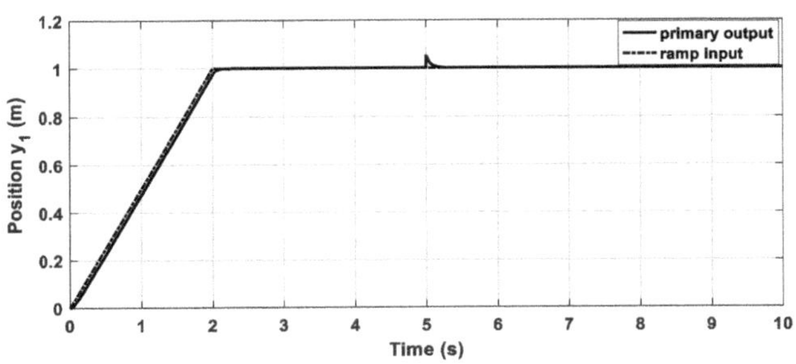

Figure 8. The translation responses to the ramp input with disturbances.

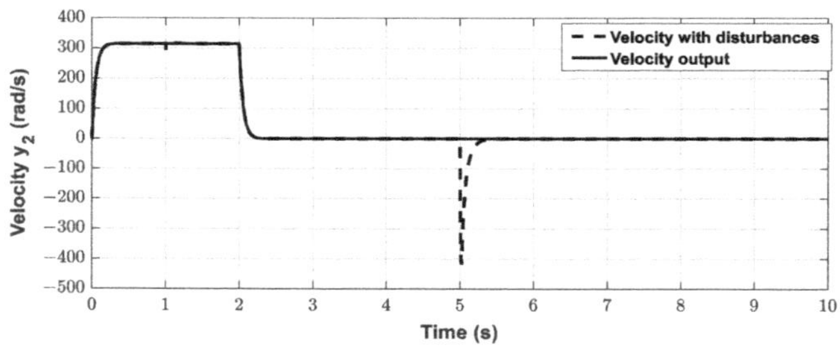

Figure 9. The velocity responses of case 1 with and without disturbances.

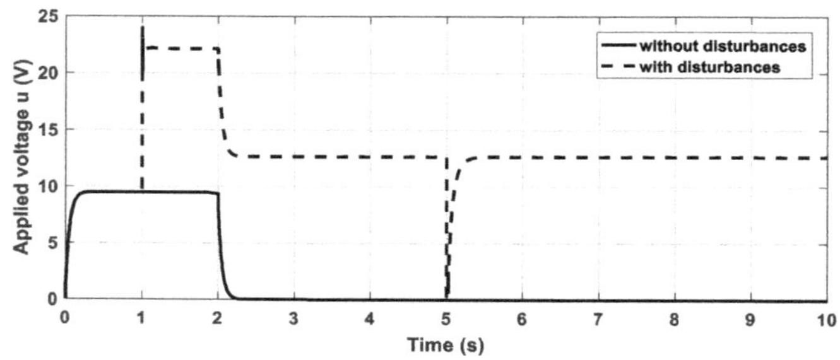

Figure 10. The applied voltages in both cases.

Moreover, the robustness of the proposed controller will be shown, as mentioned in Section 4. To obtain $l_I(\omega)$ in Equation (40), the nominal parameters of the inner transfer function are changed in the range of $\pm 50\%$. The corresponding relative errors $|(\overline{G}_{p_1} - G_{p_1})/G_{p_1}|$ are shown as functions of frequency in Figure 11a, and it can be seen that $l_I(\omega)$ is 0.4 at low frequencies and 5 at high frequencies. τ_w is chosen to guarantee that W_2 is large enough to satisfy $W_2 > l_I(\omega)$ at all frequencies. From the figure, $\tau_w = 0.0667$; therefore, the transfer function of W_2 is obtained (from Equation (41)). For the outer loop (Figure 11b), it can be performed similarly, finally also obtaining the transfer function of W_1 in this case with $\tau_w = 0.01$, $r_0 = 0.4$, and $r_\infty = 1.5$.

$$W_1 = \frac{0.01s + 0.4}{(0.01/1.5)s + 1} \tag{51}$$

$$W_2 = \frac{0.0667s + 0.4}{(0.0667/5)s + 1} \tag{52}$$

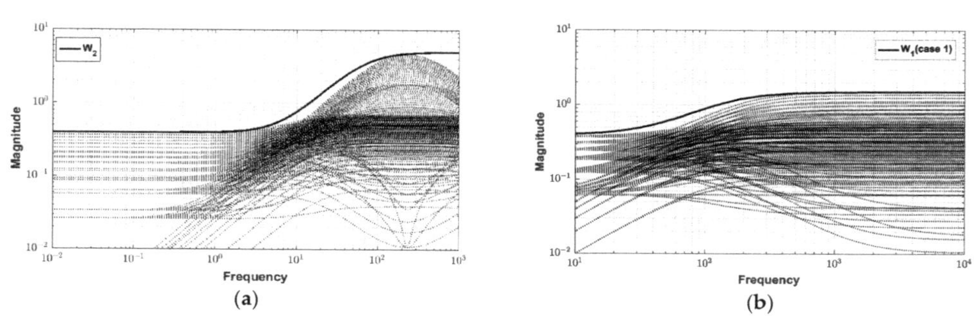

Figure 11. (**a**,**b**) The scalar multiplicative weights of both control loops.

Figure 12 illustrates the structured singular value (SSV) of the **M-Δ** structure using From Equations (43)–(46), when two sources of parametric uncertainties of both control loops affect simultaneously, two scalar weight transfer functions W_1 and W_2 are as above. As mentioned in Section 4, the condition for robust stability is that the μ value has to be always less than 1. From the figure, it is obvious that the peak of μ is 0.4 over the frequency range, which ensures the robustness of the controlled system.

$$Case\,2: \; G_{p1}(s) = \frac{2}{s(0.0014s+1)} \tag{53}$$

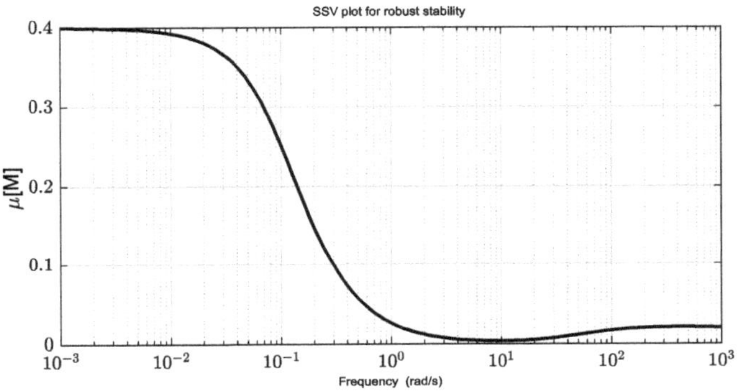

Figure 12. The SSV plot for robust stability of case 1.

In this case, $\alpha = 0.9$, $\tau_{c1} = 0.02$, and from Equation (26), $\lambda = 1.2$ is chosen. From Equations (35) and (36), the control parameters are obtained:

$$G_{c1}(s) = 8.8414 + 0.0115s^{0.9} \tag{54}$$

In this case, the position output is the rotational angle (rad), and the ramp function input is adopted to verify the position control problem. Two disturbances described in Figure 4 are illustrated by adding step signals at time t = 1 (s) and t = 5 (s). Figures 13 and 14 show the closed-loop responses at the primary output (position) and secondary output (velocity) in two cases. From the figure, it can be seen that the tracking control completely satisfies the servo performance requirements (fast response and no overshoot or oscillation). The disturbance in the inner loop has no effect on the primary output. This means that the effectiveness of the cascade scheme is clarified in this case. The primary controller plays an important part in maintaining the position in a servo manner. The control signals are shown in Figure 15. It can be seen that, in this case, the proposed method also has a smooth control signal, and only are there sudden changes in the manipulated variable due to the disturbances.

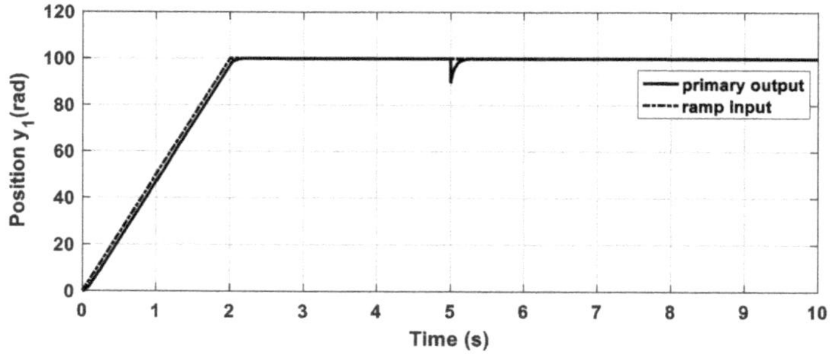

Figure 13. The rotational responses to the ramp input with disturbances.

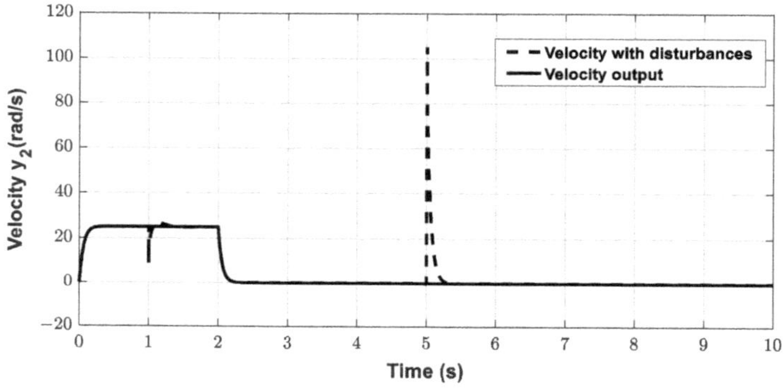

Figure 14. The velocity responses of case 2 with and without disturbances.

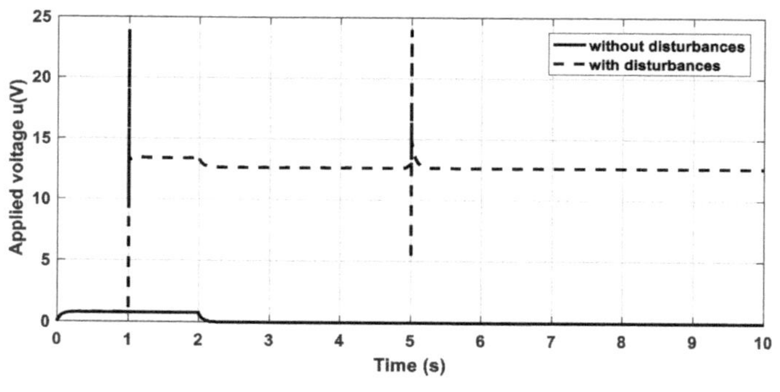

Figure 15. Applied voltages for both cases.

This is similar to case 1, to obtain $l_I(\omega)$ in Equation (40), the nominal parameters of the inner transfer function are changed in the range of $\pm 50\%$, approximately. Figure 16 illustrates the relative errors and it can be seen that $l_I(\omega)$ is 0.5 at low frequencies and 1.1 at high frequencies. τ_w is chosen as 0.002 to satisfy $W_1 > l_I(\omega)$ at all frequencies. Therefore, the transfer function of W_1 is as Equation (55). Note that the transfer function of W_2 is still the same as the one in case 1.

$$W_1 = \frac{0.002s + 0.5}{(0.002/1.1)s + 1} \tag{55}$$

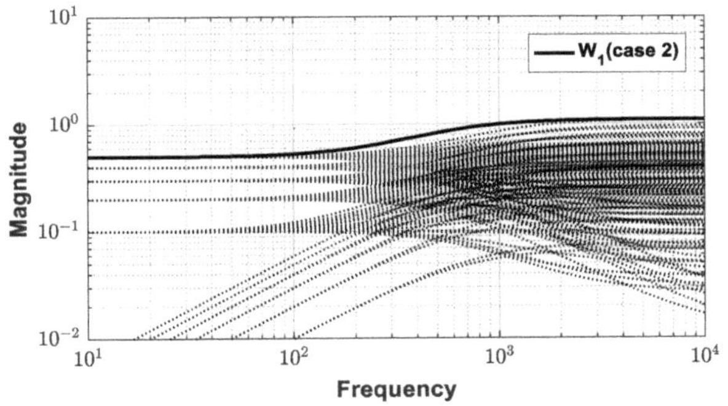

Figure 16. The scalar multiplicative weight of the primary loop of case 2.

Figure 17 illustrates the *SSV* plot of the **M-Δ** structure in this case. From the figure, it can be seen that the peak of μ is 0.52 over the frequency range, which ensures the robustness of the controlled system.

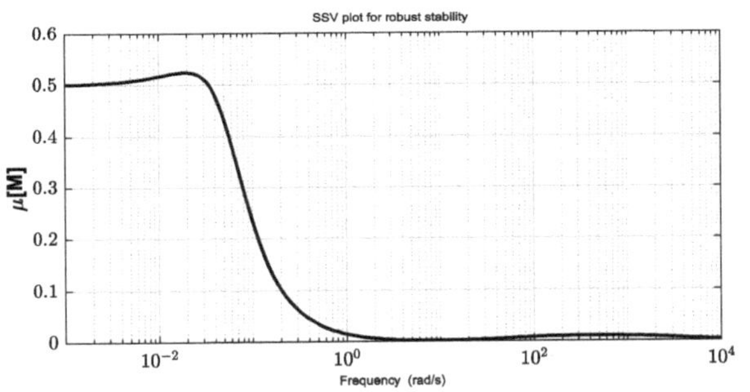

Figure 17. The *SSV* plot for robust stability of case 2.

The robust performance of the controlled systems is also evaluated by perturbing ±20% to all plant parameters. The performance indices are tabulated in Table 2 of nominal models and perturbed ones. From the data, it is obvious that the controlled systems still keep good tracking control in the presence of uncertainties, and the μ values are practically the same as the ones of the nominal models.

Table 2. The performance indices in nominal and perturbed models for both cases.

	Case 1				Case 2			
	IAE	ITAE	TV	μ (M)	IAE	ITAE	TV	μ (M)
Nominal	0.0515	0.0543	21.06	0.40	5.655	5.971	1.949	0.524
Perturbed (±20%)	0.0962	0.3693	18.63	0.42	8.801	29.883	1.543	0.5243

6. Conclusions

The fractional-order control designs for servo systems are studied in this work. To achieve the strict requirements of control performance in terms of fast response and no overshot/oscillation, the cascade control structure including two control loops, the secondary loop (inner loop) for velocity control and the primary loop (outer loop) for position control, is proposed. The FOPI and FOPD are suggested for the inner and outer, respectively, and the analytical design methods in the frequency domain are also proposed to derive the tuning rules for both controllers. In the inner loop, the maximum sensitivity function is adopted to choose the appropriate frequency to ensure the robustness of this control loop. The **M-Δ** structure is also addressed to justify the robustness of the controlled system using multiplicative output uncertainties, for which parametric uncertainties of both control loops are considered simultaneously. The simulation results indicate that the proposed method consistently satisfies the system requirements with a fast and no overshoot closed-loop time response to ramp function inputs. The performance indexes also clarify the effectiveness of the proposed method. The robust stability is also verified by the μ value, which has a peak value of less than 1 over the frequency range in all cases.

Author Contributions: Conceptualization, T.N.L.V. and L.H.G.; methodology, V.L.C. and T.N.L.V.; software, N.H.N.; validation, V.L.C. and T.N.L.V.; formal analysis, V.L.C.; investigation, V.L.C.; writing—original draft preparation, V.L.C.; writing—review and editing, V.L.C. and T.N.L.V.; visualization, V.L.C. and N.H.N.; supervision, T.N.L.V.; project administration, T.N.L.V.; funding acquisition, T.N.L.V. and L.H.G. All authors have read and agreed to the published version of the manuscript.

Funding: This research was funded by Ho Chi Minh City University of Technology and Education, grant number T2023-01NTĐ.

Data Availability Statement: Data sharing is not applicable to this article.

Conflicts of Interest: The authors declare no conflicts of interest.

References

1. Lim, H.; Seo, J.W.; Choi, C.H. Position control of XY table in CNC machining center with non-rigid ball screw. In Proceedings of the 2000 American Control Conference, ACC (IEEE Cat. No. 00CH36334), Chicago, IL, USA, 28–30 June 2000; Volume 3.
2. Franks, R.G.; Worley, C.W. Quantitative analysis of cascade control. *Ind. Eng. Chem.* **1956**, *48*, 1074–1079. [CrossRef]
3. Chen, J.S.; Huang, Y.; Cheng, K. Mechanical model and contouring analysis of high-speed ball-screw drive systems with compliance effect. *Int. J. Adv. Manuf. Technol.* **2004**, *24*, 241–250. [CrossRef]
4. Altintas, Y.; Verl, A.; Brecher, C.; Uriarte, L.; Pritschow, G. Machine feed drives. *CIRP Ann. Manuf. Technol.* **2011**, *60*, 779–796. [CrossRef]
5. Hwan, S.S.; Kang, S.K.; Chung, D.H.; Stroud, I. *Theory and Design of CNC Systems*; Springer: Cham, Switzerland, 2008; pp. 157–185.
6. Nakamura, M.; Goto, S.; Kyura, N. *Mechatronic Servo System Control*; Springer: Cham, Switzerland, 2004.
7. Sato, R.; Tsutsumi, M. Mathematical model of feed drive systems consisting of AC servo motor and linear ball guide. *J. Jpn. Soc. Precis. Eng.* **2005**, *71*, 633–638. [CrossRef]
8. Sato, R.; Tsutsumi, M. Modeling and controller tuning techniques for feed drive systems. In Proceedings of the ASME 2005 International Mechanical Engineering Congress and Exposition, Orlando, FL, USA, 5–11 November 2005; pp. 669–679.
9. Liu, Y.; Wang, Z.Z.; Wang, Y.F.; Wang, D.H.; Xu, J.F. Cascade tracking control of servo motor with robust adaptive fuzzy compensation. *Inf. Sci.* **2021**, *569*, 450–468. [CrossRef]
10. Lee, Y.; Park, S.; Lee, M. PID controller tuning to obtain desired closed loop responses for cascade control systems. *Ind. Eng. Chem. Res.* **1998**, *37*, 1859–1865. [CrossRef]
11. Rao, A.S.; Seethaladevi, S.; Uma, S.; Chidambaram, M. Enhancing the performance of parallel cascade control using Smith predictor. *ISA Trans.* **2009**, *48*, 220–227. [PubMed]
12. Uma, S.; Chidambaram, M.; Rao, A.S.; Yoo, C.K. Enhanced control of integrating cascade processes with time delays using modified Smith predictor. *Chem. Eng. Sci.* **2010**, *65*, 1065–1075. [CrossRef]
13. Padhan, D.G.; Majhi, S. An improved parallel cascade control structure for processes with time delay. *J. Process Control* **2012**, *22*, 884–898. [CrossRef]
14. Santosh, S.; Chidambaram, M. A simple method of tuning parallel cascade controllers for unstable FOPTD systems. *ISA Trans.* **2016**, *65*, 475–486. [CrossRef]
15. Raja, G.L.; Ali, A. Modified parallel cascade control strategy for stable, unstable and integrating processes. *ISA Trans.* **2016**, *65*, 394–406. [CrossRef] [PubMed]
16. Raja, G.L.; Ali, A. Smith predictor based parallel cascade control strategy for unstable and integrating processes with large time delay. *J. Process Control* **2017**, *52*, 57–65. [CrossRef]
17. Neubauer, M.; Brenner, F.; Hinze, C.; Verl, A. Cascaded sliding mode position control (SMC-PI) for an improved dynamic behavior of elastic feed drives. *Int. J. Mach. Tools Manuf.* **2021**, *169*, 103796. [CrossRef]
18. Sun, G.; Wu, L.; Kuang, Z.; Ma, Z.; Liu, J. Practical tracking control of linear motor via fractional-order sliding mode. *Automatica* **2018**, *94*, 221–235. [CrossRef]
19. Wang, J.; Shao, C.; Chen, Y.Q. Fractional order sliding mode control via disturbance observer for a class of fractional order systems with mismatched disturbance. *Mechatronics* **2018**, *53*, 8–19. [CrossRef]

20. Wang, S.; Yu, H.; Yu, J. Robust adaptive tracking control for servo mechanisms with continuous friction compensation. *Control Eng. Pract.* **2019**, *87*, 76–82. [CrossRef]
21. Xu, H.; Hinze, C.; Lechler, A.; Verl, A. Robust μ parameterization with low tuning complexity of cascaded control for feed drives. *Control Eng. Pract.* **2023**, *138*, 105607. [CrossRef]
22. Zou, M.; Yu, J.; Ma, Y.; Zhao, L.; Lin, C. Command filtering-based adaptive fuzzy control for permanent magnet synchronous motors with full-state constraints. *Inf. Sci.* **2020**, *546*, 1–12. [CrossRef]
23. Nassef, A.M.; Abdelkareem, M.A.; Maghrabie, H.M.; Baroutaji, A. Metaheuristic-based algorithms for optimizing fractional-order controllers—A recent, systematic, and comprehensive preview. *Fractal Fract.* **2023**, *7*, 553. [CrossRef]
24. Podlubny, I. Fractional-Order Systems and $PI^{\lambda}D^{\mu}$-Controllers. *IEEE Trans. Autom. Control* **1999**, *44*, 208–214. [CrossRef]
25. Padula, F.; Visioli, A. Tuning rules for optimal PID and fractional-order PID controllers. *J. Process Control* **2011**, *21*, 69–81. [CrossRef]
26. Vu, T.N.L.; Lee, M. Analytical design of fractional-order proportional-integral controllers for time-delay processes. *ISA Trans.* **2013**, *52*, 583–591. [CrossRef] [PubMed]
27. Beschi, M.; Padula, F.; Visioli, A. Fractional robust PID control of a solar furnace. *Control Eng. Pract.* **2016**, *56*, 190–199. [CrossRef]
28. Li, D.; Liu, L.; Jin, Q.; Hirasawa, K. Maximum sensitivity based fractional IMC-PID controller design for non-integer order system with time delay. *J. Process Control* **2015**, *31*, 17–29. [CrossRef]
29. Vilanova, R.; Arrieta, O.; Ponsa, P. Robust PI/PID controllers for load disturbance based on direct synthesis. *ISA Trans.* **2018**, *81*, 177–196. [CrossRef] [PubMed]
30. Yumuk, E.; Guzelkaya, M.; Eksin, I. Analytical fractional PID controller design based on Bode's ideal transfer function plus time delay. *ISA Trans.* **2019**, *91*, 196–206. [CrossRef]
31. Moradi, M. A genetic-multivariable fractional order PID control to multi-input multi-output processes. *J. Process Control* **2014**, *24*, 336–343. [CrossRef]
32. Sánchez, H.S.; Padula, F.; Visioli, A.; Vilanova, R. Tuning rules for robust FOPID controllers based on multi-objective optimization with FOPDT models. *ISA Trans.* **2017**, *66*, 334–361. [CrossRef] [PubMed]
33. Tumari, M.Z.M.; Ahmad, M.A.; Suid, M.H.; Hao, M.R. An improved marine predators algorithm-tuned fractional-order PID controller for automatic voltage regulator systems. *Fractal Fract.* **2023**, *7*, 571. [CrossRef]
34. Shalaby, R.; El-Hossainy, M.; Abo-Zalam, B.; Mahmoud, T.A. Optimal fractional-order PID controller based on fractional-order actor-critic algorithm. *Neural Comput. Appl.* **2022**, *35*, 2347–2380. [CrossRef]
35. Vu, T.N.L.; Chuong, V.L.; Truong, N.T.N.; Jung, J.H. Analytical Design of Fractional-Order PI Controller for Parallel Cascade Control Systems. *Appl. Sci.* **2022**, *12*, 2222. [CrossRef]
36. Skogestad, S.; Postlethwaithe, I. *Multivariable Feedback Control Analysis and Design*, 2nd ed.; John Wiley & Sons: Hoboken, NJ, USA, 2001; pp. 253–351.
37. Monje, C.A.; Chen, Y.; Vinagre, B.M.; Xue, D.; Feliu-Batlle, V. *Fractional-Order Systems and Controls, Fundamentals and Applications*; Springer: London, UK, 2010.
38. Choudhary, K.; Qureshi, M.S.; Singh, B. Position Control of DC Motor using Improved Sliding Mode Control Techniques. *Int. J. Appl. Eng. Res.* **2018**, *13*, 15–19.

fractal and fractional

MDPI

Article

A Fractional-Order ADRC Architecture for a PMSM Position Servo System with Improved Disturbance Rejection

Shaohua Wang [1], He Gan [1], Ying Luo [2,*], Xin Luo [1] and Yangquan Chen [3]

[1] School of Mechanical Science and Engineering, Huazhong University of Science and Technology, Wuhan 430074, China; shaohua94@hust.edu.cn (S.W.); hegan@hust.edu.cn (H.G.); mexinluo@hust.edu.cn (X.L.)

[2] School of Automation Science and Engineering, South China University and Technology, Guangzhou 510641, China

[3] The Mechatronics, Embedded Systems and Automation Lab, Department of Mechanical Engineering, School of Engineering, University of California, Merced, CA 95343, USA; ychen53@ucmerced.edu

* Correspondence: yingluosase@scut.edu.cn

Abstract: This paper proposes an active disturbance rejection control (ADRC) architecture for a permanent magnet synchronous motor (PMSM) position servo system. The presented method achieved enhanced tracking and disturbance rejection performance with a limited observer bandwidth. The model-aided extended state observer (MESO)-based ADRC was designed for the current, speed, and position loops of the PMSM position servo system. By integrating known plant information, the MESO improved disturbance estimation with a limited observer bandwidth without amplifying the noise. Additionally, a fractional-order proportional-derivative (FOPD) controller was designed as the feedback controller for the speed loop to further enhance the disturbance rejection. A simulation and experimental tests were conducted on a PMSM servo platform. The results demonstrate not only that the proposed method achieved superior tracking performance but also that the position error of the proposed strategy decreases to 2.25% when the constant disturbance was input, significantly improving the disturbance rejection performance.

Keywords: fractional-order control; active disturbance rejection control; extended state observer; PMSM position servo system; disturbance rejection.

1. Introduction

Permanent magnet synchronous motors (PMSMs) have gained widespread use in industrial applications [1]. However, the existence of nonlinearities, uncertainties, and disturbances [2] makes it challenging for a conventional proportional–integral (PI) control strategy to achieve high-precision control for the PMSM servo system [3,4]. In recent years, diverse control strategies have been developed for PMSM, aiming to enhance control effectiveness. Active disturbance rejection control (ADRC) is one of the most popular control strategies.

Proposed in the 1980s by Jingqing Han [5], ADRC was designed to address extensive uncertainties and disturbances. The fundamental concept of ADRC involves treating external disturbances and internal uncertainties collectively as the "total disturbance", which is then estimated using the extended state observer (ESO) and actively compensated for via the ESO-based feedback controller. Despite its theoretical soundness, the intricate nonlinear structure and intricate tuning process have rendered its practical application challenging. A linear version (LADRC) and bandwidth parameterization proposed by Zhiqiang Gao have greatly simplified the structures and tuning of ADRC [6]. As a result, ADRC can be quickly applied in various industrial areas, such as wind turbines [7], robot control [8], aerial vehicles [9,10], motor control [11], and so on. In recent years, the field of ADRC has undergone substantial growth.

LADRC possesses various superior properties, including a compact structure, model independence, superior disturbance rejection, and robustness [12,13]. However, there are still some challenges that necessitate additional research:

- A high observer bandwidth is required to achieve the desired disturbance estimation and rejection [6], but the high observer bandwidth will amplify the measurement noise [14].
- The LESO and the feedback controller are mutually coupled, while the tracking performance cannot be completely separated from the LESO [15,16].
- The design of LADRC based on the bandwidth parameterization method is relatively conservative, potentially leading to suboptimal control performance [17].

Various versions of ADRC have been proposed to tackle these challenges. Some studies designed ADRC to incorporate the known plant information to improve the disturbance rejection performance. For instance, the ADRC strategy with a generalized extended sta observer (GESO) was presented in [17–19], where the known plant information was incorporated into the ESO. With a GESO, ADRC shows improved control capability on non-minimum-phase, unstable, and time-delay systems. Another benefit of GESO-based ADRC demonstrated in [16] is that it satisfies the separation principle, meaning that the feedback controller design is separated from a GESO. Ref. [20] applied the ADRC with a model-aided ESO (MESO) in the current control of a PMSM, obtaining enhanced parameter robustness performance. Some studies have modified the structure of ADRC to improve the control performance. Ref. [14] presented an improved ESO to suppress the measurement noise. Recently, some intelligent algorithms have been applied in ADRC. Ref. [21] applied a diagonal recurrent neural network in ADRC for the radar position servo system facing a dead zone and friction nonlinearities. Ref. [22] introduced an ADRC method based on the deep reinforcement learning algorithm for more electric aircraft. Ref. [23] presented an optimal ADRC based on a proportional-derivative (PD) control law with particle swarm optimization to improve the dynamic and steady-state control performance.

In recent developments, integrating ADRC with fractional-order control (FOC) has garnered attention to enhance control performance. FOC is the extension of traditional integer-order control, providing more flexibility in controller design and improved control performance [24]. An application of fractional-order ADRC (FOADRC) was presented for a linear motor to achieve precise tracking performance [25]. Ref. [16] proposed an FOADRC controller with GESO to improve tracking and disturbance rejection performance for a PMSM servo system. The design of FOADRC with fractional-order ESO was investigated in [26] to suppress noise sensitivity. Ref. [27] proposed a fractional-order ESO-based ADRC method to address the trade-off between control performance and noise suppression. Ref. [28] presented a fractional-order ADRC method incorporating the fuzzy self-tuning method, which achieved an improved dynamic response and disturbance suppression capability. Ref. [29] proposed an optimal fractional-order ADRC method for the PMSM speed servo system, which met the requirement for frequency-domain indicators and achieved optimal performance in the time domain. A comparison of these versions of ADRC is presented in Table 1.

The primary objective of this paper is to improve the disturbance rejection performance with an observer bandwidth limitation for the PMSM position servo system, integrating the benefits of the model-aided ADRC and fractional-order control. The primary contributions of this paper are summarized as follows:

(1) An MESO-based ADRC architecture is introduced for the PMSM position servo system, demonstrating improved tracking and disturbance rejection under limited observer bandwidth.
(2) The MESO-based ADRC was designed for the current, speed, and position loops. Additionally, a fractional-order feedback controller was designed for the speed loop, further enhancing the disturbance rejection performance on the basis of the MESO.
(3) Simulation and experimental comparison tests on a PMSM position servo system were conducted to verify the effectiveness of the proposed method.

Table 1. Comparison of different versions of ADRC.

References	Linear/Nonlinear	ESO	Feedback Controller
[5]	Nonlinear	Nonlinear ESO	Error-Based Nonlinear Controller
[6]	Linear	LESO	State Feedback Controller
[14]	Linear	Improved ESO	State Feedback Controller
[17–19]	Linear	GESO	State Feedback Controller
[16]	Linear	GESO	Fractional-Order PD
[20]	Linear	Model-Aided ESO	Model Predictive Control
[25]	Nonlinear	Nonlinear ESO	Fractional-order PD
[29]	Linear	LESO	Fractional-Order PD
[26,27]	Linear	FOESO	PID
[28]	Linear	FOESO	Fuzzy Self-Tuning PD

The subsequent sections of this paper are organized as follows: In Section 2, the MESO-based ADRC is presented. Section 3 covers the design and tuning of the cascade ADRC strategy for the PMSM position servo system. Simulation comparisons and a discussion are presented in Section 4. Section 5 presents experimental verification. Finally, Section 6 presents the conclusion of this paper.

2. The Principle of MESO-Based ADRC

Figure 1 illustrates the structure of the MESO-based ADRC. It comprises an MESO and a feedback controller. The controlled plant is an n-th order system. The MESO estimates the lumped disturbance and plant states in real time. The feedback controller and MESO collaborate to generate the control effort. The specific details of the MESO-based ADRC are described below.

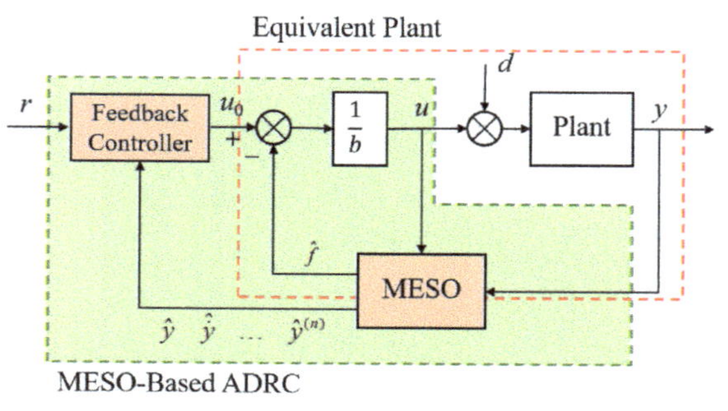

Figure 1. Block diagram of MESO-based ADRC.

2.1. MESO-Based ADRC

Consider an n-th order plant:

$$y^{(n)} + a_{n-1}y^{(n-1)} + \cdots + a_1\dot{y} + a_0 = bu + d, \tag{1}$$

where y, u, and d are the plant's output, input, and external disturbance, respectively; n is the order of the plant, and a_i $(i = 0, 1, 2, \ldots, n-1)$ and b are the plant coefficients. Its transfer function model is

$$P(s) = \frac{b}{s^n + a_{n-1}s^{n-1} + \cdots + a_1 s + a_0}. \tag{2}$$

Let f represent the lumped disturbance as

$$f = -a_{n-1}y^{(n-1)} - \cdots - a_1\dot{y} - a_0 y + d, \tag{3}$$

the plant (1) is converted to

$$y^{(n)} = f + bu. \tag{4}$$

The lumped disturbance f encompasses the known dynamics of the controlled plant and the unknown external disturbance.

When f is taken as an augmented state and $x = [y\ \dot{y}\ \dots\ y^{n-1}\ f]^T$, $h = \dot{d}$ is defined (assuming d is differentiable), the plant (4) is represented as augmented state-space form:

$$\begin{cases} \dot{x} = Ax + Bu + Eh \\ y = Cx \end{cases}, \tag{5}$$

where

$$A = \begin{bmatrix} 0 & 1 & 0 & \cdots & 0 \\ 0 & 0 & 1 & \cdots & 0 \\ \vdots & \vdots & \vdots & \ddots & \vdots \\ 0 & 0 & 0 & \cdots & 1 \\ 0 & -a_0 & -a_1 & \cdots & -a_{n-1} \end{bmatrix}, B = \begin{bmatrix} 0 \\ 0 \\ \vdots \\ b \\ -a_{n-1}b \end{bmatrix},$$
$$E = \begin{bmatrix} 0 & 0 & \cdots & 0 & 1 \end{bmatrix}^T, C = \begin{bmatrix} 1 & 0 & \cdots & 0 & 0 \end{bmatrix}.$$

The MESO can be designed based on Equation (5) as

$$\begin{cases} \dot{\hat{x}} = A\hat{x} + Bu + L(y - \hat{y}) \\ \hat{y} = C\hat{x} \end{cases}, \tag{6}$$

where $\hat{x}$ is the estimation of x, $\hat{y}$ is the estimation of y, and L is the observer gain vector,

$$L = [\beta_1\ \beta_2\ \dots\ \beta_n\ \beta_{n+1}]^T. \tag{7}$$

Remark 1. *The distinction between an MESO and an LESO lies in an MESO's utilization of all known plant information, while an LESO does not incorporate this information. The GESO proposed in [17] also utilizes known plant information, but it does not consider the known plant dynamics as part of the lumped disturbance f defined in the GESO. The GESO's disturbance compensation cannot transform the controlled plant into the ideal cascade integral plant. In contrast, an MESO can achieve cascade-integral plant equivalence because the known plant dynamics are considered as part of the lumped disturbance f in the MESO.*

The control effort is generated for the combination of the estimated disturbance $\hat{f}$ and the output of the feedback controller u_0 in real time, which is represented as

$$u = \frac{-\hat{f} + u_0}{b}, \tag{8}$$

Substitute it in (4), and the plant (1) is equivalent to

$$y^{(n)} = f - \hat{f} + u_0 \approx u_0. \tag{9}$$

If the MESO is designed appropriately, $\hat{f}$ completely compensates for the effects of the lumped disturbance f. Thus, the controlled plant is equivalent to an nth-order integrator plant. The state-feedback control law is designed as

$$u_0 = k_1(r - \hat{x}_1) - k_2\hat{x}_2 - \cdots - k_n\hat{x}_n. \tag{10}$$

In summary, the MESO-based ADRC can be expressed as

$$\begin{cases} \dot{\hat{x}} = (A - LC)\hat{x} + Bu + Ly, \\ u = K(\hat{r} - \hat{x}). \end{cases} \tag{11}$$

where $\hat{r} = [r \ 0 \ \ldots \ 0]^T$, $K = [K_0 \ 1]/b$, and $K_0 = [k_1 \ k_2 \ \ldots \ k_n]$.

2.2. Parameter Tuning

From (11), the MESO-based ADRC has two sets of gains to tune: the observer gain L and the state-feedback gain K. Many different methods can be utilized to tune L and K. The idea of the bandwidth parameterization method places all the poles of the ESO and the closed-loop system at $-\omega_o$ and $-\omega_c$ [6]. ω_o and ω_c are considered the bandwidths of the MESO and the closed-loop system.

Therefore, the characteristic equation of the MESO can be obtained using

$$|sI - (A - LC)| = (s + \omega_o)^{n+1}. \tag{12}$$

The value of β_i for L can be solved by comparing the coefficients of the two sides of (12).

Substitute (10) in (9), and the closed-loop system can be represented as

$$y^{(n)}(t) = k_1(r - \hat{x}_1) - k_2\hat{x}_2 - \cdots - k_n\hat{x}_n. \tag{13}$$

The characteristic equation of (13) can be expressed as

$$s^n + k_n s^{n-1} + \cdots + k_2 s + k_1 = (s + \omega_c)^n. \tag{14}$$

The value of k_i for K_0 can be solved from (14) as

$$k_i = C_n^{n+1-i}\omega_c^{n+1-i}. \tag{15}$$

Generally, the observer's response is faster than the closed-loop system's. Therefore, ω_o should be larger than ω_c, and an appropriate choice is $\omega_o = 2 \sim 6\omega_c$.

Remark 2. *The bandwidth parameterization method, while relatively conservative, facilitates the achievement of a non-overshoot response in the closed-loop system for setpoint tracking. In certain scenarios, there might be a need for a faster response. To address this, alternative tuning methods, including frequency specifications, loop shaping, and optimization, can be employed to attain the desired response. Moreover, various control laws, including PID, sliding mode control, and fractional-order control, can be implemented.*

2.3. Transfer Function Derivation

The MESO-based ADRC can be equivalent to a two-degrees-of-freedom closed-loop system, as shown in Figure 2, where $H(s)$ is the setpoint filter, and $C(s)$ is the feedback controller [30]. They can be obtained from the Laplace transform of (11) as follows:

$$C(s) = \frac{K[sI - (A - LC)]^{-1}L}{1 + K[sI - (A - LC)]^{-1}B}, \tag{16}$$

$$H(s) = \frac{k_1}{K[sI - (A - LC)]^{-1}L}. \tag{17}$$

From Figure 2, the transfer function from r to y is derived as follows:

$$G_{RY}(s) = \frac{H(s)C(s)P(s)}{1 + C(s)P(s)}. \tag{18}$$

It is used to derive the transfer function of the closed-loop system $G_{cl}(s)$. If $K_0 = 0$ is allowed, G_{RY} is converted to the equivalent plant $P_e(s)$.

The external disturbance transfer function $G_d(s)$ from d to y can be expressed as

$$G_{DY}(s) = \frac{P(s)}{1 + C(s)P(s)}. \tag{19}$$

It is used to analyze the disturbance rejection performance of the closed-loop system.

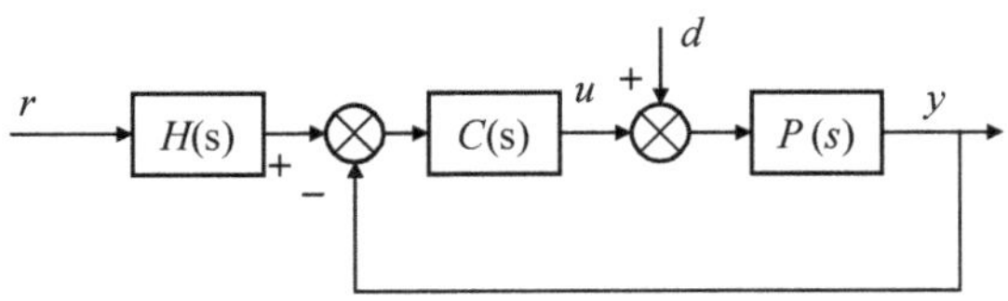

Figure 2. The transfer function form of the MESO-based ADRC

3. Cascade ADRC Strategy for a PMSM Position Servo System

The block diagram of the cascade ADRC for the PMSM position servo system is shown in Figure 3. There are three loops for the PMSM position servo system, i.e., current, speed, and position loops. The modeling of the PMSM and the MESO-based ADRC design for each loop is subsequently introduced.

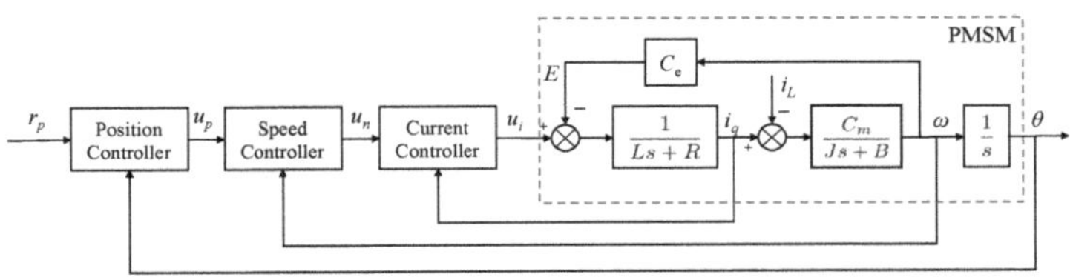

Figure 3. The structure of the cascade ADRC for a PMSM position servo system.

3.1. Modeling of PMSM

The voltage equations for a surface-mounted PMSM in d–q coordinates are expressed as follows:

$$\begin{cases} \frac{di_d}{dt} = \frac{1}{L_s}(-R_s i_d + n_p \omega L_s i_q + u_d) \\ \frac{di_q}{dt} = \frac{1}{L_s}(-R i_q - n_p \omega L_s i_d - n_p \psi_f \omega + u_q) \end{cases}, \tag{20}$$

where u_d, u_q, i_d, and i_q represent the stator voltages and currents of the d- and q-axes, respectively; L_s denotes the stator inductance, R_s is the stator resistance, ω signifies the angular velocity, n_p is the number of pole pairs, and ψ_f stands for the rotor flux linkage.

The current reference of the d-axis $i_d^* = 0$ is used to decouple the stator voltages and currents. Thus, the voltage equation represented by (20) is simplified to

$$\frac{di_q}{dt} = \frac{1}{L_s}(-R_s i_q - C_e \omega + u_q), \tag{21}$$

where $C_e = n_p \psi_f$ is the induced voltage constant.

The motion equation of the PMSM can be expressed as follows:

$$\begin{cases} \frac{d\omega}{dt} = \frac{1}{J}(C_m i_q - T_L - B_f \omega), \\ \frac{d\theta}{dt} = \omega. \end{cases} \tag{22}$$

where J is the moment of inertia, $C_m = 1.5 n_p \psi_f i_q$ is the torque coefficient, T_L is the load torque, B_f is the viscous friction coefficient, and θ is the mechanical angle of the rotor.

The transfer function model of the PMSM is obtained by applying the Laplace transform to Equations (21) and (22) under zero initial conditions as follows:

$$G_i(s) = \frac{1/L_s}{s + R_s/L_s}, \tag{23}$$

$$G_n(s) = \frac{C_m/J}{s + B_f/J}, \tag{24}$$

$$G_p(s) = \frac{1}{s}. \tag{25}$$

3.2. Current Loop Design

Taking the EMF as part of the lumped disturbance, the controlled plant for the current loop is $G_i(s)$, as in (23). According to (11), a first-order MESO-based ADRC can be designed for the current loop with $a_0 = R_s/L_s$ and $b = 1/L_s$.

The bandwidth parameterization method is adopted for the current loop. Given the observer bandwidth ω_{oi}, the observer gain $L_i = [\beta_1 \;\; \beta_2]^T$ is calculated using (12) as follows:

$$\begin{cases} \beta_1 = -a_0 + 2\omega_{oi}, \\ \beta_2 = (a_0 - \omega_{oi})^2. \end{cases} \tag{26}$$

Given the controller bandwidth ω_{ci}, the state feedback gain can be obtained using (14) as follows:

$$k_1 = \omega_{ci}. \tag{27}$$

The current closed loop can be obtained via (18) as follows:

$$G_{cli}(s) = \frac{\omega_{ci}}{s + \omega_{ci}}. \tag{28}$$

3.3. Speed Loop Design

The plant for the speed loop consists of the current closed loop (28) and the speed motion part (24), which can be represented as

$$P_n(s) = G_{cli}(s)G_n(s) = \frac{b}{s^2 + a_1 s + a_0}, \tag{29}$$

where $b = \frac{\omega_{ci}C_m}{J}$, $a_1 = \frac{\omega_{ci}J + B_f}{J}$, and $a_0 = \frac{\omega_{ci}B_f}{J}$.

The MESO for the speed loop can be designed based on (29). Given the observer bandwidth ω_{on}, the observer gain can be obtained via (12) as follows:

$$\begin{cases} \beta_1 = -a_1 + 3\omega_{on}, \\ \beta_2 = -a_0 + a_1^2 - 3a_1\omega_{on} + 3\omega_{on}^2, \\ \beta_3 = 2a_0a_1 - a_1^3 - 3a_0\omega_{on} + 3a_1^2\omega_{on} - 3a_1\omega_{on}^2 + \omega_{on}^3. \end{cases} \tag{30}$$

The equivalent plant for the speed loop can be obtained via (18) as follows:

$$P_{en}(s) = \frac{1}{s^2}. \tag{31}$$

Remark 3. *The equivalent plant is an ideal double-integrator plant equivalence, which means $P_{en}(s)$ is independent of the MESO. The design of the feedback controller is entirely decoupled from the MESO. This characteristic is also preserved in the designed ADRC for current and position loops based on the MESO. This separation enables the independent adjustment of disturbance rejection performance via the MESO without impacting the setpoint tracking determined according to the feedback controller. In contrast, the traditional LESO does not possess this property, as the equivalent plant using the LESO is always coupled with the LESO.*

A fractional-order PD (FOPD) controller was designed as the feedback controller, which can be expressed as follows:

$$u_0 = k_p(r - \hat{x}_1) - k_d D^{\alpha-1}\hat{x}_2, \tag{32}$$

where $D^{\alpha-1}(\cdot)$ represents the fractional-order differential, and the order of the derivative is $\alpha - 1$. The Caputo definition of the fractional-order derivative was adopted in this paper. The nominal control system of the speed loop is shown in Figure 4. The parameters of FOPD were designed based on this nominal control system.

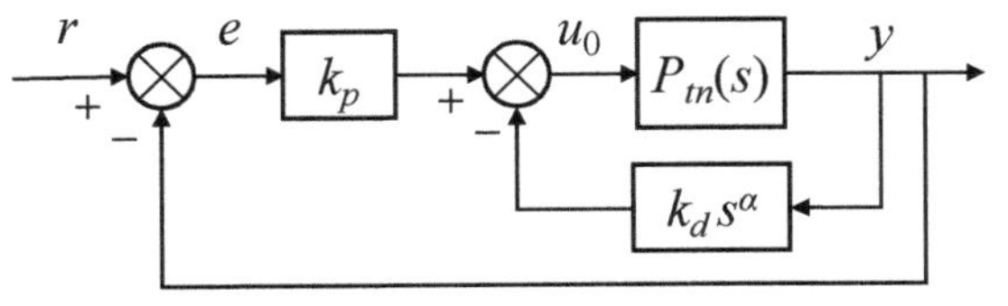

Figure 4. The nominal control system of the speed loop.

Per Figure 4, the open-loop transfer function is

$$G_{ol}(s) = \frac{k_p}{s^2 + k_d s^\alpha} = \frac{k_p}{s^\alpha (s^{2-\alpha} + k_d)}, \tag{33}$$

There is a fractional-order integrator in (33), and α should satisfy $\alpha \geq 1$ to ensure the steady-state error. Given the gain crossover frequency ω_{cn} and phase margin ϕ_m, the equations can be obtained as follows:

$$\begin{cases} |G_{on}(j\omega_{cn})| = \frac{k_p}{\sqrt{\omega_{cn}^4 + k_d^2 \omega_{cn}^{2\alpha} - 2k_d \omega_{cn}^{2+\alpha} \cos(\alpha\pi/2)}} = 1 \\ \arg[G_{on}(j\omega_{cn})] = -\tan^{-1} \frac{\omega_{cn}^2 \sin(\alpha\pi/2)}{k_d \omega_{cn}^\alpha - \omega_{cn}^2 \cos(\alpha\pi/2)} - \frac{\alpha\pi}{2} = -\pi + \phi_m \end{cases}. \tag{34}$$

Given a specific α, the parameters of k_p and k_d for (32) can be obtained as follows:

$$\begin{cases} k_p = \frac{\omega_{cn}^2 \sin(\alpha\pi/2)}{\sin(\phi_m + \alpha\pi/2)} \\ k_d = \frac{\omega_{cn}^{2-\alpha} \sin(\phi_m)}{\sin(\phi_m + \alpha\pi/2)} \end{cases}. \tag{35}$$

The choice of α is bounded as $\alpha \in [1,\ 2(\pi - \phi_m)/\pi)$ for a specific phase margin, ϕ_m. The choice of α is a trade-off between disturbance rejection and noise suppression.

The complementary sensitivity function of the nominal control system is

$$T_n(s) = \frac{G_{ol}(s)}{1 + G_{ol}(s)} = \frac{k_p}{s^2 + k_d s^\alpha + k_p}. \tag{36}$$

It is also the closed-loop transfer function of the nominal control system. The noise suppression specification defined by $T_n(s)$ determines the value of α, which is represented as

$$|T_n(j\omega_t)| \leq A_T \text{ dB}, \tag{37}$$

where

$$|T_n(j\omega_t)| = \frac{k_p}{\sqrt{k_d^2 \omega_t^{2\alpha} + (k_p - \omega_t^2)^2 + D_T}},$$

$$D_T = 2k_d \omega_t^\alpha (k_p - \omega_t^2) \cos(\alpha\pi/2).$$

Given the frequency ω_t, $|T_n(j\omega_t)|$ is calculated using (37) and (35) by sweeping α in $1 < \alpha < 2(\pi - \phi_m)/\pi$. Then, α is obtained from the curve of $|T_n(s)|$ with reference to α satisfying the gain limit of A_T dB. A flow chart of the design procedure for the speed loop is presented in Figure 5.

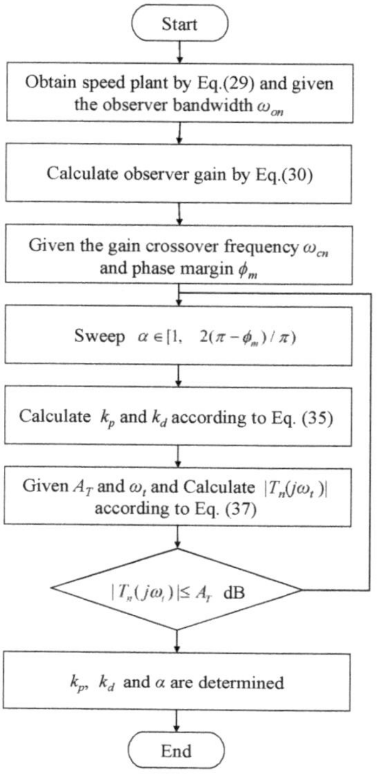

Figure 5. Flow chart of the design procedure for the speed loop.

The speed closed-loop transfer function can be obtained using (18) with $K_0 = [k_p \ k_d s^{1-\alpha} \ 1]$ as follows:

$$G_{cln}(s) = \frac{k_p}{s^2 + k_d s^\alpha + k_p}. \tag{38}$$

It is the same as the closed-loop transfer function of the nominal control system using (36).

3.4. Position Loop Design

Neglecting the fractional order, the designed speed closed-loop was approximated as an integer-order system. The difference in order was treated as part of the lumped disturbance. Therefore, according to (38) and (25), the controlled plant of the position loop can be represented as

$$P_p(s) = \frac{b}{s^3 + a_2 s^2 + a_1 s + a_0}, \tag{39}$$

where $b = k_p$, $a_2 = k_d$, $a_1 = k_p$, and $a_0 = 0$.

The MESO for the position loop can be designed based on (39), according to (5) and (6). Given the observer bandwidth ω_{op} , the observer gain can be obtained using (12) as follows:

$$\begin{cases} \beta_1 = & -a_2 + 4\omega_{op}, \\ \beta_2 = & -a_1 + a_2^2 - 4a_2\omega_{op} + 6\omega_{op}^2, \\ \beta_3 = & -a_0 + 2a_1a_2 - a_2^3 - 4a_1\omega_{op} + 4a_2^2\omega_{op} - 6a_2\omega_{op}^2 + 4\omega_{op}^3, \\ \beta_4 = & a_1^2 + 2a_0a_2 - 3a_1a_2^2 + a_2^4 - 4a_0\omega_{op} + 8a_1a_2\omega_{op} - 4a_2^3\omega_{op} \\ & -6a_1\omega_{op}^2 + 6a_2^2\omega_{op}^2 - 4a2\omega_{op}^3 + \omega_{op}^4. \end{cases} \tag{40}$$

The bandwidth parameterization method was used to tune the feedback controller for the position loop. Placing all the poles of the closed loop at $-\omega_{cp}$ allowed the parameter's value to be solved from (14) as follows:

$$k_1 = \omega_{cp}^3, k_2 = 3\omega_{cp}^2, k_3 = 3\omega_{cp}. \tag{41}$$

4. Simulation and Discussion

The mathematical models of the experimental PMSM were identified as

$$G_i(s) = \frac{403.48}{s + 153.57}, \tag{42}$$

$$G_n(s) = \frac{333.85}{s + 0.4889}. \tag{43}$$

Such a PMSM device was used for the subsequent simulation and analysis.

Three control strategies are presented for comparison: the traditional LESO-based ADRC [6], the proposed MESO-based ADRC with a PD feedback controller, and the proposed MESO-based ADRC with an FOPD feedback controller. The control strategies designed for comparison are listed in Table 2.

Table 2. Control strategies designed for comparison.

	Current Loop	Speed Loop	Position Loop
LESO-based	P-LESO	PD-LESO	PD-LESO
MESO-based	P-MESO	PD-MESO	PD-MESO
FOC- and MESO-based	P-MESO	FOPD-MESO	PD-MESO

4.1. Current Loop

The plant parameters can be determined as $a_0 = 153.57$ and $b = 403.48$ via (42). Given $\omega_{ci} = 1000$ rad/s and $\omega_{oi} = 5000$ rad/s, the parameters of the MESO-based ADRC for the current loop can be designed as $\beta_1 = 9846.43$, $\beta_2 = 2.3488 \times 10^7$, and $k_1 = 1000$. The parameters designed for the LESO-based ADRC are $\beta_1 = 10,000$, $\beta_2 = 2.5 \times 10^7$, and $k_1 = 1000$.

4.2. Speed Loop

The controlled plant of the speed loop is represented by (29), with $a_0 = 488.9$, $a_1 = 1000.49$, and $b = 3.34 \times 10^5$. The design specifications for the speed loop included $\omega_{cn} = 100$ rad/s, $\omega_{on} = 500$ rad/s, $\phi_m = 70°$, $\omega_t = 1000$ rad/s, and $A_T = -24.8$ dB. Subsequently, the MESO for the speed loop could be designed.

Considering $\phi_m = 70°$, the fractional order bound for FOPD can be calculated as $\alpha \in [1, 1.22)$. By sweeping α within this range, the parameters k_p and k_d can be determined using (35). The parameter curve for FOPD is illustrated in Figure 6a. As α approaches its upper bound, k_p increases rapidly. The gain of $T_n(s)$ at ω_t can be calculated using (37). The curve of $|T_n(j\omega_t)|$ with respect to α is depicted in Figure 6b. It indicates that the gain in $T_n(s)$ increases with a rising α, leading to a gradual decrease in the noise suppression ability. For $\alpha \in (1, 1.18]$, the constraint in (37) is satisfied. Hence, $\alpha = 1.18$ was selected to achieve better disturbance rejection performance.

The PD-LESO and PD-MESO controllers were also designed for comparison using the same design specifications. The parameters of the design controllers for the speed loop are shown in Table 3.

(1) Disturbance rejection and noise suppression for the ESO: When the estimation error is defined as $e_x = x - \hat{x}$, the dynamics of the estimation error can be obtained using (5) and (6) as follows:

$$\dot{e}_x = (A - LC)e_x + E\dot{d} - Ln. \tag{44}$$

where n is the sensor noise. By applying the Laplace transform to (44), the transfer function from d to e_{x3} ($e_{x3} = f - \hat{f}$) can be obtained as follows:

$$G_{ed}(s) = C_f[sI - (A - LC)]^{-1}Es, \tag{45}$$

The transfer function from n to e_{x3} can be derived as follows:

$$G_{en}(s) = C_f[sI - (A - LC)]^{-1}L, \tag{46}$$

where $C_f = [0\ 0\ \dots\ 0\ 1]_{n+1}$. G_{ed} and G_{en} are used to evaluate the MESO's disturbance rejection and noise sensitivity, respectively. Therefore, $G_{ed}(s)$ and $G_{en}(s)$ for the speed loop can be obtained as follows:

$$G_{ed} = \frac{s^2 + \beta_1 s + \beta_2}{(s + \omega_{on})^3}, \tag{47}$$

$$G_{en} = \frac{\beta_3 s^2 - a_0\beta_2 s}{(s + \omega_{on})^3}. \tag{48}$$

The Bode plots of $G_{ed}(s)$ and $G_{en}(s)$ are depicted in Figure 7a,b. With the same observer bandwidth ω_{on}, Figure 7 illustrates that the MESO exhibits superior disturbance estimation performance compared to the LESO, while both maintain the same noise suppression performance. The disturbance estimation performance of the LESO improves as ω_{on} increases, but the noise suppression performance decreases simultaneously.

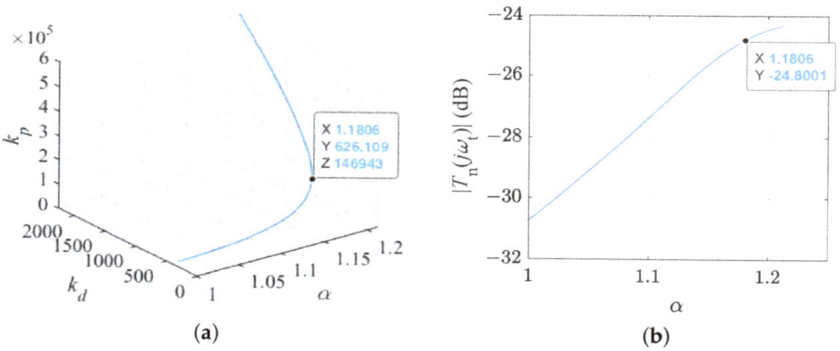

Figure 6. Curves for FOPD design: (**a**) parameters curve; (**b**) $|T_n(j\omega_t)|$ w.r.t. α.

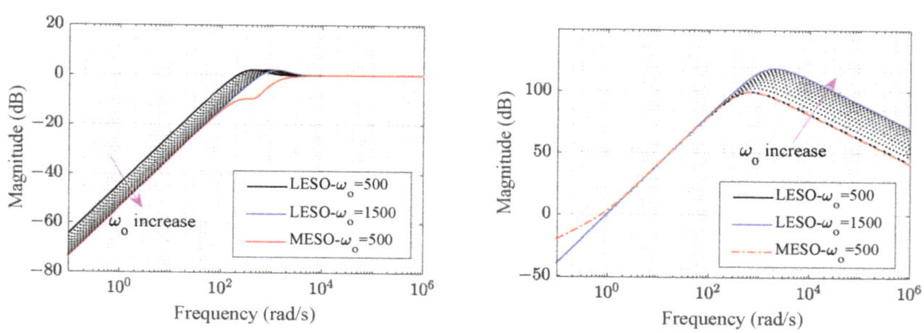

Figure 7. Frequency responses comparison of a different ESO for the speed loop: (**a**) disturbance estimation ; (**b**) noise suppression.

Table 3. Parameters of the designed controllers for the speed loop.

	k_1	k_2	α	β_1	β_2	$\beta_3(\times 10^8)$
PD-LESO	29,238.0	274.75	1	1500	750,000	1.25
PD-MESO	29,238.0	274.75	1	499.51	249,755	−1.2512
FOPD-MESO	144,897	618.93	1.18	499.51	249,755	−1.2512

Remark 4. *Since the MESO incorporates the plant information as the known part of the total disturbance, the burden of MESO tracking f is lower compared to the traditional LESO. It is well known that the selection of observer bandwidth involves a trade-off between disturbance estimation and noise sensitivity. Therefore, the MESO alleviates this conflict and achieves better disturbance estimation and noise suppression performance simultaneously compared to the LESO under a limited observer bandwidth.*

(2) The equivalent plant: The Bode plots of the equivalent plant are presented in Figure 8. It shows that the MESO-based equivalent plant is identical to the nominal plant, represented as $1/s^2$. However, the LESO-based equivalent plant differs significantly from the nominal plant. Although the differences between the equivalent plant and the nominal plant decrease as ω_{on} increases, the LESO becomes more sensitive to noise at a high ω_{on}.

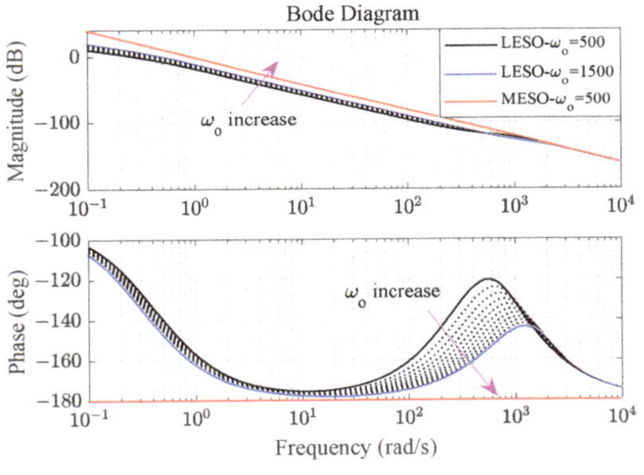

Figure 8. Bode plots of the equivalent plant for the speed loop.

(3) Open-loop Bode plots: Open-loop Bode plots are illustrated in Figure 9a. It is evident from the plots that the designed PD-MESO and FOPD-MESO met the design specifications, whereas the PD-LESO did not. Due to the significant difference between the LESO-based equivalent plant and the nominal plant, the designed PD feedback controller based on the nominal plant could not meet the design specifications in the real system. In the case of the FOPD-MESO, the gain at the low-frequency range increased with a rising α, leading to an improvement in the steady-state error. However, simultaneously, the gain in the high-frequency range also increased with a rising α, indicating a decrease in noise suppression performance.

(4) Closed-loop Bode plots: The closed-loop transfer function is identical to the complementary sensitivity function, as per (38) and (36). The Bode plots are depicted in Figure 9b. It is observed that the PD-MESO exhibited the lowest resonance peak. The resonance peak of the PD-LESO was notably higher than that of the other controllers. The resonance peak of the FOPD-MESO increased with a rising α, accompanied by an increase in the gain in the high-frequency range. The relative stability and noise suppression of the FOPD-MESO decreased as α increased.

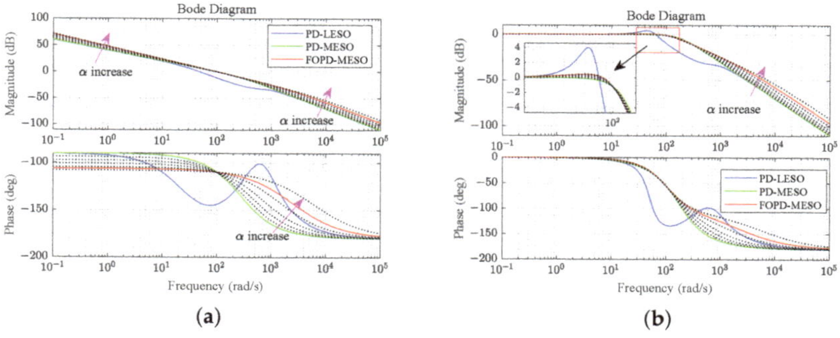

Figure 9. Bode plots of the transfer functions for the speed loop: (**a**) the open loop; (**b**) the closed loop.

(5) The disturbance rejection of the closed loop: According to (19), the disturbance transfer function of the speed loop is

$$G_{dn}(s) = \frac{b(\beta_2 + k_p)s + b(\beta_1 + s)(s^2 + k_d s^\alpha)}{(s^2 + k_d s^\alpha + k_p)(s + \omega_o)^3}. \tag{49}$$

Bode plots of the disturbance transfer function $G_{dn}(s)$ are presented in Figure 10. It is evident that the MESO-based ADRC exhibited superior disturbance rejection performance compared to the other controllers. Moreover, the disturbance rejection performance of the FOPD-MESO improved with an increasing α. Therefore, in addition to the MESO, the FOPD-MESO can enhance the disturbance rejection performance by adjusting α. The choice of α involves a trade-off between disturbance rejection and noise suppression.

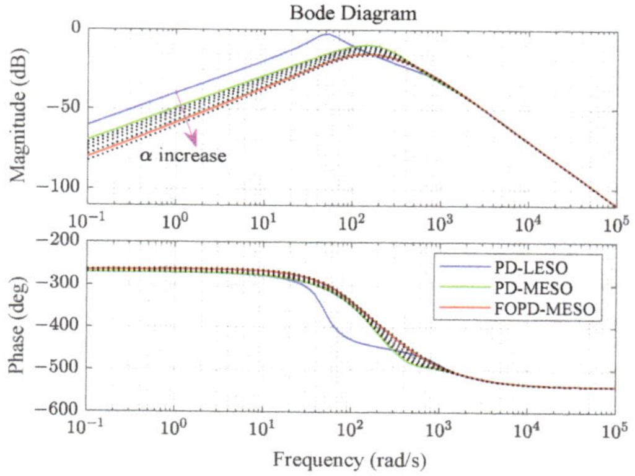

Figure 10. Bode plots of the disturbance transfer function for the speed loop.

(6) The disturbance rejection tuning with ω_{on} and α: Based on the analysis above, it can be concluded that the disturbance rejection performance of the FOPD-MESO can be simultaneously tuned by adjusting ω_{on} and α. To explore this property, the step responses with disturbance for various ω_{on} and α values are depicted in Figure 11a,b. Figure 11a demonstrates that the disturbance rejection performance improved as the value of ω_{on} increased, while the tracking performance remained unchanged. This verifies that the MESO-based ADRC satisfied the separation principle. The tracking performance was independent of the observer gain. Figure 11b illustrates that the disturbance rejection performance improved as α increased. However, the overshoot of the setpoint tracking also increased simultaneously. Therefore, the FOPD provides another tunable parameter α to enhance the disturbance rejection performance based on the MESO.

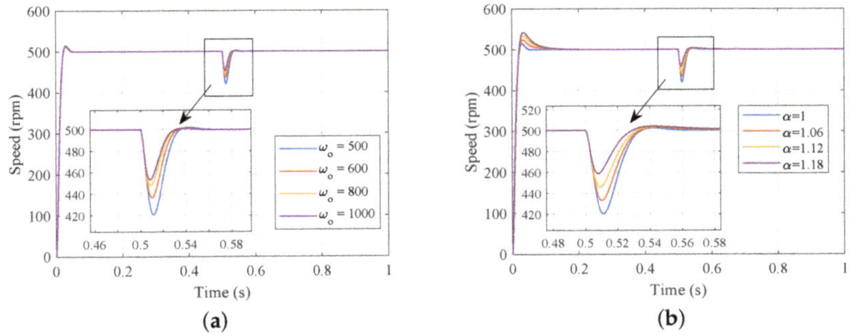

Figure 11. The step response of the MESO-based ADRC for the speed loop (simulation): (**a**) different ω_{on} values ($\alpha = 1$); (**b**) different α values ($\omega_o = 500$ rad/s).

(7) Step response comparison for different control strategies: The step response of the speed loop is depicted in Figure 12. It is evident that the MESO-based ADRC exhibited superior tracking and disturbance rejection. The PD-LESO exhibited a worse transient response compared to the other controllers. When a disturbance was applied, the speed drop of the FOPD-MESO was 8.3%, significantly less than the PD-MESO's 15.9%. Compared to the PD-MESO, the overshoot of the FOPD-MESO increased slightly from 3.0% to 8.4%. Therefore, the fractional order α can be adjusted to balance tracking and disturbance rejection performance.

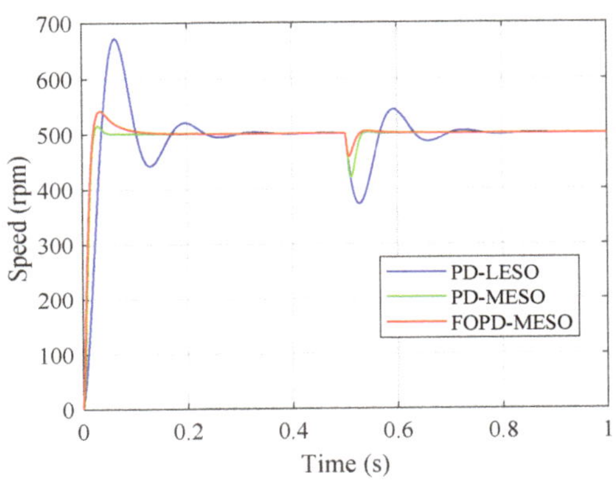

Figure 12. The step response comparison for the speed loop (simulation).

4.3. Position Loop

The approximate plant of the position loop can be determined using (39), with $b = 29{,}238.0$, $a_0 = 0$, $a_1 = 29{,}238.0$, and $a_2 = 274.747$. When the design specifications of the position loop are set as $\omega_{cp} = 50$ rad/s and $\omega_{op} = 250$ rad/s, PD-MESO and PD-LESO can be designed based on this plant. The parameters of the design controllers for the position loop are presented in Table 4.

Table 4. Parameters of the designed controllers for the position loop.

	k_1	k_2	k_2	β_1	β_2	$\beta_3(\times 10^6)$	$\beta_4(\times 10^8)$
PD-LESO	125,000	7500	150	1000	375,000	62.5	39.0625
PD-MESO	125,000	7500	150	725.252	146,500	1.04435	−6.64074

Figure 13 presents the step response of the position loop. It is evident that the LESO-based ADRC strategy exhibited a greater overshoot and position error when the disturbance was applied. This did not meet the design expectations, as the equivalent plant based on the LESO was significantly different from the nominal plant. However, the step response of the proposed MESO-based ADRC and FOC- and MESO-based ADRC could closely match the step response of the nominal system when using the same observer bandwidth. Additionally, the FOC- and MESO-based ADRC exhibited the smallest position error after disturbance input at 2.0%, compared to the MESO-based ADRC's 4.5%. Therefore, the simulation indicated the best disturbance rejection performance of the FOC- and MESO-based ADRC compared with the other controllers.

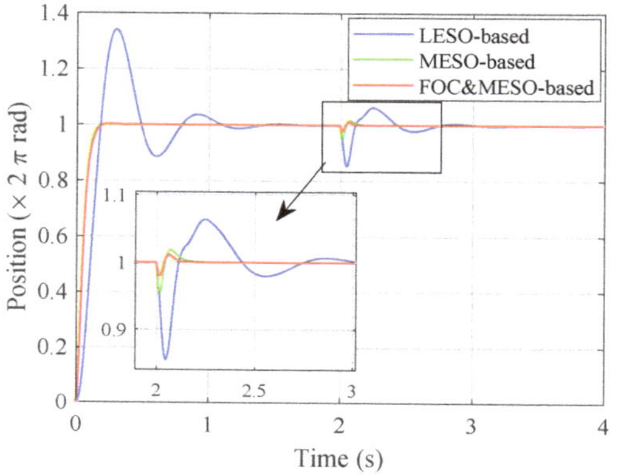

Figure 13. The step response comparison for the position loop (simulation).

5. Experimental Verification

5.1. Experimental Setup

Experimental validation was performed on a PMSM servo system, illustrated in Figure 14. The experimental setup consisted of essential components, including a PMSM, a servo drive, a DC generator, and a PC. The servo driver was powered via DSP-TMS320F28335. The structure of the PMSM position servo system based on field-oriented control is depicted in Figure 15. The control sampling frequencies for the current, speed, and position loops were 10 kHz, 5 kHz, and 2 kHz, respectively. The specifications of the PMSM are provided in detail in Table 5.

Figure 14. The experimental PMSM position servo system.

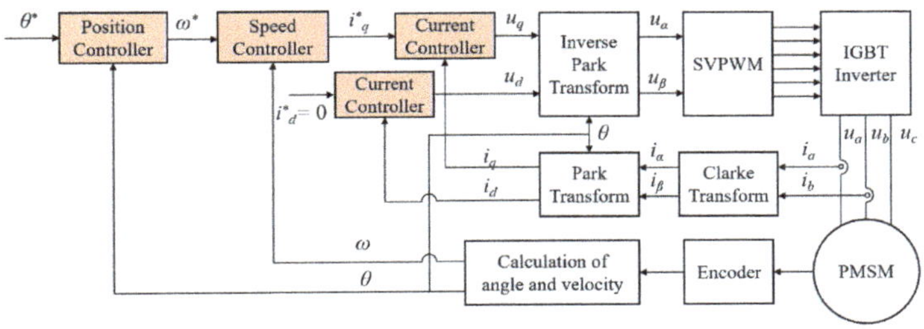

Figure 15. The structure of the PMSM position servo system based on field-oriented control.

Table 5. Specifications of the PMSM.

Parameters	Unit	Value
Rated power	kW	2.0
Rated speed	rpm	2000
Rated current	A	9.4
Line resistance	Ω	0.38
Line inductance	mH	2.48
Induced voltage coefficient	V/krpm	67.4
Moment of inertia	$\text{kg} \cdot \text{m}^2$	24.3×10^{-4}

5.2. Experimental Tests for the Speed Loop

The fractional-order differential $s^{0.18}$ was approximated using a five-order discrete transfer function obtained through the impulse response invariant discretization method of [16]:

$$s^{0.18} \approx \frac{N}{D}, \tag{50}$$

where

$$\begin{aligned} N &= z^5 - 3.05222z^4 + 3.43539z^3 - 1.71645z^2 + 0.352724z - 0.0193436, \\ D &= 0.248528z^5 - 0.708956z^4 + 0.730482z^3 - 0.321782z^2 + 0.0534573z - 0.00163956. \end{aligned} \tag{51}$$

The Bode plot of the approximated $s^{0.18}$ is shown in Figure 16. It can be seen that the approximated fractional-order operator matched the true Bode plot of $s^{0.18}$ around $\omega = 100$ rad/s, the gain crossover frequency of the speed loop, which means the approximated fractional-order ADRC could perform as the real fractional-order ADRC approximately.

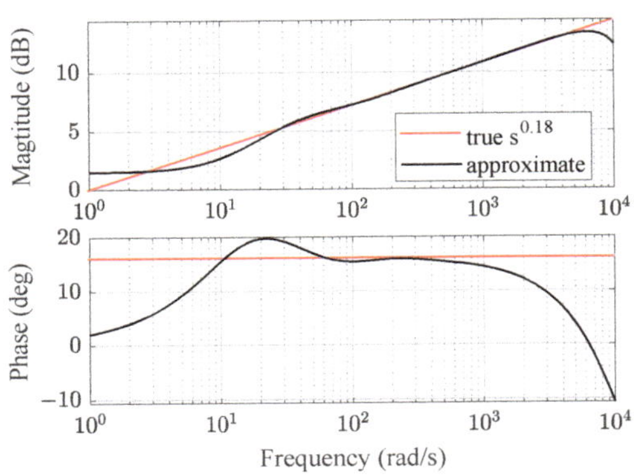

Figure 16. The Bode plot comparison of approximated and true $s^{0.18}$.

Figure 17a,b illustrates the step response of the PMSM's speed loop with varying values of ω_{on} and α. The experimental results validated the effectiveness of tuning disturbance rejection performance using ω_{on} and α, as observed in the simulation. It was observed that the tracking performance remained unaffected by the MESO bandwidth, while the disturbance rejection performance improved with an increasing ω_{on}. Furthermore, the disturbance performance improved with an increasing α, but it accompanied an increase in the overshoot.

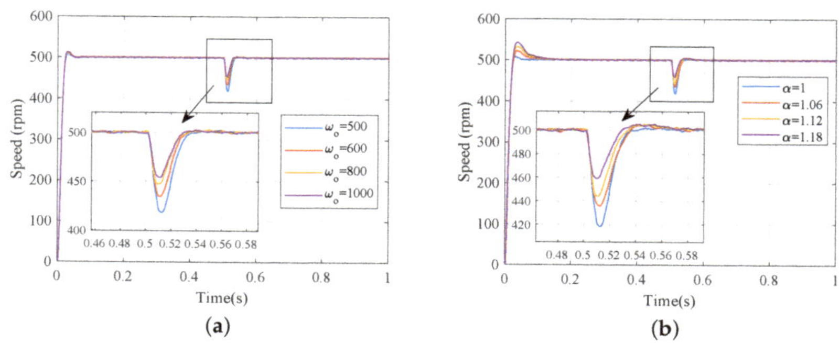

Figure 17. The step response of the MESO-based ADRC for the speed loop (experiment): (**a**) different ω_{on} values ($\alpha = 1$); (**b**) different α values ($\omega_{on} = 500$ rad/s).

Figure 18 presents the step response of the PMSM's speed loop employing different control methods. The performance indices are listed in Table 6. Consistent with the simulation results, the experimental results validated the findings. In comparison to the PD-LESO, the PD-MESO exhibited a significant improvement in both tracking and disturbance rejection performance using the same observer bandwidth. Additionally, the FOPD-MESO can enhance disturbance rejection performance by tuning the order of the feedback controller based on the MESO. The speed drop of the FOPD-MESO was 8.19%, versus the PD-MESO's 16.29%.

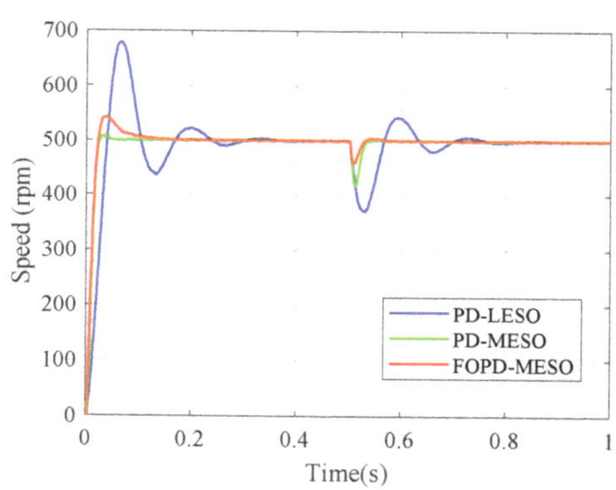

Figure 18. Step response comparison for the speed loop (experiment).

Table 6. Indices of the step responses for the speed loop.

	Overshoot	Settling Time	Speed Drop	Recovery Time
PD-LESO	35.75%	0.225 s	25.92%	0.181 s
PD-MESO	1.51%	0.0205 s	16.29%	0.029 s
FOPD-MESO	8.36%	0.085 s	8.19%	0.0238 s

5.3. Experimental Tests for the Position Loop

Figure 19 illustrates the step response of the PMSM's position loop. The legend names represent the different control strategies for comparison, which can be found in Table 2. Consistent with the simulation, the experimental results aligned with the expected outcomes. The setpoint tracking of the LESO-based ADRC exhibited a large overshoot, deviating significantly from the behavior of the nominal control system. In contrast, the setpoint tracking of the MESO-based and the FOC- and MESO-based ADRC aligned with the nominal control system. The disturbance rejection performance of the MESO-based ADRC was significantly improved compared to the LESO-based ADRC. Moreover, the FOC- and MESO-based ADRC further enhanced the disturbance rejection performance based on the MESO-based ADRC, When the disturbance was applied, the position error of the FOC- and MESO-based ADRC improved to 2.2 % from the MESO-based ADRC's 4.3%.

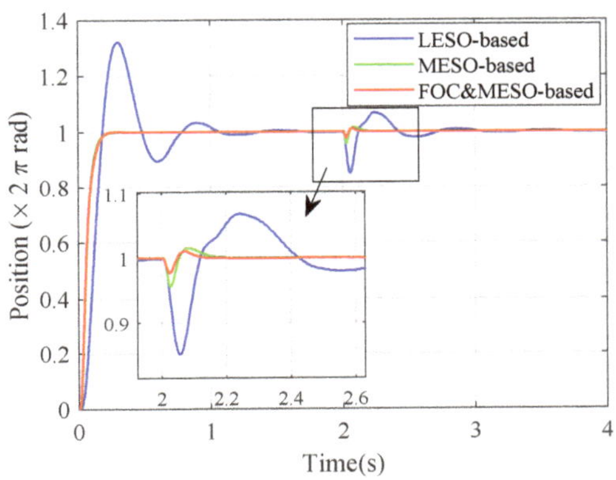

Figure 19. The step response comparison for the position loop (experiment).

6. Conclusions

This paper has proposed a cascade FOADRC architecture with an MESO for a PMSM position servo system. The MESO-based ADRC was designed for the current, speed, and position loops of the PMSM position servo system. Moreover, a fractional-order PD feedback controller was designed for the speed loop to further improve disturbance rejection performance on the basis of the MESO. The simulation and experimental verification were conducted on a PMSM servo platform. The results demonstrate that the proposed method achieved the desired tracking performance. Additionally, using the same observer bandwidth, the position error of the proposed strategy decreased to 2.25% when the disturbance was input, in contrast to the traditional LADRC's 14.9% and the MESO-based integer-order ADRC's 4.3%. The proposed method achieved superior tracking and disturbance rejection performance with a limited observer bandwidth.

Author Contributions: Conceptualization, Y.L., X.L. and Y.C.; methodology, S.W.; software, S.W. and H.G.; validation, S.W. and H.G.; formal analysis, S.W.; investigation, S.W.; writing—original draft preparation, S.W.; writing—review and editing, Y.L., X.L. and Y.C.; visualization, S.W.; supervision, Y.C.; project administration, Y.L. and X.L.; funding acquisition, Y.L. All authors have read and agreed to the published version of the manuscript.

Funding: This research was funded by the National Natural Science Foundation of China (grant number: 51975234).

Data Availability Statement: No new data were created or analyzed in this study.

Conflicts of Interest: The authors declare no conflicts of interest.

Abbreviations

PMSM	Permanent magnet synchronous motor
ADRC	Active disturbance rejection control
LADRC	Linear ADRC
PID	Proportional-integral-derivative
FOC	Fractional-order control
FOPD	Fractional-order proportional-derivative
FOADRC	Fractional-order ADRC
ESO	Extended state observer
GESO	Generalized ESO
MESO	Model-aided ESO

References

1. Wang, G.; Liu, R.; Zhao, N.; Ding, D.; Xu, D. Enhanced linear ADRC strategy for HF pulse voltage signal injection-based sensorless IPMSM drives. *IEEE Trans. Power Electron.* **2019**, *34*, 514–525. [CrossRef]
2. Chen, W.; Yang, J.; Guo, L.; Li, S. Disturbance-observer-based control and related methods —An overview. *IEEE Trans. Ind. Electron.* **2016**, *63*, 1083–1095. [CrossRef]
3. Liu, H.; Li, S. Speed control for PMSM servo system using predictive functional control and extended state observer. *IEEE Trans. Ind. Electron.* **2011**, *59*, 1171–1183. [CrossRef]
4. Alcántara, S.; Vilanova, R.; Pedret, C. PID control in terms of robustness/performance and servo/regulator trade-offs: A unifying approach to balanced autotuning. *J. Process. Control* **2013**, *23*, 527–542. [CrossRef]
5. Han, J. From PID to Active Disturbance Rejection Control. *IEEE Trans. Ind. Electron.* **2009**, *56*, 900–906. [CrossRef]
6. Gao, Z. Scaling and Bandwidth-Parameterization Based Controller Tuning. In Proceedings of the The 2003 American Control Conference, Denver, CO, USA, 4–6 June 2003; Volume 6, pp. 4989–4996.
7. Ren, L.; Mao, C.; Song, Z.; Liu, F. Study on Active Disturbance Rejection Control with Actuator Saturation to Reduce the Load of a Driving Chain in Wind Turbines. *Renew. Energy* **2019**, *133*, 268–274. [CrossRef]
8. Fareh, R.; Khadraoui, S.; Abdallah, M.Y.; Baziyad, M.; Bettayeb, M. Active Disturbance Rejection Control for Robotic Systems: A Review. *Mechatronics* **2021**, *80*, 102671. [CrossRef]
9. Sun, H.; Sun, Q.; Wu, W.; Chen, Z.; Tao, J. Altitude Control for Flexible Wing Unmanned Aerial Vehicle Based on Active Disturbance Rejection Control and Feedforward Compensation. *Int. J. Robust Nonlinear Control* **2020**, *30*, 222–245. [CrossRef]
10. Liu, J.; Sun, M.; Chen, Z.; Sun, Q. Output Feedback Control for Aircraft at High Angle of Attack Based upon Fixed-Time Extended State Observer. *Aerosp. Sci. Technol.* **2019**, *95*, 105468. [CrossRef]
11. Meng, Y.; Liu, B.; Wang, L. Speed Control of PMSM Based on an Optimized ADRC Controller. *Math. Probl. Eng.* **2019**, *2019*, 1074702. [CrossRef]
12. Wu, Z.; Zhou, H.; Guo, B.; Deng, F. Review and New Theoretical Perspectives on Active Disturbance Rejection Control for Uncertain Finite-Dimensional and Infinite-Dimensional Systems. *Nonlinear Dyn.* **2020**, *101*, 935–959. [CrossRef]
13. Li, D.; Ding, P.; Gao, Z. Fractional Active Disturbance Rejection Control. *ISA Trans.* **2016**, *62*, 109–119. [CrossRef] [PubMed]
14. Du, Y.; Cao, W.; She, J. Analysis and Design of Active Disturbance Rejection Control With an Improved Extended State Observer for Systems With Measurement Noise. *IEEE Trans. Ind. Electron.* **2023**, *70*, 855–865. [CrossRef]
15. Guo, B.; Bacha, S.; Alamir, M.; Hably, A.; Boudinet, C. Generalized integrator-extended state observer with applications to grid-connected converters in the presence of disturbances. *IEEE Trans. Control Syst. Technol.* **2020**, *29*, 744–755. [CrossRef]
16. Chen, P.; Luo, Y. A two-degree-of-freedom controller design satisfying separation principle with fractional-order PD and generalized ESO. *IEEE/ASME Trans. Mechatron.* **2021**, *27*, 137–148. [CrossRef]
17. Fu, C.; Tan, W. Tuning of linear ADRC with known plant information. *ISA Trans.* **2016**, *65*, 384–393. [CrossRef] [PubMed]
18. Fu, C.; Tan, W. Control of Unstable Processes with Time Delays via ADRC. *ISA Trans.* **2017**, *71*, 530–541. [CrossRef]
19. Zhou, R.; Tan, W. Analysis and Tuning of General Linear Active Disturbance Rejection Controllers. *IEEE Trans. Ind. Electron.* **2019**, *66*, 5497–5507. [CrossRef]
20. Liu, H.; Lin, W.; Liu, Z.; Buccella, C.; Cecati, C. Model Predictive Current Control With Model-Aid Extended State Observer Compensation for PMSM Drive. *IEEE Trans. Power Electron.* **2023**, *38*, 3152–3162. [CrossRef]
21. Cui, S.; Zhu, G.; Zhao, T. Linear Active Disturbance Rejection Control-Based Diagonal Recurrent Neural Network for Radar Position Servo Systems with Dead Zone and Friction. *Appl. Sci.* **2022**, *12*, 12839. [CrossRef]
22. Wang, Y.; Fang, S.; Hu, J. Active Disturbance Rejection Control Based on Deep Reinforcement Learning of PMSM for More Electric Aircraft. *IEEE Trans. Power Electron.* **2023**, *38*, 406–416. [CrossRef]
23. Zhang, Z.; Cheng, J.; Guo, Y. PD-Based Optimal ADRC with Improved Linear Extended State Observer. *Entropy* **2021**, *23*, 888. [CrossRef] [PubMed]
24. Dastjerdi, A.A.; Vinagre, B.M.; Chen, Y.; HosseinNia, S.H. Linear fractional order controllers; A survey in the frequency domain. *Annu. Rev. Control* **2019**, *47*, 51–70. [CrossRef]
25. Shi, X.; Chen, Y.; Huang, J. Application of Fractional-Order Active Disturbance Rejection Controller on Linear Motion System. *Control Eng. Pract.* **2018**, *81*, 207–214. [CrossRef]

26. Wang, S.; Gan, H.; Luo, Y.; Wang, X.; Gao, Z. Active Disturbance Rejection Control with Fractional-Order Model-Aided Extended State Observer. *ISA Trans.* **2023**, *142*, 527–537. [CrossRef]
27. Zhao, M.; Hu, Y.; Song, J. Improved Fractional-Order Extended State Observer-Based Hypersonic Vehicle Active Disturbance Rejection Control. *Mathematics* **2022**, *10*, 4414. [CrossRef]
28. Zhao, J.; Zhao, T.; Liu, N. Fractional-Order Active Disturbance Rejection Control with Fuzzy Self-Tuning for Precision Stabilized Platform. *Entropy* **2022**, *24*, 1681. [CrossRef]
29. Chen, P.; Luo, Y.; Peng, Y.; Chen, Y. Optimal Fractional-Order Active Disturbance Rejection Controller Design for PMSM Speed Servo System. *Entropy* **2021**, *23*, 262. [CrossRef]
30. Tian, G.; Gao, Z. Frequency Response Analysis of Active Disturbance Rejection Based Control System. In Proceedings of the 2007 IEEE International Conference on Control Applications, Singapore, 1–3 October 2007; pp. 1595–1599.

fractal and fractional

MDPI

Article

Fractional Control of a Class of Underdamped Fractional Systems with Time Delay—Application to a Teleoperated Robot with a Flexible Link

Saddam Gharab [1] and Vicente Feliu Batlle [2,*]

[1] Instituto de Investigaciones Energéticas y Aplicaciones Industriales, 13005 Ciudad Real, Spain; saddam.gharab@uclm.es

[2] Escuela Técnica Superior de Ingeniería Industrial, Universidad de Castilla La Mancha, 13071 Ciudad Real, Spain

* Correspondence: vicente.feliu@uclm.es

Abstract: This work addresses the robust control of processes of the form $G(s) = K \cdot e^{-\tau \cdot s}/(1 + T \cdot s^{\lambda})$ with $1 < \lambda \leq 2$. A new method for tuning fractional-order PI and PD controllers is developed. The stability is assessed based on the frequency domain tuning of the regulators to control such delayed fractional-order underdamped processes. In order to analyze the closed-loop stability and robustness, the new concept of *Robust High-Frequency Condition* is introduced. The analysis based on that demonstrates that each controller has a different region of feasible frequency specifications, and, in all cases, they depend on their fractional integral or derivative actions. Finally, an application example, the position control of a teleoperated manipulator with a flexible link, is presented. Simulations and experiments illustrate that the region of feasible frequency specifications defined at low and high frequencies allows us to obtain robust controllers that fulfill frequency requirements.

Keywords: time delay; fractional order systems; flexible robots; teleoperated robots; fractional order control

1. Introduction

Fractional calculus has indeed experienced significant advancements in theory, methods, and applications over the past few decades. It provides a powerful mathematical framework for describing and analyzing systems with memory effects and hereditary properties. Using the fractional-order framework in modeling is crucial for describing real-world applications such as in the field of forecasting and predictions [1], or physical domain such as modeling the nonlinear water waves that integer-order models cannot accurately represent [2]. Thereby, unlike traditional integer-order modeling, fractional order is important because it offers a more accurate and versatile way to describe various real-world phenomena and systems.The development of fractional calculus has led to a deeper understanding of complex systems exhibiting non-local and non-Markovian behavior. It has found applications in diverse fields such as physics [3], engineering [4], biology [5], finance [6], signal processing [7], and control systems [8]. Viscoelastic materials are a class of materials that exhibit mechanical behavior between pure elasticity and pure viscosity. They possess both elastic and viscous characteristics, resulting in unique mechanical properties that have found wide applications in various fields [9]. Fractional calculus has been successfully employed in modeling the constitutive relations of viscoelastic behaviors [10,11]. Several fractional-order viscoelastic models have been established and validated by experimental data, such as the fractional Poynting-Thomson model [12], the fractional Maxwell model [11], and the fractional Kelvin-Voigt model [13].

In addition, the fractional-order PID (Proportional-Integral-Derivative) controller was defined as the generalization of the classical PID controller that incorporates fractional calculus principles. It utilizes fractional derivatives and integrals to capture memory effects and hereditary properties, providing improved control performance in systems with non-local and non-Markovian behaviors. The design of such a controller based on frequency specifications is a fairly broadly used approach [14]. While the traditional phase margin, gain margin, gain crossover frequency, and phase crossover frequency are commonly used in the design of classical PID controllers, their application to fractional-order PID controllers is less straightforward [15]. This is because fractional-order PID controllers introduce additional complexity due to the fractional calculus operators involved. Those specifications have a twofold purpose in the design of control systems, including both classical and fractional-order PID controllers. Frequency specifications relate to both time-domain performance and robustness considerations. An interesting feature of these techniques is, therefore, that they allow the design of the dynamic response of closed-loop control systems, and also permit the design of robustness properties for the controller [16]. Meanwhile, achieving robustness properties in both classical and fractional-order PID controllers becomes more complex when using time domain or Laplace domain-based design techniques than when using frequency domain approaches [17].

The presence of time delay in controlled systems can produce a detrimental effect on the closed-loop performance if it is not considered during the controller design process. In the worst-case scenario, it can even result in instability. For example, time delays can arise from the physical and operational characteristics of a system. Friction-induced delays [18,19] and the nature of certain manufacturing processes such as milling [20] or metal cutting [21] can contribute to these delays. Delays can also be attributed to the mechatronic instrumentation used in real-time experimental systems, where sensors, actuators, and communication networks serve as the primary sources of such delays [22]. Over the years, numerous prominent researchers have addressed this issue by proposing various control solutions. These include integer and fractional-order PID control [23], model predictive control [24], Smith predictor [25,26], communication disturbance observer [27,28], sliding mode control [29], and switching control [30].

From the literature, one can notice that the control of delayed fractional-order monotonic systems has received more investigation and attention than the delayed fractional-order underdamped systems . This is principally due to the complexity of underdamped systems, which exhibit an oscillatory behavior that generally is more complex to analyze and control [31]. The dominance of delayed monotonic systems in applications is another reason for giving them more attention. Those systems exhibit steady-state convergence without oscillations and are more prevalent in many practical applications [32]. In this sense, we study a kind of delayed, underdamped system in this paper.

On the other hand, Ref. [33] proposed a new method for non-integer order integration based on a Multi-layer Perceptron modelization, which allows for an easy circuit design and a wide range of integration orders. Later, Caponetto discussed the design of an auto-tuning fractional-order proportional-integral-derivative (PID) controller and its implementation on a hardware-in-the-loop simulator for real-time control of unknown plants [34]. The proposed procedure is applicable to systems with delay and order greater than one, provided that specifications on cross-over frequency and phase margin are available.

In this study, we focus on the PI^{α} and PD^{μ} controllers with the aim of deriving expressions that simplify their design process for the case of underdamped fractional-order delayed systems with a single oscillation mode. The objective is to design controllers that fulfill either: (a) specific frequency requirements or (b) certain time specifications indirectly by achieving some corresponding frequency specifications. Unlike previous works, we take a different approach by defining the region of attainable frequency specifications rather than the region of stabilizing controllers. Consequently, we obtain sets of reachable frequency specifications instead of sets of controller parameters. We demonstrate that these regions, which are more restrictive than stability regions, not only provide insights into the set of controllers but also offer information on the achievable outcomes with a controller of a given structure. We address these particular structures instead of the general one, which is the $PI^{\alpha}D^{\mu}$ controller structure, because (1) we are interested in achieving only the three specifications of steady-state error, gain crossover frequency and phase margin for the nominal plant, which can be achieved—in principle—by the proposed structures, that have only three parameters to be tuned, (2) the $PI^{\alpha}D^{\mu}$ structure has five parameters to be tuned which makes much more complicated to obtain theorems and analytic functions that approximate the specification regions, such as the ones obtained in this paper, and (3) specification plots and robustness regions that will be expressed in this paper in two or three dimension graphics, cannot be graphically represented in the case of the $PI^{\alpha}D^{\mu}$ structure because they would require plots of four or five dimensions.

As an application example, we focus on the air table FLR (Flexible Link Robot) located at the University of Castilla-La Mancha. In this particular case, we have deliberately introduced an artificial time delay into the system's outer loop. The system is subsequently fine-tuned around a specific operating point by utilizing a delayed fractional underdamped model. Following this adjustment, a comparative analysis is conducted to evaluate the performance of PI^{α} and PD^{μ} controllers in positioning the payload of the robot.

2. Frequency Domain Analysis

In general, the frequency response has to be obtained by the direct evaluation of the irrational-order transfer function of the fractional-order system along the imaginary axis for $s = j\omega$, $\omega \in [0, \infty]$. However, for the commensurate order systems, we can obtain Bode-like plots. In other words, the frequency response can be obtained by the addition of the individual contributions of the terms of order λ resulting from the factorization of the function. For each of these terms, referred to as $(s^{\lambda} + \sigma)^{\pm 1}$, the magnitude curve will have a slope that starts at zero and tends to $\pm\lambda \cdot 20dB/dec$ for higher frequencies, and the phase plot will go from 0 to $\pm\lambda \cdot \pi/2$. If λ were a real number, it would give more degrees of freedom than in the case of standard rational models, in which λ is integer $\lambda = (1, 2, 3, \ldots)$ [35]. In addition, there will be resonances for $\lambda > 1$, which is the case of our considered underdamped system. Figure 1 shows the magnitudes of the frequency responses of fractional-order underdamped, integer-order underdamped, and integer-order undamped systems, whose transfer functions are, respectively:

$$G(s) = \frac{K}{T \cdot s^{\lambda} + 1} \cdot e^{-\tau \cdot s} \tag{1}$$

$$G(s) = \frac{K}{s^2 + 2 \cdot \zeta \cdot s + \omega_0^2} \cdot e^{-\tau \cdot s} \tag{2}$$

$$G(s) = \frac{K}{s^2 + \omega_0^2} \cdot e^{-\tau \cdot s} \tag{3}$$

where

- $K = 1$ is the static gain;
- ω_0 = 1.64 (rad/s) is the fundamental frequency of vibration of the beam;
- $\zeta \in [0, 1]$ is the damping coefficient;
- $T = 0.645$ s is the time constant of the model (1), and $\lambda \in [1\ 2]$ is the order of the fractional order model;
- $\tau = 0.5$ s is the time delay, which is unique for the three models.

In Figure 1, the green plot represents the magnitude of model (1). It is evident that this model exhibits a wider distribution, particularly in the high-frequency domain, compared to the other two integer models.

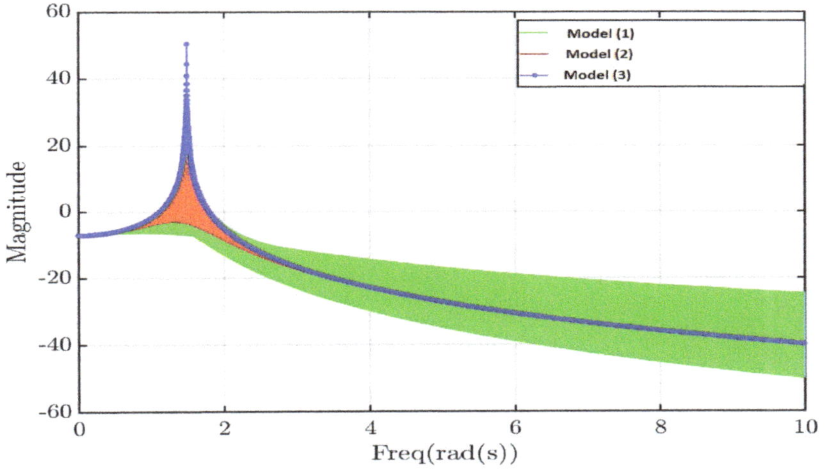

Figure 1. Frequency-responses: Model (1) vs. Model (2) vs. Model (3).

3. Region of Feasible Frequency Specifications: PI^{α} Controller

The proposed method focuses on controlling fractional-order delayed processes of the form (1) such that $(1 < \lambda \leqslant 2)$ using fractional order PI^{α} controller $(0 < \alpha \leqslant 1)$. The stability and the robustness of the controlled system are studied at low and high-frequency domains. Moreover, for values of $\lambda > 1$, the step response of the process is underdamped, and it is stable if $T > 0$.

We consider the negative unity feedback control scheme shown in Figure 2. Our aim is to design a control system for the process (4) that verifies certain typical frequency specifications: (a) desired gain crossover frequency, ω_c, which is related to the desired nominal speed of the closed loop system response, (b) desired phase margin, ϕ_m, which provides the desired damping and robustness to changes in time delay and (c) zero steady-state error to a step command, which means that the controller must include an integral term (of integer or fractional order), according to the Final Value Theorem.

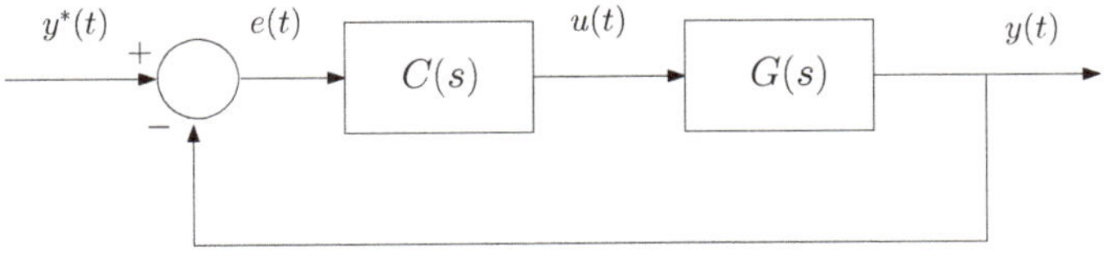

Figure 2. Block diagram of the unity feedback control system.

The aim is to develop a general tuning method based on a set of frequency specifications. In order to obtain general results, we carry out a normalization procedure. The time vector t is scaled by $T^{1/\lambda}$ $(t_n = t/T^{1/\lambda})$ and the process output $y(t)$ is scaled by K $(y_n(t) = y(t)/K)$. This yields the new normalized process:

$$G_n(s) = \frac{1}{s^{\lambda}+1}e^{-\tau_n \cdot s} \tag{4}$$

where τ_n is the normalized time delay $(\tau_n = \tau/T^{\lambda})$. Following the normalization process, it is necessary to normalize the gain crossover frequency by applying the formula $\omega_{cn} = \omega_c T^{1/\lambda}$. The phase margin does not change because of the normalization process.

The new normalized specifications (ω_{cn}, ϕ_m) and the zero steady error to step commands can then be attained by using a normalized controller $C_n(s)$ with a pole at the origin of fractional or integer order and at least two parameters to be tuned. We propose the ideal PI^{α} controller characterized by the following transfer function:

$$C_n(s) = \frac{K_{in}}{s^{\alpha}} + K_{pn} \tag{5}$$

The condition of having given phase margin ϕ_m and gain crossover frequency ω_{cn} can be expressed in a compact form using complex numbers:

$$C_n(j \cdot \omega_{cn}) \cdot G_n(j \cdot \omega_{cn}) = -e^{j \cdot \phi_m} \tag{6}$$

Let us denote $Z_r = \Re[-e^{j \cdot \phi_m} / G_n(j \cdot \omega_{cn})]$ and $Z_i = \Im[-e^{j \cdot \phi_m} / G_n(j \cdot \omega_{cn})]$, where $\Re$ and $\Im$ are, respectively, the real and imaginary parts of a complex number. Then the controller can be tuned using Equation (6):

$$C_n(j \cdot \omega_{cn}) = Z_r + jZ_i \tag{7}$$

Operating $-e^{j \cdot \phi_m} / G_n(j \cdot \omega_{cn})$ yields:

$$Z_r = -cos(\phi_m + \tau_n \cdot \omega_{cn}) \cdot \left(\omega_{cn}^{\lambda} \cdot cos\left(\frac{\lambda \cdot \pi}{2}\right) + 1\right) + \omega_{cn}^{\lambda} \cdot sin(\phi_m + \tau_n \cdot \omega_{cn}) \cdot sin\left(\frac{\lambda \cdot \pi}{2}\right) \tag{8}$$

$$Z_i = -\omega_{cn}^{\lambda} \cdot cos(\phi_m + \tau_n \cdot \omega_{cn}) \cdot sin\left(\frac{\lambda \cdot \pi}{2}\right) - \left(\omega_{cn}^{\lambda} \cdot cos\left(\frac{\lambda \cdot \pi}{2}\right) + 1\right) \cdot sin(\phi_m + \tau_n \cdot \omega_{cn}) \tag{9}$$

The normalized time delay τ_n as it appears in the above expressions is always linked in the same manner to ϕ_m. Let us then define the normalized phase margin:

$$\phi_{mn} = \phi_m + \tau_n \cdot \omega_{cn} \tag{10}$$

Expressions (8) and (9) become:

$$Z_r = -cos(\phi_{mn}) \cdot \left(\omega_{cn}^{\lambda} \cdot cos\left(\frac{\lambda \cdot \pi}{2}\right) + 1\right) + \omega_{cn}^{\lambda} \cdot sin(\phi_{mn}) \cdot sin\left(\frac{\lambda \cdot \pi}{2}\right) \tag{11}$$

$$Z_i = -\omega_{cn}^{\lambda} \cdot cos(\phi_{mn}) \cdot sin\left(\frac{\lambda \cdot \pi}{2}\right) - \left(\omega_{cn}^{\lambda} \cdot cos\left(\frac{\lambda \cdot \pi}{2}\right) + 1\right) \cdot sin(\phi_{mn}) \tag{12}$$

Since the magnitude Bode plot of $G_n(s) \cdot C_n(s)$ is independent of the time delay of the process, the gain crossover frequency is independent of the value of the time delay. However, the time delay shifts the Bode phase plot, which is the reason for the previous phase margin normalization. Expressions (8), (9), (11) and (12) suggest that there are two manners to design the controller $C_n(s)$:

(a): The controller must fulfill specifications (ϕ_m, ω_{cn}) with the process $G_n(s)$ of (4).
(b): The controller must fulfill specifications (ϕ_{mn}, ω_{cn}) with the process $G_n^*(s)$:

$$G_n^*(s) = \frac{1}{s^{\lambda} + 1} \tag{13}$$

For the sake of simplicity, we decided to use procedure (b) considering process (13) and the normalized phase margin defined in (10), and taking into account the real phase margin, the normalized delay, and the normalized gain crossover frequency. Accordingly, the real controller is obtained from the normalized one as

$$C(s) = \frac{1}{K} C_n(s^*)\big|_{s^* = sT^{\frac{1}{\lambda}}} \tag{14}$$

For the sake of simplicity, from now on we will use s instead of s^* in $C_n(s^*)$. Substituting Expressions (11) and (12) in Equation (7), the parameters of the fractional order controller can be determined in function of the system characteristics, the normalized phase margin, and the normalized gain crossover frequency:

$$K_{in} = \frac{\omega_{cn}^{\alpha}}{sin(\alpha \cdot \pi/2)} \cdot \left[\omega_{cn}^{\lambda} \cdot sin\left(\phi_{mn} + \frac{\lambda \cdot \pi}{2}\right) + sin(\phi_{mn})\right] \tag{15}$$

$$K_{pn} = \frac{-1}{sin(\alpha \cdot \pi/2)} \cdot \left[\omega_{cn}^{\lambda} \cdot sin\left(\phi_{mn} + \frac{(\lambda + \alpha) \cdot \pi}{2}\right) + sin(\phi_{mn} + \frac{\alpha \cdot \pi}{2})\right] \tag{16}$$

It is obtained from (10) that $\phi_m = \phi_{mn} - \tau_n \cdot \omega_{cn}$, which shows that the phase margin that can be obtained diminishes as τ_n and ω_{cn} increase. Accordingly, there is a limit on the time delay for each pair of specifications (ω_{cn}, ϕ_{mn}), which is valid for integer order systems too.

$$\tau_{n,max} = \frac{\phi_{mn}}{\omega_{cn}} \tag{17}$$

3.1. Low Frequency Condition

The Nyquist criterion of stability applied to process (13) with $1 < \lambda \leqslant 2$ using controller (5) imposes that the plot must not encircle the limit point of stability $(-1, 0)$. Since $C_n(s) \cdot G_n^*(s)$ exhibits a pole at the origin of order equal to α, this implies that:

$$\lim_{\omega \to 0^+} C_n(j\omega) \cdot G_n^*(j\omega)$$

must be in the third or fourth quadrant of the complex plane. Otherwise, the gain of the integral term of the controller defines an asymptotic behavior of the controller at low frequencies, and it must verify the following condition in order to achieve closed-loop stability:

$$K_{in} > 0 \tag{18}$$

3D low frequency region: The three-dimensional region that defines the low-frequency condition for the normalized fractional-order process (13), controlled by a PI^{α} controller (5), is presented in Figure 3. The x-axis represents the normalized gain crossover frequency ω_{cn}, the y-axis represents the normalized phase margin ϕ_{mn}, and the z-axis represents the order of the process λ. The region does not depend on the fractional order α, which means that all PI^{α} controllers (including the standard PI) share the same low-frequency region for a given process.

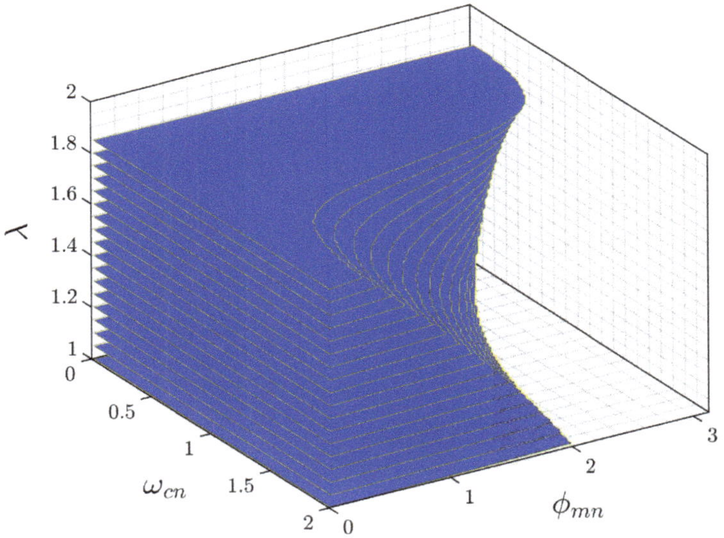

Figure 3. Low frequency condition for $1 < \lambda \leqslant 2$.

The base of the volume corresponds to the combinations of ω_{cn} and ϕ_{mn} that verify the above stability condition at low frequencies, for the lowest order λ (slightly higher than 1). The volume starts decreasing its horizontal section, which reaches its minimum at $\lambda = 2$.

We refer to this region as the "Low-Frequency Condition Region", and it is challenging to find analytical formulae that guarantee this condition for the process (13) combined with controllers (5). In this subsection, we have developed a mathematical model that describes this region for all processes where $1 < \lambda \leqslant 2$.

It is challenging to find analytical formulae that guarantee the feasible frequency conditions for process Equation (13) combined with controllers (5) for all the combinations of fractional order operators α and λ. However, in Theorems 1 and 2, we will develop some useful partial results that fit the curves of the low and robust high-frequency conditions using mathematical models.

Theorem 1. *The region of frequency specifications that verify (18) for a process $G_n^*(s)$ of the form (13) and a controller $C_n(s)$ of the form (5) is given by:*

1. *If $\alpha \in (0,\ 2]$, and $\lambda \in (1,\ 1.3]$*

$$\phi_{mn} < m_1 \cdot e^{m_2 \cdot \omega_{cn}} + m_3 \tag{19}$$

where:

$$m_1 = 0.008 \cdot e^{5.806 \cdot \lambda} + 0.424 \cdot e^{0.598 \cdot \lambda}$$

$$m_2 = 4.4 \cdot \lambda + 0.4$$

$$m_3 = -0.0085 \cdot e^{3.377 \cdot \lambda} + 2.446 \cdot e^{-0.049 \cdot \lambda}$$

2. *If $\alpha \in (0, 2]$, and $\lambda \in (1.3, 2]$,*

$$\phi_{mn} < \frac{\sum_{k=0}^{5} a_k \cdot \omega_{cn}^k}{\omega_{cn}^6 + \sum_{k=0}^{5} b_k \cdot \omega_{cn}^k} \tag{20}$$

$$\begin{cases} a_0 = 0.642 \cdot e^{3.167 \cdot \lambda} - 0.591 \cdot e^{0.262 \cdot \lambda} \\ a_1 = 6.029 \cdot e^{1.649 \cdot \lambda} + 3.738 \cdot e^{-2.01 \cdot \lambda} \\ a_2 = -0.441 \cdot e^{2.9 \cdot \lambda} - 0.176 \cdot e^{1.375 \cdot \lambda} \\ a_3 = 0.004 \cdot e^{0.722 \cdot \lambda} - 1.189 \cdot e^{1.11 \cdot \lambda} \\ a_4 = 1.889 \cdot e^{4.177 \cdot \lambda} + 2.203 \cdot e^{0.096 \cdot \lambda} \\ a_5 = 9.266 \cdot e^{0.133 \cdot \lambda} - 3.118 \cdot e^{1.205 \cdot \lambda} \end{cases}$$

$$\begin{cases} b_0 = -6.79 \cdot e^{-1.646 \cdot \lambda} + 0.001 \cdot e^{9.060 \cdot \lambda} \\ b_1 = 0.316 \cdot e^{4.075 \cdot \lambda} + 1.142 \cdot e^{-6.485 \cdot \lambda} \\ b_2 = -2.48 \cdot e^{-1.074 \cdot \lambda} - 3.571 \cdot e^{-0.55 \cdot \lambda} \\ b_3 = -1.69 \cdot e^{1.168 \cdot \lambda} + 3.749 \cdot e^{1.876 \cdot \lambda} \\ b_4 = -0.74 \cdot e^{-0.161 \cdot \lambda} + 0.701 \cdot e^{4.075 \cdot \lambda} \\ b_5 = 4.397 \cdot e^{0.018 \cdot \lambda} + 0.401 \cdot e^{1.754 \cdot \lambda} \end{cases}$$

Figure 4 represents the adjusted mathematical model of the Low-frequency condition for several values of λ.

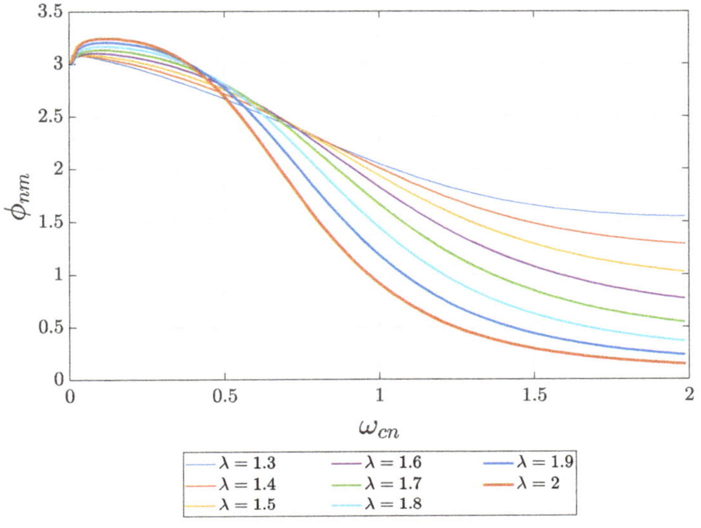

Figure 4. Low frequency region for $1 < \lambda \leqslant 2$.

3.2. Robust High-Frequency Condition

3.2.1. Motivation

Figure 5 presents typical Nyquist plots of two systems: $L_{1n}(j \cdot \omega) = C_{1n}(j \cdot \omega) \cdot G_n(j \cdot \omega)$ and $L_{2n}(j \cdot \omega) = C_{2n}(j \cdot \omega) \cdot G_n(j \cdot \omega)$. Both systems satisfy the same phase margin and gain crossover frequency conditions. Moreover, $L_{1n}(j \cdot \omega)$ satisfies the condition

$$|L_{1n}(j \cdot \omega)| < 1, \; \forall \omega > \omega_{cn} \tag{21}$$

proposed in [17] to guarantee the fulfillment of the previous two frequency specifications. However, $L_{2n}(j \cdot \omega)$ verifies the condition:

$$\frac{d|L_{2n}(j \cdot \omega)|}{d\omega} < 0, \; \forall \omega > \omega_{cn} \tag{22}$$

This last condition, referred to as the '**Robust High-Frequency Condition:** $RHFC$' not only guarantees the fulfillment of the two frequency specifications but is also more restrictive than the first one.

The plots of Figure 5 demonstrate that the stability of the two systems can be determined by the common phase margin ϕ_{mn}, which is obtained by evaluating $\pi + \angle L_n(j\omega)$ at the point on the Nyquist plot where $|L_n(j\omega)| = 1$ with the highest possible value of ω (referred to as ω_{cn}). A change in the loop gain results in a scaling of the Nyquist plot, which alters the phase margin of both systems. This behavior is illustrated in Figure 6a,b. In the first case in which Condition (21) is verified, the phase margin of $L_{1n}(j\omega)$ exhibits a significant and sudden reduction due to a small variation in the loop gain. Consequently, there is a substantial increase in the gain crossover frequency. The new phase margin, measured at ω^*_{cn1} (where $\omega^*_{cn1} > \omega_{cn1}$), corresponds to a negative value. This sudden instability corresponds to a sharp increase in the gain crossover frequency of the system. However, this phenomenon does not occur in the case in which Condition (22) is verified because the magnitude of $L_{2n}(j\omega)$ strictly decreases with ω.

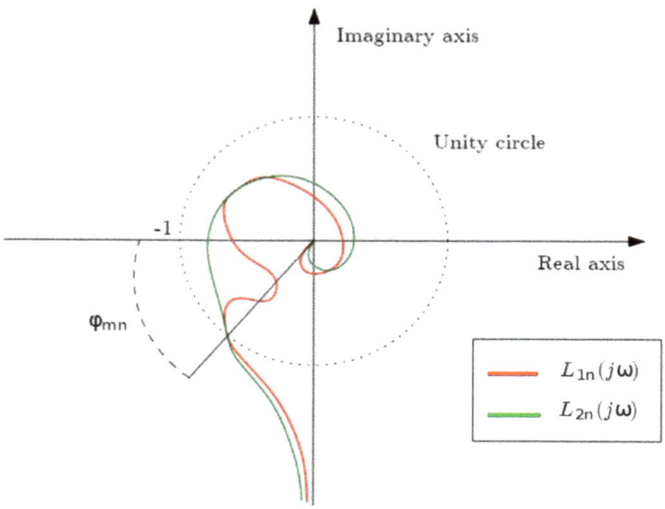

Figure 5. Nyquist-diagrams: $L_{1n}(j\omega)$ vs. $L_{2n}(j\omega)$.

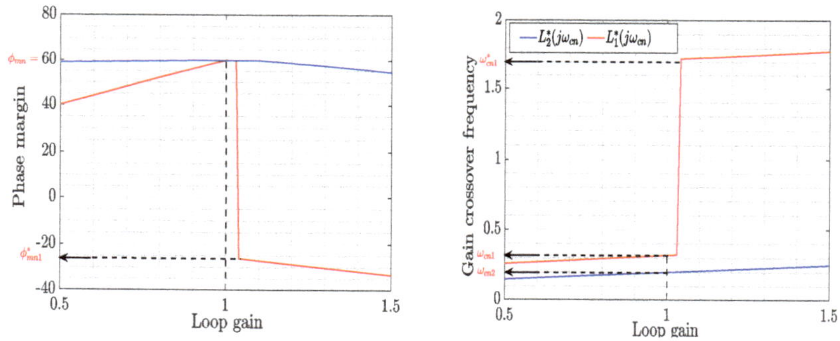

(**a**) Phase margin vs. loop gain changes (**b**) Gain crossover frequency vs. loop gain changes

Figure 6. Problem-illustration.

3.2.2. Analytical Global Model

Theorem 2. *The region of frequency specifications that verify the defined RHFC condition for a process $G_n^*(s)$ of the form (13) and a controller $C_n(s)$ of the form (5) for $\alpha \in (0\ 1]$, $\lambda \in (1\ 2]$ is given by:*

1. *If $\alpha \in (0,\ 0.5]$, and $\lambda \in (1,\ 1.5]$*

$$\phi_{mn} < \Sigma_1 \cdot e^{\Sigma_2 \cdot \omega_{cn}} + \Sigma_3 \cdot e^{\Sigma_4 \cdot \omega_{cn}} \tag{23}$$

where:

$$\Sigma_1 = 0.322 \cdot e^{f_1(\lambda)\cdot\alpha}$$

$$f_1(\lambda) = -14.25 \cdot e^{\frac{0.0007-\lambda}{0.0684}} - 239.63 \cdot e^{\frac{\lambda-0.0924}{0.135}} - 1.047 \cdot e^{\frac{11.22-\lambda}{2.415}}$$

$$\Sigma_2 = 347.18 \cdot e^{-9.86\cdot\lambda} - 2.14 \cdot e^{0.54\cdot\lambda} \cdot e^{\frac{\alpha-0.268\cdot e^{2.441\cdot\lambda}}{7.04\cdot e^{0.039\cdot\lambda}}}$$

$$\Sigma_3 = 2.81 \cdot e^{-0.91\cdot\lambda} + 0.24 \cdot e^{1.71\cdot\lambda} \cdot e^{\frac{\alpha-0.0185\cdot e^{-0.51\cdot\lambda}}{0.54\cdot e^{1.083\cdot\lambda}}}$$

$$\Sigma_4 = -1.93 \cdot e^{\left(\frac{0.87-\lambda}{0.49}\right)^2} \cdot \left(e^{\left(-1.75e^3\cdot\alpha e^{(-16.77\cdot\lambda)^2}\right)} + e^{\left(19.9\cdot\alpha e^{(-2.6\cdot\lambda)^2}\right)}\right)$$

2. *If* $\alpha \in (0.5,\ 1]$, *and* $\lambda \in (1,\ 1.5]$

$$\phi_1 < \phi_{mn} < \phi_2 \tag{24}$$

$$\phi_1 = \sum_{i=1}^{4}\sum_{j=1}^{6} p_{ij} \cdot \alpha^{j-1} \cdot \omega_{cn}^{i-1} \tag{25}$$

$$\begin{bmatrix} p_{11} \\ p_{12} \\ p_{13} \\ p_{14} \\ p_{15} \\ p_{16} \end{bmatrix} = \begin{bmatrix} -0.945 & 1.472 & -1.127 & -4.056 \\ 4.518 & -1.014 & -193.87 & 175.004 \\ 9.617 & 0.005 & -0.015 & 0.0172 \\ -0.785 & 1.105 & -0.843 & 1.462 \\ -4.436 & 0.739 & 80.818 & -122.67 \\ 4.858 & -0.038 & 0.031 & -0.063 \end{bmatrix} \cdot \begin{bmatrix} \lambda^3 \\ \lambda^2 \\ \lambda \\ 1 \end{bmatrix}$$

$$\begin{bmatrix} p_{21} \\ p_{22} \\ p_{23} \\ p_{24} \\ p_{25} \\ p_{26} \end{bmatrix} = \begin{bmatrix} -1.057 & 0.509 & 0.046 & 1.58 \\ -12.72 & 0.725 & 382.07 & -234.49 \\ -20.42 & -0.008 & -0.030 & -0.030 \\ 0.021 & 0.333 & -0.157 & -0.206 \\ 1.267 & 0.159 & -26.36 & 13.518 \\ -5.97 & 0.008 & -0.069 & 0.035 \end{bmatrix} \cdot \begin{bmatrix} \lambda^3 \\ \lambda^2 \\ \lambda \\ 1 \end{bmatrix}$$

$$\begin{bmatrix} p_{31} \\ p_{32} \\ p_{33} \\ p_{34} \\ p_{35} \\ p_{36} \end{bmatrix} = \begin{bmatrix} -0.333 & 0.293 & -0.181 & 0.4174 \\ 2.2071 & -1.7171 & -240.8178 & 127.0347 \\ 3.3305 & -0.010 & 0.003 & 0.007 \\ 0.022 & -0.378 & -0.044 & 1.126 \\ -8.485 & -0.576 & 59.70 & -23.38 \\ 2.080 & 0.004 & -0.040 & -0.021 \end{bmatrix} \cdot \begin{bmatrix} \lambda^3 \\ \lambda^2 \\ \lambda \\ 1 \end{bmatrix}$$

$$\begin{bmatrix} p_{41} \\ p_{42} \\ p_{43} \\ p_{44} \\ p_{45} \\ p_{46} \end{bmatrix} = \begin{bmatrix} 0.008 & -0.102 & 0.047 & 0.489 \\ -3.038 & 11.785 & -19.24 & 8.826 \\ 1.056 & 0.009 & -0.020 & 0.005 \\ 0.022 & 0.109 & -0.112 & -0.483 \\ 2.914 & -1.439 & -0.455 & 4.388 \\ -2.59 & 0.003 & -0.002 & -0.005 \end{bmatrix} \cdot \begin{bmatrix} \lambda^3 \\ \lambda^2 \\ \lambda \\ 1 \end{bmatrix}$$

$$\begin{aligned} \phi_2 = {} & M_2(1,1) \cdot \alpha^2 \cdot e^{M_2(2,1)\cdot\alpha^2\cdot\omega_{cn}} + M_2(3,1) \cdot \alpha^2 \cdot e^{M_2(4,1)\cdot\alpha^2\cdot\omega_{cn}} \\ & + M_2(1,2) \cdot \alpha \cdot e^{M_2(2,2)\cdot\alpha\cdot\omega_{cn}} + M_2(3,2) \cdot \alpha \cdot e^{M_2(4,2)\cdot\alpha\cdot\omega_{cn}} \\ & + M_2(1,3) \cdot e^{M_2(2,3)\cdot\omega_{cn}} + M_2(3,3) \cdot e^{M_2(4,3)\cdot\omega_{cn}} \end{aligned} \tag{26}$$

$$M_2 = \begin{bmatrix} 573.7 \cdot e^{478.17\cdot\lambda} & -712.7 \cdot e^{541.24\cdot\lambda} & 142.9 \cdot e^{509.11\cdot\lambda} \\ -341.7 \cdot e^{-588.18\cdot\lambda} & 1.82 \cdot e^{541.24\cdot\lambda} & 0.02 \cdot e^{-31.58\cdot\lambda} \\ -148.6 \cdot e^{-333.25\cdot\lambda} & -249.4 \cdot e^{563.71\cdot\lambda} & -565.6 \cdot e^{307\cdot\lambda} \\ -3.4 \cdot e^{-168.96\cdot\lambda} & 110.2 \cdot e^{-354.09\cdot\lambda} & -2.6 \cdot e^{-27.7\cdot\lambda} \end{bmatrix}$$

3. *If* $\alpha \in (0,\ 1]$, *and* $\lambda \in (1.5,\ 2]$

$$\omega_{cn} > \sum_{i=0}^{3} r_i \cdot \phi_{mn}^{i} \tag{27}$$

where:

$$\begin{bmatrix} r_0 \\ r_1 \\ r_2 \\ r_3 \end{bmatrix} = R \cdot \begin{bmatrix} \alpha^3 \\ \alpha^2 \\ \alpha \\ 1 \end{bmatrix} \tag{28}$$

and

$$R = R_1 \cdot \lambda^2 + R_2 \cdot \lambda + R_3$$

$$R_1 = \begin{bmatrix} 8.59 & 4.45 & -10.17 & 2.85 \\ -34.04 & 23.91 & 4.276 & -2.76 \\ -1.47 & 15.37 & -17.77 & 2.63 \\ 7.325 & -13.68 & 6.55 & -1.50 \end{bmatrix}$$

$$R_2 = \begin{bmatrix} -26.37 & -18.47 & 35.97 & -9.763 \\ 107.3 & -67.75 & -21.68 & 10.13 \\ 6.69 & -56.36 & 63.12 & -9.06 \\ -23.85 & 45.17 & -21.43 & 4.77 \end{bmatrix}$$

$$R_3 = \begin{bmatrix} 20.27 & 18.32 & -31.59 & 8.337 \\ -84.33 & 46.81 & 23.95 & -9.097 \\ -7.06 & 51.05 & -55.59 & 7.69 \\ 19.4 & -37.26 & 17.43 & -2.68 \end{bmatrix}$$

Figure 7 shows the curves that represent the borders of the $RHFC$ of the normalized process (13) controlled by the fractional order controller (5). The mentioned condition depends on the variation of the four variables $(\lambda, \alpha, \omega_{cn}, \phi_{mn})$. Figure 7 plots the before mentioned curves for different values of α in the interval $[0,\ 0.5]$ and for values of $\lambda = 1.1$ and $\lambda = 1.3$. These curves with the before defined low-frequency condition zoomed around the interference zone with the $RHFC$ define the feasible frequency specification regions for the considered PI^{α} controller.

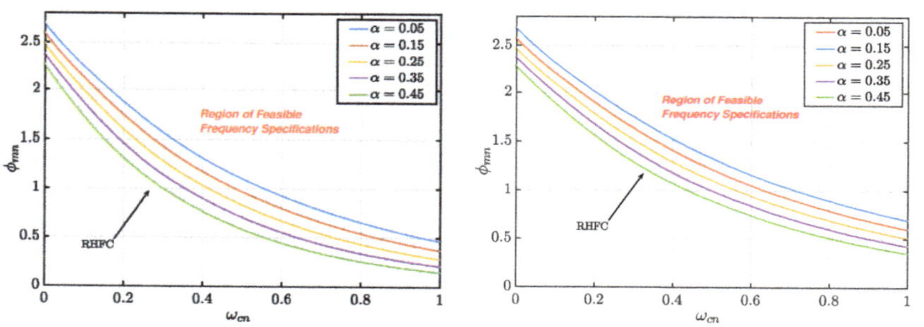

(a) $\lambda = 1.1$ **(b)** $\lambda = 1.3$

Figure 7. $RHFC$: $\alpha \in [0,\ 0.5]$ and $\lambda \in [1,\ 1.5]$.

Figure 8 shows the $RHFC$ of the PI^{α} controller as a function of α and λ, where $\alpha \in [0.5, 1]$ and $\lambda = 1.1$ and $\lambda = 1.3$. For $\alpha \in [0.5, 1]$, it can be observed that the $RHFC$ modifies the curves corresponding to the low-frequency condition (the upper curves of the subplots in Figure 8 zoomed in the interval $[0\ 1]$). However, this is not the case for $\alpha \in [0, 0.5]$, where the two conditions do not interfere.

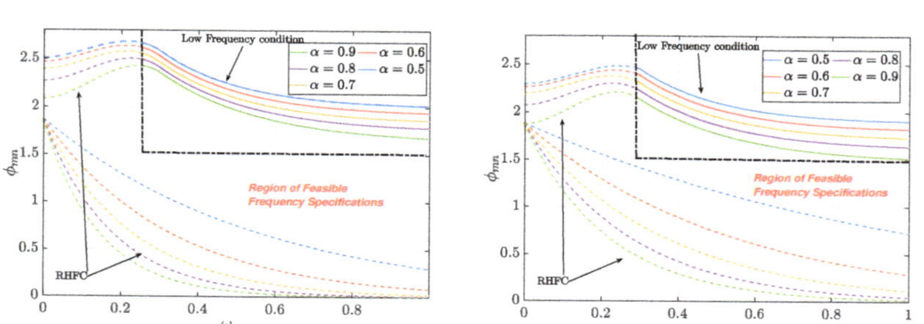

(a) $\lambda = 1.1$ **(b)** $\lambda = 1.3$

Figure 8. $RHFC$: $\alpha \in [0.5,\ 1]$ and $\lambda \in [1,\ 1.5]$.

Figure 9 shows the $RHFC$ of the PI^{α} controller in function of α and λ for $\alpha \in [0.5,\ 1]$ and $\lambda \in [1.5\ 2]$. Figures 7–9 demonstrate that the low-frequency condition becomes more and more restrictive for higher values of λ. Accordingly, the $RHFC$ region decreases for $\lambda > 1.5$ to reach its minimum at $\lambda = 2$.

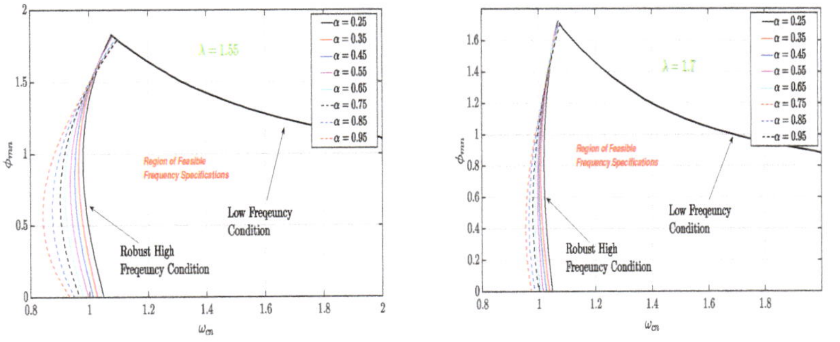

(a) $\lambda = 1.55$ **(b)** $\lambda = 1.7$

Figure 9. $RHFC$: $\alpha \in [0,\ 1]$ and $\lambda \in [1.5,\ 2]$.

4. Fractional Order PD^{μ} Controller

In this section, we will examine the underdamped system represented by the transfer Function (1) and the fractional-order proportional-derivative controller

$$C_{PD}(s) = K_p + K_d \cdot s^{\mu},\ 0 < \mu < 2 \tag{29}$$

PD controllers are often used in mechatronics. In this sense, we have an interest in analyzing the fractional version of such a controller and calculating its feasible frequency specifications. In this sense, the condition of having a desired phase margin ϕ_m and gaining a crossover frequency ω_c is expressed by

$$C_{PD}(j \cdot \omega_c) \cdot G(j \cdot \omega_c) = -e^{j \cdot \phi_m} \tag{30}$$

From Equation (30), the normalized parameters of the PD^{μ} controller are expressed as a function of the normalized specifications ϕ_{mn} and ω_{cn} by:

$$K_{dn} = \frac{Z_i}{\omega_{cn}^{\mu} \cdot sin\left(\frac{\mu \cdot \pi}{2}\right)} \tag{31}$$

$$K_{pn} = Z_r - K_d \cdot \omega_{cn}^{\mu} \cdot cos\left(\frac{\mu \cdot \pi}{2}\right) \tag{32}$$

where:

$$Z_i = \Im\left(-e^{j \cdot \phi_{mn}} / G_n(j \cdot \omega_{cn}))\right) \tag{33}$$

$$Z_r = \Re\left(-e^{j \cdot \phi_{mn}} / G_n(j \cdot \omega_{cn}))\right) \tag{34}$$

In the case of the PD^{μ} control of a fractional-order process of the form (13), the resulting open-loop system does not have a pole at the origin. As a result, there is no stability constraint to adhere to the low-frequency behavior.

The phase margin is determined at the highest frequency where the magnitude of the open-loop system $C_{PD_n}(j \cdot \omega) \cdot G_n^{*}(j \cdot \omega)$ is equal to unity. Here, $C_{PD_n}(j \cdot \omega)$ represents the frequency response of the normalized PD^{μ} controller.

4.1. Robust High-Frequency Condition

The robust high-frequency condition stipulates that the magnitude of the frequency response should strictly decrease within the unit circle. This condition can be expressed as:

$$\left.\frac{d|C_{PD_n}(j\cdot\omega)\cdot G_n^*(j\cdot\omega)|}{d\omega}\right|_{\omega>\omega_{cn}} < 0 \quad (35)$$

It is a challenging task to find analytical formulae that guarantee the fulfillment of this condition for the normalized process (13) combined with the normalized PD^{μ} controller across all combinations of fractional-order operators μ and λ. Therefore, Figure 10 illustrates the three-dimensional region that defines the $RHFC$ in terms of the normalized specifications ω_{cn} and ϕ_{mn} and the fractional order of the process λ. Moreover, these are the points that guarantee the $RHFC$ for any value $\mu \in (0, 2]$.

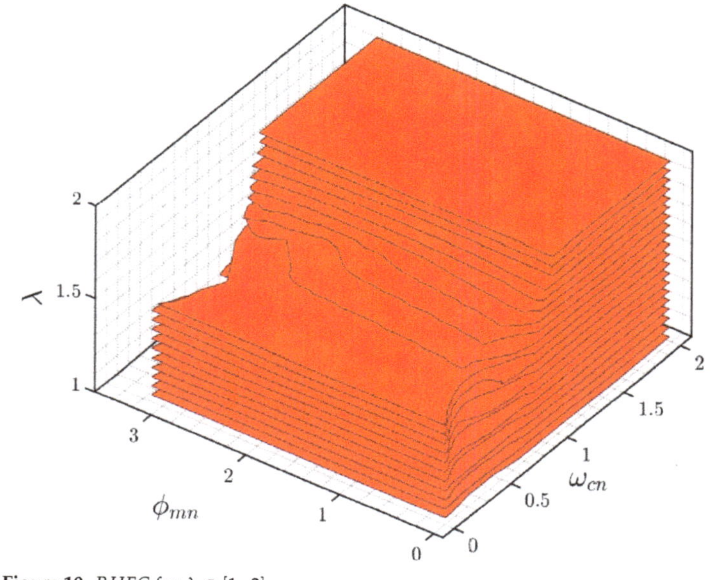

Figure 10. $RHFC$ for $\lambda \in [1, 2]$.

Since the PD^{μ} controller does not have a low-frequency condition, the red surfaces in Figure 10 correspond to the $RHFC$. Unlike the PI^{α} controller, this figure displays broad zones, even for process orders close to 2.

4.2. Modified PD^{μ} Controller

When analyzing the feasible frequency conditions obtained for each of the two controllers PI^{α} and PD^{μ} for the case of a delayed fractional order system with an underdamped step response, it is evident that:

For the case of PI^{α}:

1. It eliminates the steady-state error with the integral action.
2. It presents two drawbacks: (a) very reduced regions for λ close to 2, which clearly limits the design operation, (b) the relative stability is reduced.

For the case of PD^{μ}:

1. The regions of feasible frequency specifications are very extended compared to the case of a PI^{α}.
2. It increases relative stability.

3. The main drawbacks are: (a) it increases the sensitivity to noise and (b) it cannot eliminate the steady-state error.

Therefore, in this section, we propose a new alternative that combines the advantages of both controllers. We modify the standard control scheme by incorporating feedback from the system's output in such a way that we impose a fictitious pole at the origin of our fractional order model. This approach allows us to eliminate the static error. Subsequently, we design the PD^{μ} controller while ensuring that it meets the desired specifications.

Let us consider the positive feedback loop of Figure 11 applied to the process (1). Its closed-loop transfer function is

$$F_{cl}(s) = \frac{G(s)}{1-\delta \cdot G(s)} = \frac{K \cdot e^{-\tau \cdot s}}{1+T \cdot s^{\lambda} - \delta \cdot K \cdot e^{-\tau \cdot s}} \tag{36}$$

If we impose $\delta = \frac{1}{K}$ in the previous equation we obtain:

$$F_{cl}(s) = \frac{K \cdot e^{-\tau \cdot s}}{T \cdot s^{\lambda} + (1-e^{-\tau \cdot s})} \tag{37}$$

In addition, using the Taylor series expansion, the term $(1-e^{-\tau s})$ can be approximated as $\tau s - \frac{\tau^2}{2}s^2 + \ldots$ Therefore, considering that $\lambda > 1$, the resulting model features a pole at the origin of order 1 which enables us to eliminate the static error.

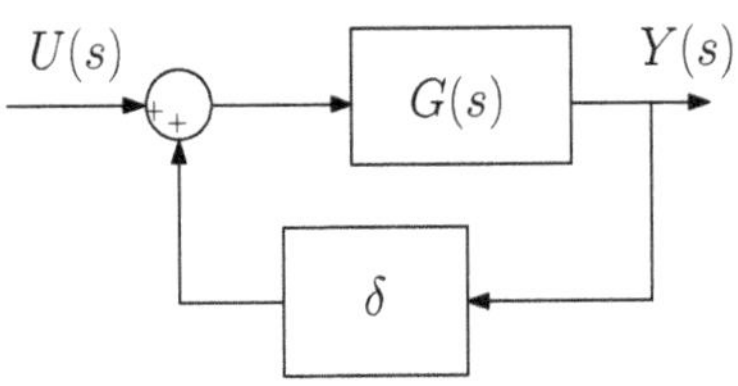

Figure 11. Modified dynamic system.

The resulting closed-loop scheme is shown in Figure 12, in which disturbances z_1 and z_2 have been included at the input of the process and the output measurement, respectively. The closed-loop transfer function assuming no disturbances, is:

$$Y(s) = \frac{K \cdot e^{-\tau \cdot s} \cdot C(s)}{1 - K \cdot \delta \cdot e^{-\tau \cdot s} + K \cdot e^{-\tau \cdot s} \cdot C(s) + T \cdot s^{\lambda}} \cdot Y^{*}(s) \tag{38}$$

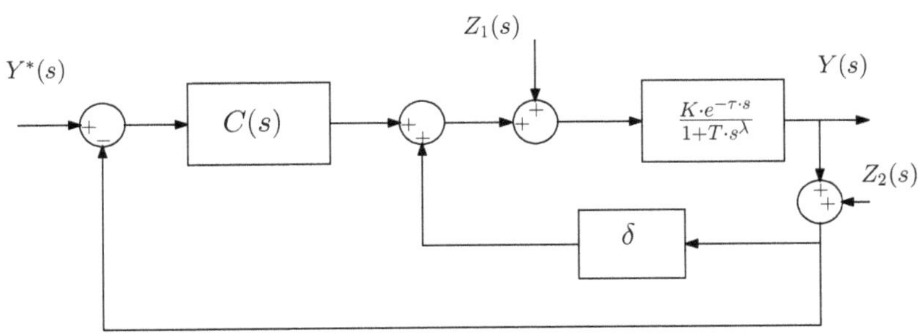

Figure 12. Closed-loop control scheme considering the modified dynamic system.

If $K \cdot \delta = 1$, Expression (38) becomes

$$Y(s) = \frac{K \cdot e^{-\tau \cdot s} \cdot C(s)}{1 - e^{-\tau \cdot s} + K \cdot e^{-\tau \cdot s} \cdot C(s) + T \cdot s^{\lambda}} \cdot Y^{*}(s) \tag{39}$$

It is easy to verify, by applying the Final Value Theorem, that the response of this system to a step command $Y^*(s)$ has zero steady-state error.

Let us study the characteristic equation of the transfer Function (39). It is expressed by

$$1 - e^{-\tau \cdot s} + Ke^{-\tau \cdot s}C(s) + T \cdot s^{\lambda} = 0 \Rightarrow (1 + T \cdot s^{\lambda}) + Ke^{-\tau \cdot s}\left(C(s) - \tfrac{1}{K}\right) = 0$$
$$\Rightarrow 1 + \tfrac{Ke^{-\tau \cdot s}}{1+T \cdot s^{\lambda}}(C(s) - \delta) = 0 \tag{40}$$

If we denote $\hat{C}(s) = C(s) - \delta$, this equation becomes $1 + G(s) \cdot \hat{C}(s) = 0$. Then the closed-loop specifications of this modified control system can be achieved using the *RHFC* region of Figure 10 to design controller $\hat{C}(s) = K_p + K_d \cdot s^{\mu}$. After, the PD^{μ} controller $C(s)$ of Figure 12 is obtained from

$$C(s) = \hat{C}(s) + \delta = K_p + \delta + K_d \cdot s^{\mu} \tag{41}$$

4.3. Disturbance Rejection Analysis

Consider Figure 12 and step disturbances in $z_1(t)$ and $z_2(t)$.

4.3.1. Case 1: Variable Step Disturbance at the Input of the Process

In this case $Z_1(s) = \frac{a}{s}$, where a is the amplitude of the step, and $Z_2(s) = 0$. Simplifying the scheme of Figure 12, $Y(s)$ is expressed as a function of the reference $Y^*(s)$ and the disturbance $Z_1(s)$:

$$Y(s) = \frac{K \cdot e^{-\tau \cdot s} \cdot Z_1(s) + K \cdot e^{-\tau \cdot s} \cdot C(s) \cdot Y^*(s)}{1 - K \cdot \delta \cdot e^{-\tau \cdot s} + K \cdot e^{-\tau \cdot s} \cdot C(s) + T \cdot s^{\lambda}} \tag{42}$$

If $K \cdot \delta = 1$ then

$$Y(s) = \frac{K \cdot e^{-\tau \cdot s} \cdot Z_1(s) + K \cdot e^{-\tau \cdot s} \cdot C(s) \cdot Y^*(s)}{1 - e^{-\tau \cdot s} + K \cdot e^{-\tau \cdot s} \cdot C(s) + T \cdot s^{\lambda}} \tag{43}$$

which has zero steady-state error to a step command $Y^*(s)$ and a steady state output $y_d = a/C(0) = a/(K_p + \delta)$ to the step disturbance $Z_1(s)$. Then the PD^{μ} cannot completely remove the steady-state error caused by disturbance $Z_1(s)$.

4.3.2. Case 2: Variable Step Disturbance at the Measurement of the Output of the Process

In this case $Z_1(s) = 0$ and $Z_2(s) = \frac{a}{s}$ where a is the amplitude of the step. Simplifying the scheme of Figure 12, $Y(s)$ is expressed as a function of the reference $Y^*(s)$ and the disturbance $Z_2(s)$:

$$Y(s) = \frac{-K \cdot e^{-\tau \cdot s} \cdot (C(s) - \delta) \cdot Z_2(s) + K \cdot e^{-\tau \cdot s} \cdot C(s) \cdot Y^*(s)}{1 - K \cdot \delta \cdot e^{-\tau \cdot s} + K \cdot e^{-\tau \cdot s} \cdot C(s) + T \cdot s^{\lambda}} \tag{44}$$

If $K \cdot \delta = 1$ then

$$Y(s) = \frac{-K \cdot e^{-\tau \cdot s} \cdot \left(K_p + K_d \cdot s^{\mu}\right) \cdot Z_2(s) + K \cdot e^{-\tau \cdot s} \cdot C(s) \cdot Y^*(s)}{1 - e^{-\tau \cdot s} + K \cdot e^{-\tau \cdot s} \cdot C(s) + T \cdot s^{\lambda}} \tag{45}$$

which has zero steady-state error to a step command $Y^*(s)$ and a steady state output $y_d = -a \cdot (C(0) - \delta)/C(0) = -a \cdot K_p/(K_p + \delta)$ to the step disturbance $Z_2(s)$. Then the PD^{μ} can completely remove the steady state error caused by disturbance $Z_2(s)$ if $K_p = 0$, i.e., if $C(s) = \delta + K_d \cdot s^{\mu}$.

5. Application to the Control of Flexible Manipulator

5.1. Presentation of the Platform

Figure 13 provides a detailed depiction of the experimental platform, including all its components. The platform comprises a three-legged metallic structure that supports a Harmonic

Drive mini servo DC motor, specifically the model $RH-8D-6006-E036AL-SP(N)$. This motor features a reduction ratio of $n = 25$.

This frame enables stable and unrestricted rotation of the motor in the horizontal plane around the vertical axis of the platform. The motor is characterized by the following parameter values: inertia $J = 6.87 \times 10^{-5}$ kg $\cdot$ m^2, viscous friction $\nu = 1.041 \times 10^{-3}$ N $\cdot$ m $\cdot$ s, and electromechanical constant $k = 0.21$ N $\cdot$ m/V.

The servo-amplifier is designed to accept control inputs from the computer within the range of $[-10, 10]$ volts. The flexible link is connected to the motor, and a load (disc) is positioned above the surface of an air table in a manner that nullifies the effects of friction and gravity. The disc is attached at its center to the end of the beam with a freely pivoted pin joint. The flexible link possesses the following characteristics: length $L_b = 0.5$ m, cross-sectional diameter $d = 3 \times 10^{-3}$ m, and flexural rigidity $EI = 0.260$ N$\cdot$m^2.

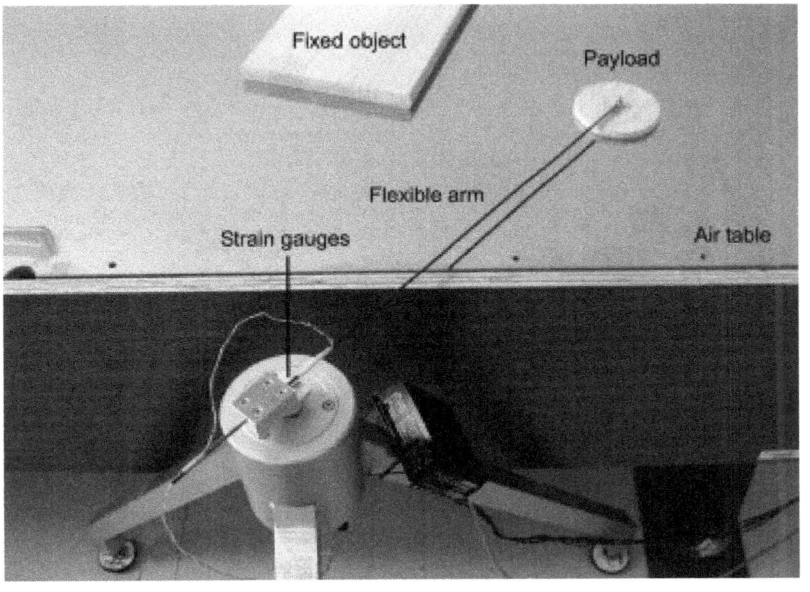

Figure 13. Experimental platform: flexible manipulator robot.

The nominal tip load is a wood disc of mass $m_0 = 0.09$ kg. Because the mass of the link is small compared to that of the disc and because the pinned joint prevents the generation of torques at the end of the beam, this mechanical system has a single vibration mode. Then the nominal vibration angular velocity of the link is characterized as $\omega_0 = 5.8434$ rad/s ($f_0 = 0.93$ Hz).

The sensor system consists of an encoder embedded in the motor—which allows us to know the motor angular position $\theta_m(t)$ with a precision of $7 \cdot 10^{-5}$ rad—and a pair of strain gauges with gauge factor 2.16 and resistance 120.2 Ω that give the moment at the base of the link $\Gamma(t)$, i.e., the coupling torque between the motor and the link. The sampling time in the signals processing is $T_s = 2$ ms.

5.2. The Dynamic Model

The dynamics of the described system, composed by the attachment of the flexible beam and the payload disc, was studied and experimentally identified in [36] yielding an underdamped fractional model of order $\lambda = 1.92$:

$$G_b(s) = \frac{\theta_d(s)}{\theta_m(s)} = \frac{\omega_0^2}{s^\lambda + \omega_0^2} \tag{46}$$

and

$$G_{\Gamma}(s) = \frac{\Gamma(s)}{\theta_m(s)} = \frac{c \cdot s^{\lambda}}{s^{\lambda} + \omega_0^2} e^{-\tau \cdot s} \tag{47}$$

where θ_d is the robot tip (disc) angle and c is an identified coefficient that characterizes the stiffness of the beam and the calibration of the strain gauge sensor.

5.3. Control Design and Analysis of the Results

The disc angle $\theta_d(t)$ can be estimated from the measurements using expression [37]:

$$\theta_d(t) = \theta_m(t) - \frac{\Gamma(t)}{c} \tag{48}$$

Then, a two-nested control loop scheme can be implemented. The inner loop controls the motor angular position feeding back θ_m measurements, and the outer loop controls the tip angular position by feeding back θ_d estimated by (48).

In the inner loop, the motor angle θ_m is corrected using a robust 2 DOF PID controller designed based on a pole placement technique and including compensation terms of the Coulomb friction and the motor-beam coupling torque. Besides removing disturbance effects, this inner loop is designed to be as fast as possible without saturating the actuator. Then a feedback loop with high-gain controllers is implemented. Details of this servo motor system can be found in [38]. The equivalent transfer function of the inner loop is

$$M(s) = \frac{\theta_m(s)}{\theta_m^*(s)} = \frac{1}{(1 + \epsilon \cdot s)^2} \tag{49}$$

where θ_m^* is the reference to be followed by the output θ_m of the inner loop and ϵ is made as small as possible.

The inner loop is implemented "in situ" while the outer control loop is assumed to be implemented in a remote station. Then the time delay is implemented between the output of the outer loop controller $u(t)$ and the input of the inner loop control system θ_m^*. Consequently, the complete dynamic model of the robot, including the inner loop dynamics and the time delay, is

$$G(s) = \frac{\theta_d(s)}{u(s)} = \frac{\omega_0^2}{s^{\lambda} + \omega_0^2} \cdot M(s) \cdot e^{-\tau \cdot s} \tag{50}$$

We have artificially implemented a time delay in the output of the outer loop controller in order to simulate a teleoperated robotic system. Previous works on the teleoperation of flexible manipulators report time delays between one-tenth and the value of the period of the main vibration mode (the one with the lowest frequency of vibration) [39]. This period in the case of the nominal disc is 1.0753 s. We have therefore implemented a time delay of $\tau = 0.1613$ s, which is 15% of the period of the main vibration mode of the flexible link.

Moreover, the inner loop has been designed such that $\epsilon = 0.0125$, i.e., the double pole of $M(s)$, is located at -80. Then the absolute value of this pole is more than ten times bigger than the vibration angular velocity of the beam, allowing us to assume that the inner loop dynamics are much faster than the dynamics involved in the outer loop. We can therefore assume that $M(s) \approx 1$ and that

$$G(s) \approx \frac{\omega_0^2}{s^{\lambda} + \omega_0^2} \cdot e^{-\tau \cdot s} = \frac{1}{1 + \frac{1}{\omega_0^2} \cdot s^{\lambda}} \cdot e^{-\tau \cdot s} \tag{51}$$

where the right member of the equation expresses $G(s)$ in the standarized form (1).

Figure 14a shows the detailed control scheme using a PI^{α} controller in the outer loop. It can be observed the place where the artificial time delay has been implemented: the reference signal of the inner loop is the delayed output of the fractional-order controller used in the outer loop. Furthermore, a disturbance step signal d is incorporated, representing the offset of the strain gauge sensor. Figure 14b shows a simplified control scheme that is expressed in terms of the equivalent transfer functions of the subsystems involved in Figure 14a. Note that $\hat{d} = d/c$ is introduced in Figure 14b.

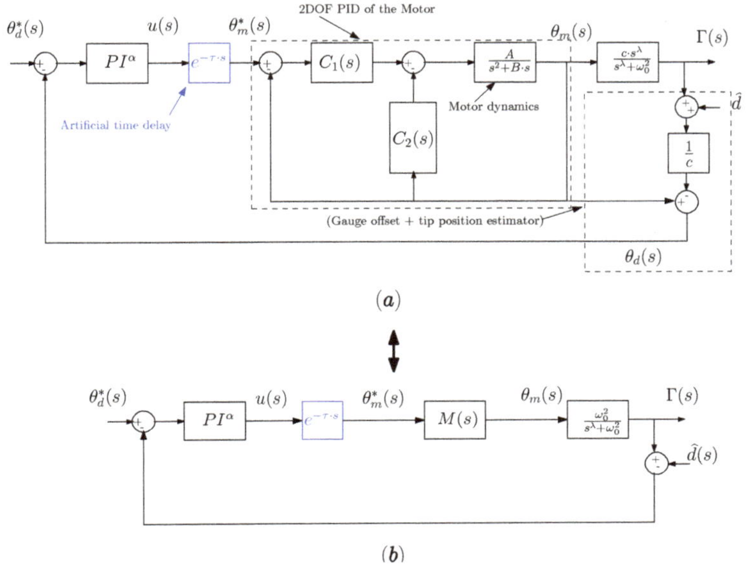

Figure 14. PI^{α} based control scheme. ((**a**): General control scheme with the inner loop of the motor. (**b**): Simplified control scheme).

Figure 15 depicts the control scheme using a PD^{μ} controller in the outer loop. It is also a simplified scheme version expressed in terms of the equivalent transfer functions of the involved subsystems. In this figure, we have implemented the modified dynamic system by providing feedback on the disc angular position with a positive sign which, based on the analysis conducted in the preceding section, incorporates a fictitious pole at the origin. In our case, taking into account the right-hand side of Expression (51), we obtain that $\delta = 1$.

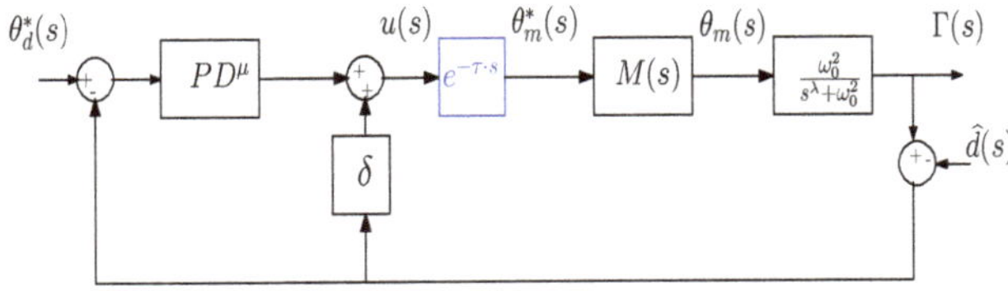

Figure 15. PD^{μ} based control scheme.

For our teleoperated robot, we have conducted a design study for the PI^{α} and PD^{μ} controllers using the tuning method described in the previous sections. Figure 16 displays the $RHFC$ regions of both controllers, as well as the low-frequency condition, considering $\alpha = \mu = 0.88$ (arbitrarily chosen) . For the case of the PI^{α} controller (Figure 16a), the red region represents the overlap between the $RHFC$ condition and the low-frequency condition. Regarding the PD^{μ} controller (Figure 16b), as previously mentioned, there is no low-frequency condition as the controller does not have a pole at the origin. However, there is a secondary stability constraint referred to as the 'Stability condition', which requires positive values for the proportional parameter $K_p + \delta$ of the controller represented here by the blue region. The intersection of this last with the $RHFC$ is presented in red color and presents then the feasible frequency region of the PD^{μ} controller.

To carry out a fair comparison between the PI^{α} controller and the PD^{μ} controller, including the positive feedback proposed in this paper, we have imposed identical specifications for both controllers: a gain crossover frequency $\omega_c = 6.86$ rad/s and a phase margin $\phi_m = 0.25$ rad.

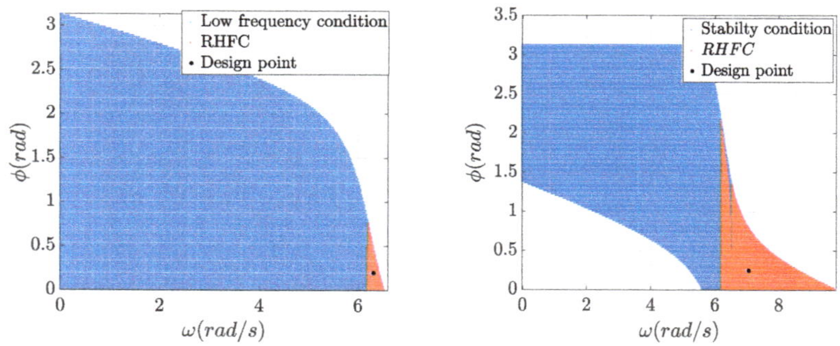

(**a**) PI^{α} controller (**b**) PD^{μ} controller

Figure 16. RHFC region.

The numerical expressions of the designed controllers are:

$$C_{PI^{\alpha}} = 0.1219 + \frac{0.1255}{s^{0.88}} \tag{52}$$

$$C_{PD^{\mu}} = 0.0293 + 0.0256 \cdot s^{0.88} \tag{53}$$

Note that these are the real controllers $C(s)$ and not the normalized ones $C_n(s*)$ (see (14)). In the case of PD^{μ} controller, the proportional parameter K_p of the normalized controller should be 1, according to Section 4.3.2, in order to eliminate the steady state error caused by the strain gauges offset. This gain becomes $(1/\omega_0^2)$ in the real controller, which is the value expressed in (53).

Figure 17 illustrates the Nyquist plots for both controllers. It is evident that both controllers fulfill the desired specifications. Furthermore, it is notable that the PD^{μ} controller provides an extra margin of robustness for the system compared to the PI^{α} controller: the minimum distance of the Nyquist plot to the point $(-1, 0)$ is bigger using the PD^{μ} than using the PI^{α}.

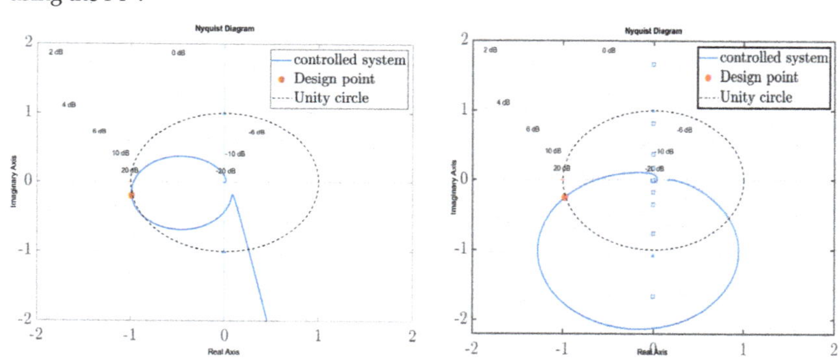

(**a**) PI^{α} (**b**) PD^{μ}

Figure 17. Nyquist-plots.

The flexible manipulator is controlled using a LABVIEW graphical interface. We have devised a second-order, smoothly differentiable trajectory with an amplitude of 0.5 radians to avoid motor saturation. The duration of the trajectory, denoted as L_t, is adjustable. The fractional-order operators $s^{-\alpha}$ and s^{λ} have been implemented using the Grunwald-Letnikov approximation and the principle of short memory [40].

As mentioned in the platform description, a pair of strain gauges was utilized to measure the deformation of a flexible link and determine the coupling torque at the base of the robot. However, it is worth noting that strain gauge sensors can sometimes display a noticeable offset. This offset refers to an output reading that deviates from zero even in the absence of applied strain.

In the specific case of the employed sensory system, despite implementing a calibration procedure, the offset signal persisted throughout the movement until the final position of the disc. This led to a step disturbance characterized by a variable value instead of a consistent offset signal. Figure 18 illustrates the controlled final position of the disc when a proportional controller with $K_p = 1$ is used. The experiment is repeated multiple times to demonstrate the non-constancy of the offset value.

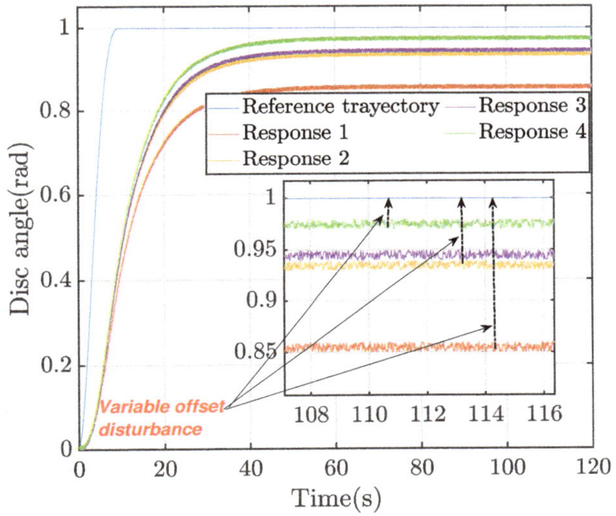

Figure 18. Experimental demonstration of the offset of the Strain gauge sensory system.

During the experimentation, we recorded the position θ_d of the disc for various values of L_t. In Figure 19, it can be observed that both controllers exhibit stable behavior when operating with slow trajectories. However, the response of the PI^{α} controller is considerably slower due to the integration process. Additionally, the last three sub-figures illustrate the system's response when $L_t < 0.05$, which can be regarded as a step command. It is evident that the PI^{α} controller leads to a critically stable behavior, whereas the PD^{μ} controller demonstrates a strictly stable response.

The responses of Figure 19 prove that it is possible to eliminate the offset disturbance of the strain gauges in flexible manipulators without resorting to controllers with an integral term, which could decrease the relative stability of the closed-loop system and its robustness to payload variations. Additionally, differentiating the strain gauge signal to remove the offset would introduce noise to the closed-loop control and compromise the compensation of steady-state tip position errors.

Observing Figure 16, it becomes evident that the feasible frequency specification region for the PD^{μ} controller is more extensive than that of the PI^{α} controller. The latter exhibits the advantage of eradicating steady-state errors, a significant issue inherent in the strain gauge sensors employed in the studied application. The newly formulated control scheme takes all these constraints into account by introducing a fictitious pole at the origin, effectively eliminating static errors (as shown in Figure 19). Consequently, the system's response becomes more rapid, capitalizing on the broader feasible frequency region offered by the PD^{μ} controller.

Finally, we mention that the phase margin specification used in this comparative study is small. It is caused by the small $RHFC$ region that we obtain if we use a PI^{α} controller (see Figure 16a). Figure 16b shows that the PD^{μ} controller yields a bigger $RHFC$ region that would allow larger phase margins.

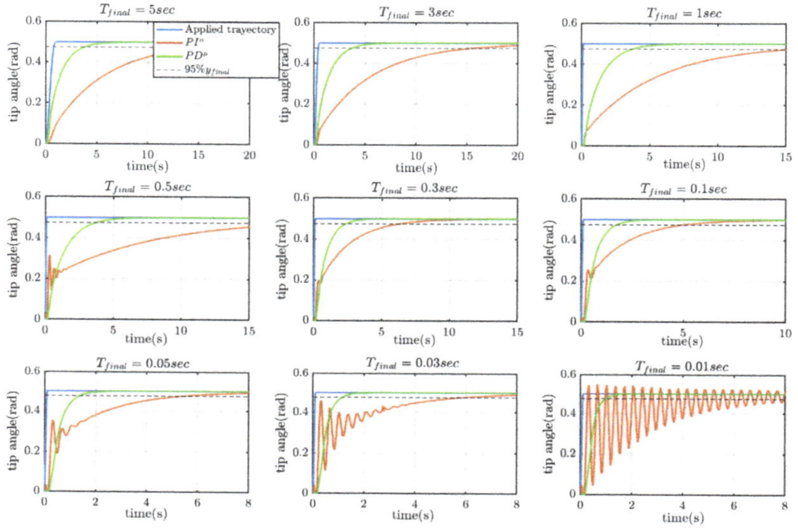

Figure 19. Experimental- results.

6. Conclusions

In this paper, feasible frequency specification regions have been computed for both PI^{α} and PD^{μ} fractional-order controllers. However, the study has two limitations. Firstly, the analysis has been conducted specifically for delayed underdamped systems with a range of $1 < \lambda \leqslant 2$. This choice was motivated by the fact that previous research has primarily focused on controlling delayed fractional-order monotonic systems with $0 < \lambda \leqslant 1$, while neglecting the case where $\lambda > 1$.

The second limitation of our study is that we have focused solely on the gain crossover frequency and phase margin as design specifications. We could have considered alternative pairs of specifications, such as gain margin and phase margin, which are more suitable for robustness analyses. However, our primary objective in this study was to explore the enhancements achievable in the closed-loop response of a system by designing frequency specifications that simultaneously improve time response specifications and robustness properties. To achieve this, we employed frequency domain techniques, which are better suited for fractional-order controllers than time domain techniques, and we selected the above frequency specifications because they are closely related to time specifications, such as damping and speed of response, while also providing indications of robustness properties.

Regions of feasible pairs of specifications (ϕ_m, ω_c) were determined for each of the considered controllers, taking into account all the combinations of fractional-order operators for the model and the controller. These regions were obtained to ensure that the Bode plot of the closed-loop system passes through the designated design point of the complex plane (the desired point of the unity circumference). Additionally, these regions guarantee robust stability of the closed-loop system by enforcing that the magnitude of the frequency response of the open-loop controlled system be a strictly decreasing function after the Nyquist plot has entered the unity circle. By using a design method based on these regions, we ensure that:

1. The design specifications reflect the actual behavior of the closed-loop system, i.e., there are no frequencies beyond the gain crossover frequency at which the magnitude of the frequency response could be one, which could lead to different phase margin and gain crossover frequency specifications than the desired ones. In essence, the design method ensures that the selected specifications represent the true characteristics of the closed-loop system.
2. Changes in the gain of the process (particularly increments of the gain) do not produce sharp changes in the phase margin and the gain crossover frequency that could suddenly unstabilize the closed-loop system. Note that this robustness concept is different from the well-known isophase margin condition. This one guarantees a local constant phase margin when the gain changes, while our condition guarantees that the phase margin changes smoothly in a broad range of gain variations. In particular, this smooth change is guaranteed if the gain grows with respect to the nominal value, which is the case that most likely can produce instability.

An application example involving the position control of a teleoperated flexible manipulator was developed in this study. Both PI^{α} and PD^{μ} controllers were designed to meet the same specifications for the system. The main challenge was to achieve the fastest and most precise response while eliminating the static error caused by the offset of the strain gauge sensor. To address this challenge, a modification was made to the system by implementing positive feedback on the angular position of the disc. This modification allowed for the introduction of a fictitious pole at the origin, which enabled the PD^{μ} controller to effectively eliminate the static error. As a result, the closed-loop system exhibited improved behavior and performance. Experimental results validated the controllers developed in this paper.

Our future research will address the obtention of Low-Frequency Conditions and Robust High-Frequency Conditions for $PI^{\alpha}D^{\mu}$ controllers.

Author Contributions: Conceptualization, V.F.B. and S.G.; methodology, V.F.B. and S.G.; software, S.G.; validation, S.G.; formal analysis, V.F.B.; investigation, S.G. and V.F.B.; resources, V.F.B.; writing—original draft preparation, S.G.; writing—review and editing, V.F.B.; visualization, S.G.; supervision, V.F.B.; project administration, V.F.B.; funding acquisition, V.F.B. All authors have read and agreed to the published version of the manuscript.

Funding: This research was funded in part by the Spanish University of Castilla-La Mancha under Grant 2019-PREDUCLM-10960 and in part by the Grant PID2019-111278RB-C21 funded by MCIN/AEI/10.13039/501100011033 and "ERDF A way of making Europe".

Data Availability Statement: Not applicable.

Conflicts of Interest: The authors declare no conflict of interest. The funders had no role in the design of the study; in the collection, analyses, or interpretation of data; in the writing of the manuscript, or in the decision to publish the results.

Abbreviations

The following abbreviations are used in this manuscript:

DC	Direct Current
DOF	Degree Of Freedom
$RHFC$	Robust High Frequency Condition
FLR	Flexible Link Robot
PD^{μ}	Fractional Order Proportional-Derivative controller
PI^{α}	Fractional Order Proportional-Integral controller
PID	Proportional-Integral-Derivative controller

References

1. Abdulwasaa, M.A.; Abdo, M.S.; Shah, K.; Nofal, T.A.; Panchal, S.K.; Kawale, S.V.; Abdel-Aty, A.H. Fractal-fractional mathematical modeling and forecasting of new cases and deaths of COVID-19 epidemic outbreaks in India. *Results Phys.* **2021**, *20*, 103702. [CrossRef]
2. AlBaidani, M.M.; Ganie, A.H.; Aljuaydi, F.; Khan, A. Application of Analytical Techniques for Solving Fractional Physical Models Arising in Applied Sciences. *Fractal Fract.* **2023**, *7*, 584. [CrossRef]
3. Ahmad, I.; Ahmad, H.; Thounthong, P.; Chu, Y.M.; Cesarano, C. Solution of multi-term time-fractional PDE models arising in mathematical biology and physics by local meshless method. *Symmetry* **2020**, *12*, 1195. [CrossRef]
4. Chen, W.; Sun, H.; Li, X. *Fractional Derivative Modeling in Mechanics and Engineering*; Springer: Berlin/Heidelberg, Germany, 2022.
5. Srivastava, H.; Dubey, V.; Kumar, R.; Singh, J.; Kumar, D.; Baleanu, D. An efficient computational approach for a fractional-order biological population model with carrying capacity. *Chaos Solitons Fractals* **2020**, *138*, 109880. [CrossRef]
6. Chen, S.B.; Jahanshahi, H.; Abba, O.A.; Solís-Pérez, J.; Bekiros, S.; Gómez-Aguilar, J.; Yousefpour, A.; Chu, Y.M. The effect of market confidence on a financial system from the perspective of fractional calculus: Numerical investigation and circuit realization. *Chaos Solitons Fractals* **2020**, *140*, 110223. [CrossRef]
7. Shen, K.; Yu, W. Fractional programming for communication systems—Part I: Power control and beamforming. *IEEE Trans. Signal Process.* **2018**, *66*, 2616–2630. [CrossRef]
8. Hekimoğlu, B. Optimal tuning of fractional order PID controller for DC motor speed control via chaotic atom search optimization algorithm. *IEEE Access* **2019**, *7*, 38100–38114. [CrossRef]
9. Peng, X.; Shang, Y.; Zheng, X. Lower bounds for the blow-up time to a nonlinear viscoelastic wave equation with strong damping. *Appl. Math. Lett.* **2018**, *76*, 66–73. [CrossRef]
10. Lewandowski, R.; Wielentejczyk, P. Nonlinear vibration of viscoelastic beams described using fractional order derivatives. *J. Sound Vib.* **2017**, *399*, 228–243. [CrossRef]
11. Ding, X.; Zhang, G.; Zhao, B.; Wang, Y. Unexpected viscoelastic deformation of tight sandstone: Insights and predictions from the fractional Maxwell model. *Sci. Rep.* **2017**, *7*, 11336. [CrossRef]
12. Xu, H.; Jiang, X. Creep constitutive models for viscoelastic materials based on fractional derivatives. *Comput. Math. Appl.* **2017**, *73*, 1377–1384. [CrossRef]
13. Eldred, L.B.; Baker, W.P.; Palazotto, A.N. Kelvin-Voigt versus fractional derivative model as constitutive relations for viscoelastic materials. *AIAA J.* **1995**, *33*, 547–550. [CrossRef]
14. Ogata, K. *Modern Control Engineering*; Prentice Hall: Upper Saddle River, NJ, USA, 2010; Volume 5, p. 894.
15. Sanchis, R.; Romero, J.A.; Balaguer, P. Tuning of PID controllers based on simplified single parameter optimisation. *Int. J. Control* **2010**, *83*, 1785–1798. [CrossRef]
16. Hamamci, S.E.; Tan, N. Design of PI controllers for achieving time and frequency domain specifications simultaneously. *ISA Trans.* **2006**, *45*, 529–543. [CrossRef] [PubMed]
17. Castillo-Garcia, F.; Feliu-Batlle, V.; Rivas-Perez, R. Frequency specifications regions of fractional-order PI controllers for first order plus time delay processes. *J. Process Control* **2013**, *23*, 598–612. [CrossRef]
18. Ariyatanapol, R.; Xiong, Y.P.; Ouyang, H. Partial pole assignment with time delays for asymmetric systems. *Acta Mech.* **2018**, *229*, 2619–2629. [CrossRef]
19. Sinou, J.J.; Chomette, B. Active vibration control and stability analysis of a time-delay system subjected to friction-induced vibration. *J. Sound Vib.* **2021**, *500*, 116013. [CrossRef]
20. Olgac, N.; Sipahi, R. Dynamics and stability of variable-pitch milling. *J. Vib. Control* **2007**, *13*, 1031–1043. [CrossRef]
21. Gu, K.; Niculescu, S.I. Survey on recent results in the stability and control of time-delay systems. *J. Dyn. Syst. Meas. Control* **2003**, *125*, 158–165. [CrossRef]
22. Li, S.; Zhu, C.; Mao, Q.; Su, J.; Li, J. Active disturbance rejection vibration control for an all-clamped piezoelectric plate with delay. *Control Eng. Pract.* **2021**, *108*, 104719. [CrossRef]
23. Birs, I.; Muresan, C.; Nascu, I.; Ionescu, C. A survey of recent advances in fractional order control for time delay systems. *IEEE Access* **2019**, *7*, 30951–30965. [CrossRef]
24. Shi, T.; Shi, P.; Wang, S. Robust sampled-data model predictive control for networked systems with time-varying delay. *Int. J. Robust Nonlinear Control* **2019**, *29*, 1758–1768. [CrossRef]
25. Wu, M.; Cheng, J.; Lu, C.; Chen, L.; Chen, X.; Cao, W.; Lai, X. Disturbance estimator and smith predictor-based active rejection of stick–slip vibrations in drill-string systems. *Int. J. Syst. Sci.* **2020**, *51*, 826–838. [CrossRef]
26. Araujo, J.M.; Santos, T.L. Control of second-order asymmetric systems with time delay: Smith predictor approach. *Mech. Syst. Signal Process.* **2020**, *137*, 106355. [CrossRef]
27. Natori, K.; Oboe, R.; Ohnishi, K. Stability analysis and practical design procedure of time delayed control systems with communication disturbance observer. *IEEE Trans. Ind. Inform.* **2008**, *4*, 185–197. [CrossRef]
28. Natori, K.; Oboe, R.; Ohnishi, K. Robustness on model error of time delayed control systems with communication disturbance observer. *IEEJ Trans. Ind. Appl.* **2008**, *128*, 709–717. [CrossRef]
29. Zhou, W.; Wang, Y.; Liang, Y. Sliding mode control for networked control systems: A brief survey. *ISA Trans.* **2022**, *124*, 249–259. [CrossRef] [PubMed]
30. Nian, F.; Shen, S.; Zhang, C.; Lv, G. Robust Switching Control for Force-reflecting Telerobotic with Time-varying Communication Delays. In Proceedings of the 2020 Chinese Control and Decision Conference (CCDC), Hefei, China, 22–24 August 2020; pp. 1714–1719.

31. Rogers, G. *Power System Oscillations*; Springer Science & Business Media: Berlin/Heidelberg, Germany, 2012.
32. Muresan, C.I.; Dutta, A.; Dulf, E.H.; Pinar, Z.; Maxim, A.; Ionescu, C.M. Tuning algorithms for fractional order internal model controllers for time delay processes. *Int. J. Control* **2016**, *89*, 579–593. [CrossRef]
33. Abbisso, S.; Caponetto, R.; Diamante, O.; Fortuna, L.; Porto, D. Non-Integer Order Integration by Using Neural Networks. In Proceedings of the IEEE International Symposium on Circuits and Systems (Cat. No. 01CH37196), Sydney, Australia, 6–9 May 2001; Volume 3, pp. 688–691.
34. Caponetto, R.; Dongola, G.; Pappalardo, F.; Tomasello, V. Auto-tuning and fractional order controller implementation on hardware in the loop system. *J. Optim. Theory Appl.* **2013**, *156*, 141–152. [CrossRef]
35. Monje, C.A.; Chen, Y.; Vinagre, B.M.; Xue, D.; Feliu-Batlle, V. *Fractional-Order Systems and Controls: Fundamentals and Applications*; Springer Science & Business Media: Berlin/Heidelberg, Germany, 2010.
36. Benftima, S.; Gharab, S.; Feliu-Batlle, V. Fractional Modeling and Control of Lightweight 1 DOF Flexible Robots Robust to Sensor Disturbances and Payload Changes. *Fractal Fract.* **2023**, *7*, 504. [CrossRef]
37. Mamani, G.; Becedas, J.; Feliu-Batlle, V. Sliding mode tracking control of a very lightweight single-link flexible robot robust to payload changes and motor friction. *J. Vib. Control* **2011**, *18*, 1141–1155. [CrossRef]
38. Gharab, S.; Benftima, S.; Batlle, V.F. Fractional Control of a Lightweight Single Link Flexible Robot Robust to Strain Gauge Sensor Disturbances and Payload Changes. *Actuators* **2021**, *10*, 317. [CrossRef]
39. Yaryan, M.; Naraghi, M.; Rezaei, S.; Zareinejad, M.; Ghafarirad, H. Bilateral Nonlinear Teleoperation for Flexible Link Surgical Robot with Vibration Control. In Proceedings of the 2012 19th Iranian Conference of Biomedical Engineering (ICBME), Tehran, Iran, 20–21 December 2012; pp. 101–106.
40. Podlubny, I. Fractional differential equations. *Math. Sci. Eng.* **1999**, *198*, 7–35.

fractal and fractional

MDPI

Article

Fractional-Order Phase Lead Compensation Multirate Repetitive Control for Grid-Tied Inverters

Fen Liang [1,2], Ho-Joon Lee [1,*] and Hongwei Zhang [3]

1 Department of Electrical & Control Engineering, Cheongju University, Cheongju 28503, Republic of Korea; lf01023354@cju.ac.kr
2 School of Mechanical and Electrical Engineering, Henan Industry and Trade Vocational College, Zhengzhou 451191, China
3 School of Electronic and Information, Zhongyuan University of Technology, Zhengzhou 451191, China; zhanghongwei@zut.edu.cn
* Correspondence: hjlee@cju.ac.kr; Tel.: +82-043-229-8436

Abstract: To reduce computational load and memory consumption, multirate repetitive control (MRC) with downsampling rates provides a flexible and efficient design for proportional-integral multi-resonant repetitive control (PIMR-RC) systems for grid-tied inverters. However, in MRC systems, repetitive controllers with low sampling rates produce low delay periods, and integer-order phase lead compensation may cause undercompensation or overcompensation. These imprecise linear phase lead compensations may result in deteriorated control performance. To address these problems, based on an infinite impulse response (IIR) filter, a fractional-order phase lead proportional-integral multi-resonant multirate repetitive control (FPL-PIMR-MRC) is proposed for grid-tied inverters in this paper. The proposed method can provide a suitable fractional phase lead step to achieve a wide stability region, minor tracking errors, and low hardware costs. The IIR fractional-order lead filter design, stability analysis, and the step-by-step parameter tuning of the FPL-PIMR-MRC system are derived in detail. Finally, simulation performed confirms the feasibility and effectiveness of the proposed scheme.

Keywords: fractional-order phase lead compensation; grid-tied inverters; multirate repetitive control

1. Introduction

With the development of renewable energy generation systems, pulse-width modulated (PWM) grid-tied inverters, serving as a crucial interface connecting distributed generation systems to the grid, have attracted widespread and increasing attention [1,2]. Nevertheless, owing to uncertainties in the parameters, nonlinear loads, and other factors, when the inverter works, it generates a lot of harmonics, resulting in poor control performance and high total harmonic distortion (THD) [3,4]. Repetitive control (RC) effectively achieves zero steady-state error tracking for any periodic signals and provides an excellent harmonic compensation capability for grid-tied inverters [5–7]. Owing to the inherent delay in RC, it has poor dynamic performance [8]. Therefore, other feedback controllers, including proportional-integral (PI) or proportional resonance (PR), are combined with RC to form a composite RC scheme [9,10]. The composite RC, consisting of RC and PI control, typically exists in two structural forms, either in series or in parallel, and is widely used in grid-tied inverter control systems. For instance, ref. [11] introduces a proportional-integral multi-resonant RC (PIMR-RC) scheme with outstanding harmonic suppression performance. This scheme incorporates an enhanced RC and a proportional gain in parallel.

Generally, to cut down the cost of power devices and improve the grid current quality, raising the grid-tied inverter's switching frequency is an effective method [12]. Hence, the switching frequency of power devices can reach 10 kHz or even higher in grid-tied inverters. Furthermore, in a traditional single-rate RC (SRC) system, the sampling rate of the RC is generally the same as the switching frequency of the grid-tied inverter. A higher switching frequency implies an elevated sampling frequency. Nevertheless, as the sampling rate of the RC increases, so does the digital controller's memory consumption and computational load [13], and the stability of the system will also deteriorate. Therefore, a multirate RC (MRC) scheme for PWM inverter control has been proposed [14]. In MRC systems, the RC controller works at a reduced sampling rate (RC rate), whereas the inverter system operates at a high sampling rate (feedback rate). Typically, the feedback rate is an integer multiple of the RC rate. Hence, as the RC loop has a reduced sampling rate, the computational load and the memory consumption will significantly reduce. A low sampling frequency will reduce the switching loss and cut down hardware costs effectively. Furthermore, previously proposed MRCs, normally with down-sampling schemes, have been successfully applied to active power filters (APF) [15], PWM inverters [16,17], motion control [18], and in other application fields.

However, in MRC systems, the low sampling rate will result in low delay periods N_m, which leads to imprecise linear phase lead compensation. For instance, assuming the phase lead compensator is ${z_m}^k$, this introduces a phase lead angle $\theta = k \times m \times (\omega/\omega_N) \times 180^\circ$ at ω, where k, ω, and ω_N represent the phase lead step, the angular frequency, and the Nyquist frequency, respectively. m is termed the "sampling ratio" [19], which is equal to the ratio of the feedback rate to the RC rate. Obviously, as m increases, this can easily result in lower phase compensation resolution. Because larger m leads to a lower sampling rate of RC, this will decrease the phase compensation precision [20] or even make the system unstable. Thus, selecting a suitable phase lead step k is necessary to achieve better phase lead compensation, a wider stability region, and more minor tracking errors. Considering the above issues, a fractional-order phase lead RC is proposed, utilizing a finite impulse response fractional lead (FIR-FL) filter [21,22] or an infinite impulse response fractional lead (IIR-FL) filter [23]. Most proposed schemes can achieve better control performance using an FIR-FL filter or an IIR-FL filter approximating the item ${z_m}^k$ with a fractional k. Furthermore, fractional-order phase lead compensation has also been successfully applied to MRC systems [24]. Nevertheless, most of these MRC schemes are based on plug-in RC systems and an FIR-FL filter, and there are no MRC schemes based on a PIMR-RC system and IIR-FL filter in the published literature. Furthermore, in practical applications, it is necessary to further research fractional-order MRCs to find a better universal MRC design [25]. There are also many studies on the application of fractional-order control (FOC) in inverters [26,27] and in some other fields [28,29], and many methods have also emerged for solving and analyzing fractional-order models [30–32].

In view of this, a fractional-order phase lead proportional-integral multi-resonant multirate repetitive control (FPL-PIMR-MRC) scheme is proposed for grid-tied inverters. The proposed scheme, characterized by a rapid error convergence rate and low THD, can attain precise fractional-order phase lead compensation, reduced computational load, and diminished memory consumption. The primary contributions of this work can be succinctly summarized as follows:

(1) A PIMR-MRC controller is introduced to enhance control performance, offering reduced computational load and lower memory consumption for grid-tied inverters. It not only provides a wider stability region but also maintains excellent harmonic suppression performance.
(2) Using an IIR-FL filter to approximate the fractional part of the phase lead step, precise compensation for the phase lag problem in the PIMR-MRC system is achieved.
(3) The proposed FPL-PIMR-MRC scheme not only improves the system stability and the quality of the grid-injected current but also has a fast error convergence rate, because it can accommodate a larger RC gain. A comprehensive analysis of FPL-PIMR-MRC based on an IIR-FL filter, and an FIR-FL filter, respectively, is also provided.

2. Modeling the Single-Phase PWM Inverter

A model of the single-phase LCL-type grid-tied PWM inverter is illustrated in Figure 1 [33,34]. It mainly comprises an LCL filter, a single-phase full-bridge inverter, a current controller $G_i(s)$, and a phase-locked loop (PLL). In the LCL filter, L_1 and L_2 are the inductors of the inverter side and the grid side, respectively, and C is the filter capacitor. i_1 and i_c are the currents of the inductor L_1 and the capacitor C, respectively. R_1 and R_2 are the parasitic resistors of the LCL filter. R_1 and R_2 are the parasitic resistors of the LCL filter. R_c is a passive damping resistor used to suppress the resonant peak of the LCL filter. u_{inv} is the output voltage of the inverter; it contains a large number of harmonics near the switching frequency, which can be filtered by LCL filters to reduce pollution to the grid. i_g, I_{ref}, and u_g are the grid current, the reference current amplitude, and the grid voltage, respectively. u_{PCC} is the voltage at the point of common coupling (PCC) for the grid. E_{dc} and ZOH are the DC bus voltage and a zero-order holder.

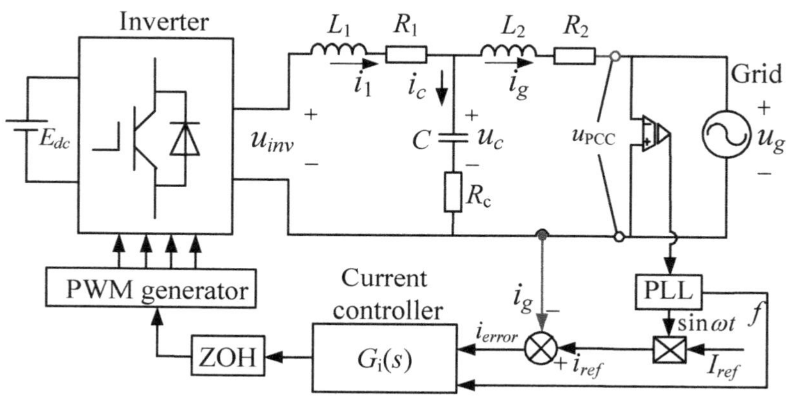

Figure 1. Structure model of the single-phase LCL-type grid-tied PWM inverter.

The inverter converts DC into AC, and the LCL low-pass filter is used to filter out higher-order harmonics in the system. The current controller plays a pivotal role in achieving high-performance control in the grid-tied inverter system, and it is also the primary focus of investigation in this paper. The PLL collects the grid voltage phase and provides the sampling frequency f for the current controller. The current error i_{error} is produced by the grid feedback current i_g to the reference current i_{ref}. The current error signal i_{error} is fed into the current controller, and its output serves as the PWM inverter control signal. Then, a closed-loop current control system with grid current negative feedback is established.

Ignoring the parasitic resistors of the L_1 and L_2, and assuming the state variable $x = \begin{bmatrix} i_1 & i_g & u_c \end{bmatrix}^T$, the system input $u = \begin{bmatrix} u_{inv} & u_g \end{bmatrix}^T$, and the system output $y = i_g$. According to Kirchhoff's theorem, the balance equations for LCL filters are:

$$\begin{cases} u_{inv} - u_c = L_1 \dfrac{di_1}{dt} + C\dfrac{du_c}{dt}R_c, \\ u_c - u_g = L_2 \dfrac{di_g}{dt} - C\dfrac{du_c}{dt}R_c, \\ i_1 - i_g = C\dfrac{du_c}{dt}. \end{cases} \quad (1)$$

By Laplace transformation on Equation (1), the equation in the s-domain can be obtained as follows:

$$\begin{cases} I_1(s) = \dfrac{U_{inv}(s) - U_c(s) - R_c C s U_c(s)}{sL_1}, \\ I_g(s) = \dfrac{U_c(s) - U_g(s) + R_c C s U_c(s)}{sL_2}, \\ U_c(s) = \dfrac{I_1(s) - I_g(s)}{sC}. \end{cases} \tag{2}$$

Accordin to Equations (1) and (2), the transfer function of the u_{inv} to i_g can be derived as:

$$P(s) = \frac{R_c C s + 1}{L_1 L_2 C s^3 + (L_1 + L_2) R_c C s^2 + (L_1 + L_2) s}. \tag{3}$$

where $P(s)$ is the controlled plant of the current controller in the s-domain.

Grid-tied inverters typically use high-frequency sinusoidal PWM (SPWM) modulation technology with a switching frequency exceeding 10 kHz. Consequently, the inverter module can be considered a unit with a gain of one. Therefore, a diagram of the grid current closed-loop control system can be represented as in Figure 2.

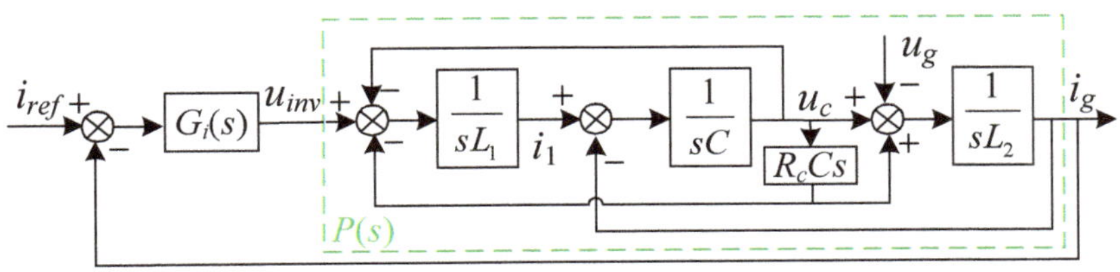

Figure 2. Structural diagram of the grid current closed-loop control system.

Substituting the parameter values (shown in Table 1) into the expression (3) and discretizing at the sampling frequency of 10 kHz with the ZOH discretization method, the controlled plant of the current controller in the z-domain can be derived as

$$P(z) = \frac{0.005908z^2 + 0.004191z - 0.002328}{z^3 - 2.024z^2 + 1.521z - 0.4976}. \tag{4}$$

Table 1. Parameters for the grid-tied inverter.

Parameters	Symbols	Values
DC-link voltage	E_{dc}	380 V
RMS value of grid voltage	U_g	220 V
Inverter side inductor	L_1	3.8 mH
Grid side inductor	L_2	2.3 mH
Capacitor	C	10 µF
L_1 equivalent resistance	R_1	0.48 Ω
L_2 equivalent resistance	R_2	0.32 Ω
Passive damping resistance	R_c	10 Ω
Grid frequency	f_g	50 Hz
Switching frequency	f_{sw}	10 kHz
Sampling frequency	f_s	10 kHz
Switching dead time	–	3 µs

3. Multirate Repetitive Control

3.1. Conventional Repetitive Control

Figure 3 illustrates the structural diagram of the conventional repetitive control (CRC).

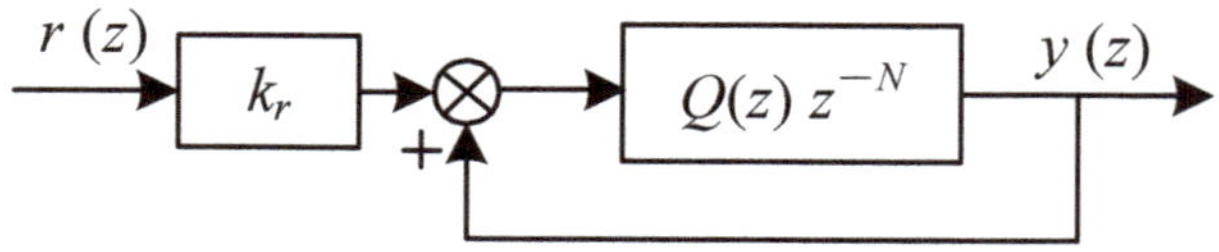

Figure 3. Structural diagram of CRC.

From Figure 3, $r(z)$ is the input signal, and $y(z)$ is the output signal. The transfer function expression from the input to the output of CRC in the z-domain is expressed as

$$G_{crc}(z) = \frac{y(z)}{r(z)} = \frac{k_r Q(z) z^{-N}}{1 - Q(z) z^{-N}}, \tag{5}$$

where k_r is the RC gain, which affects the amplitude of the RC output signal, and the value of k_r reflects the magnitude of the control effect of RC in the system. The smaller the k_r, the more conducive it is for system stability, but at the same time, it weakens the role of the repetitive controller. $Q(z)$ is an internal mode filter or a constant less than 1 to ensure stable convergence of the system and improve system stability. Typically, selecting $Q(z)$ as a zero-phase low-pass filter provides greater stability compared to selecting a constant. N is the number of samples in a period of the repetitive signal. The larger the N, the higher the error-tracking accuracy of RC [35].

3.2. Proportional-Integral Multi-Resonant Repetitive Control

Based on CRC, a PIMR-RC system with excellent dynamic response and a fast convergence rate is proposed and described in detail in [11]. Figure 4 shows the structural diagram of the PIMR-RC system.

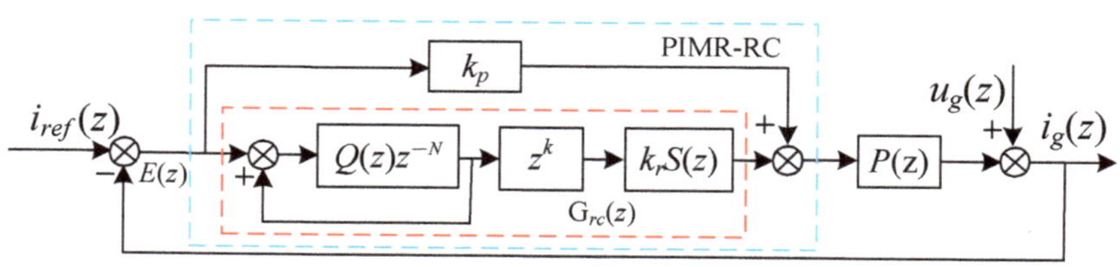

Figure 4. Structural diagram of the PIMR-RC system.

From Figure 4, the proportional gain, k_p, is employed to enhance the dynamic response of the system. $S(z)$ is a compensator normally selected as a low-pass filter, which is employed to decay the high-frequency harmonics. $i_{ref}(z)$ is the reference current, $i_g(z)$ is the output current, $u_g(z)$ is a disturbance signal, $P(z)$ is the plant model, and $E(z)$ is the tracking error. $G_{rc}(z)$ is the RC controller, and its discrete transfer function expression is expressed as

$$G_{rc}(z) = \frac{Q(z) z^{-N}}{1 - Q(z) z^{-N}} z^k k_r S(z). \tag{6}$$

3.3. Proportional-Integral Multi-Resonant Multirate Repetitive Control

The MRC system with downsampling rates can reduce the computational load and memory consumption due to the reduced sampling frequency of the system. In the MRC system, the feedback rate of the inverter feedback control loop is assumed to be f_s, the RC rate is f_m, and the corresponding sampling periods are T_s and T_m, respectively. z^{-N_m} is the period delay of the MRC system; N_m is equivalent to the ratio of f_m to the fundamental frequency f_g. In practical applications, to simplify analysis and reduce losses, m is typically an integer. Therefore, only the downsampling rate MRC with $m \geq 1$ is discussed in this paper. The two sampling rates in the MRC system can be summarized as

$$\begin{cases} f_s = mf_m, T_m = mT_s, N_m = f_m/f_g, \\ z = e^{sT_s}, z_m = e^{sT_m} = e^{smT_s} = z^m. \end{cases} \tag{7}$$

In the MRC system, for example, m = 2, the parameters of the inverter feedback control circuit are f_s = 10 kHz, f_g = 50 Hz, and the sampling period $T_s = 1/f_s$ = 0.0001 s. Then, in the repetitive controller, the RC rate $f_m = f_s/m$ = 5000 Hz, and the corresponding sampling period $T_m = mT_s$ = 0.0002 s. The number of samples per cycle $N_m = f_m/f_g$ = 100. z is the operator under the feedback rate, and z_m is the operator under the RC rate.

Based on the PIMR-RC system, a proportional-integral multi-resonant type multirate repetitive control (PIMR-MRC) is designed. Figure 5 displays the structural diagram of the PIMR-MRC system.

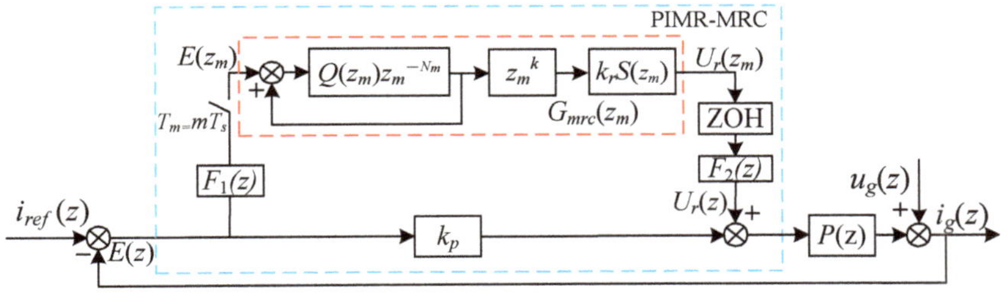

Figure 5. Structural diagram of the PIMR-MRC system.

Similar to the function expression (6) of $G_{rc}(z)$ in Figure 4, the expression of $G_{mrc}(z_m)$ can be described as follows:

$$G_{mrc}(z_m) = \frac{Q(z_m)z_m^{-N_m}}{1 - Q(z_m)z_m^{-N_m}} z_m^k k_r S(z_m), \tag{8}$$

where $Q(z_m)$ is an internal mode filter, k_r is the MRC gain, N_m is the number of samples in one period, z_m^k is the phase lead compensator, and $S(z_m)$ is a low-pass filter used to improve the system stability.

As shown in Figure 5, the proportional control k_p and a multirate RC controller $G_{mrc}(z_m)$ are connected in parallel to form a PIMR-MRC composite controller. i_{ref} is the input reference grid current signal, i_g is the grid current feedback signal, $P(z)$ is the controlled pant model, u_g is the disturbance signal, and $E(z)$ is the tracking error signal. The $G_{mrc}(z_m)$ has a reduced sampling rate f_m, while the other grid current feedback system has a high sampling rate f_s. The $G_{mrc}(z_m)$ works at a reduced sampling rate and can reduce the computational load and memory consumption. Accordingly, the PIMR-MRC scheme is an improved PIMR-RC.

$F_1(z)$ is the anti-spectral aliasing filter, which intercepts signals with frequency $|\omega| > \pi/T_m$. $E(z_m)$ is a down-sampled tracking error signal, and $U_r(z_m)$ is the output of $G_{mrc}(z_m)$. Finally, after up-sampling by a ZOH, the $U_r(z)$ can be obtained by an anti-imaging filter $F_2(z)$ with a cut-off frequency less than π/T_m.

To simplify analysis, the multirate system needs to be equivalent to a single-rate system. Figure 6 shows the equivalent single sampling rate control system structure for PIMR-MRC; the sampling rate of the whole system is the RC rate. Clearly, the CRC system is a special MRC when $m = 1$.

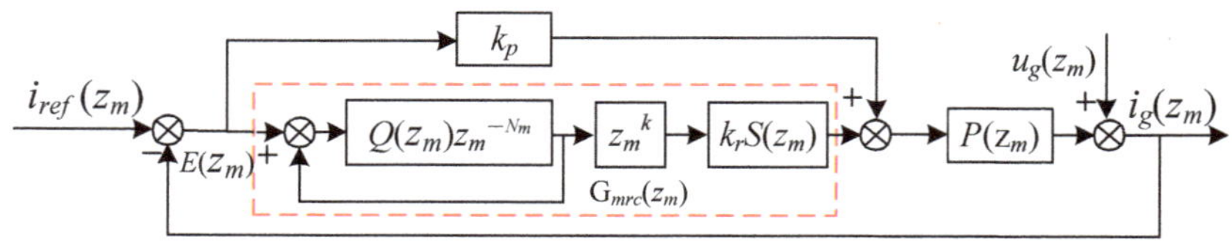

Figure 6. Structural diagram of equivalent single sampling rate system.

According to Figure 6, $P(z_m)$ is the plant model at the RC rate. If the sampling frequency of the system f_s = 10 kHz, m = 2, then $f_m = f_s/2$ = 5 kHz. $P(z_m)$, with a low sampling frequency f_m in the z-domain, is derived as

$$P(z_m) = \frac{0.02205z_m^2 + 0.01975z_m - 0.002614}{z_m^3 - 1.052z_m^2 + 0.3z_m - 0.2476}. \quad (9)$$

z_m^k not only addresses the phase lag induced by $P(z_m)$ and $S(z_m)$ but also compensates for the phase lag introduced by $F_1(z)$, $F_2(z)$, and ZOH. When f_s is 10 kHz and m = 1, 2, 3, 4, the corresponding RC rates f_m are 10 kHz, 5 kHz, 10/3 kHz, and 2.5 kHz, respectively. The phase lead compensation θ provided by z_m^k at the angular frequency ω with different m is shown in Figure 7.

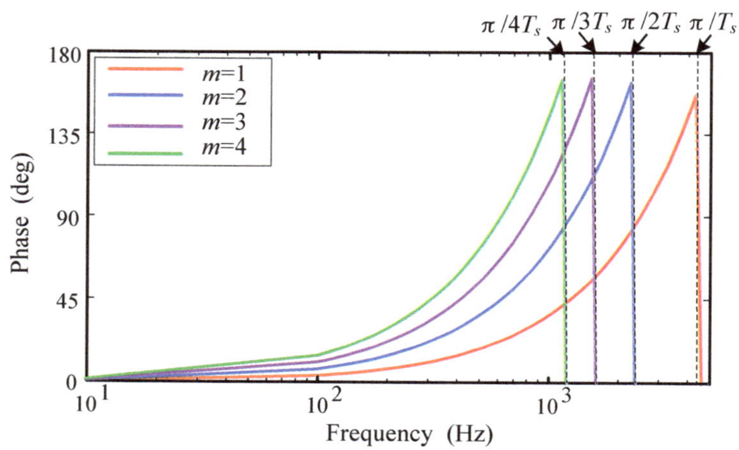

Figure 7. Phase lead compensation angle θ of z_m^k with the sampling ratio m = 1, 2, 3, and 4.

As shown in Figure 7, the phase lead compensation angle θ grows more and more rapidly as the value of the sampling ratio m increases. It can be inferred that the accuracy of the phase lead compensation decreases at a low sampling rate when m is too large. Thus, the item z_m^k with an integer k may result in overcompensation or undercompensation, leading to deterioration of the current quality and even system instability. Therefore, to compensate for the phase delay precisely and to fully utilize the advantages of the MRC, an FPL-PIMR-MRC scheme is proposed.

4. Fractional-Order Phase Lead Compensation Multirate Repetitive Control

4.1. Design of Fractional-Order Phase Lead Compensation Filter

Based on the Thiran method [36], the IIR filter requires less order, less computational complexity, faster speed, and lower CPU performance requirements compared to the FIR filter with the same performance. Especially for the IIR filter based on an all-pass structure, the amplitude-frequency characteristics are all 0 dB within the Nyquist frequency band range. Thus, during the filter design process, only the phase frequency characteristics of the filter need to be adjusted, reducing the difficulty of filter design significantly [37].

In the MRC system, when the value of the phase lead step k is a fraction, that is

$$k = k_i + d \tag{10}$$

where k_i, d are the integer and the fractional part of k, respectively, then,

$$z_m^k = z_m^{k_i} z_m^d. \tag{11}$$

The fractional-order phase lead item can be derived by substituting a delay operator z^{-1} with a lead operator z [22]. If the fractional item z_m^d is approximated by an M-order all-pass IIR filter, the discrete domain expression is

$$z_m^d \approx H(z_m) = \frac{1 + a_1 z_m^{-1} + \ldots + a_{M-1} z_m^{-(M-1)} + a_M z_m^{-M}}{a_M + a_{M-1} z_m^{-1} + \ldots + a_1 z_m^{-(M-1)} + z_m^{-M}}. \tag{12}$$

The calculation formula for the coefficient a_k of the filter is

$$a_k = (-1)^k \binom{M}{k} \prod_{n=0}^{M} \frac{d - M + n}{d - M + k + n}, k = 1, 2, \ldots, M, \tag{13}$$

where M is the order of the all-pass filter. When M = 1, 2, 3, 4, and 5, the coefficients a_k can be calculated according to the formulas in Table 2.

Table 2. Coefficients of the IIR filter when M = 1, 2, 3, 4, and 5.

	$M = 1$	$M = 2$	$M = 3$	$M = 4$	$M = 5$
a_1	$\frac{1-d}{1+d}$	$\frac{-2(d-2)}{d+1}$	$\frac{-3(d-3)}{d+1}$	$\frac{-4(d-4)}{d+1}$	$\frac{-5(d-5)}{d+1}$
a_2		$\frac{(d-1)(d-2)}{(d+1)(d+2)}$	$\frac{3(d-2)(d-3)}{(d+1)(d+2)}$	$\frac{6(d-3)(d-4)}{(d+1)(d+2)}$	$\frac{10(d-4)(d-5)}{(d+1)(d+2)}$
a_3			$-\frac{(d-1)(d-2)(d-3)}{(d+1)(d+2)(d+3)}$	$-\frac{4(d-2)(d-3)(d-4)}{(d+1)(d+2)(d+3)}$	$-\frac{10(d-3)(d-4)(d-5)}{(d+1)(d+2)(d+3)}$
a_4				$\frac{(d-1)(d-2)(d-3)(d-4)}{(d+1)(d+2)(d+3)(d+4)}$	$\frac{5(d-2)(d-3)(d-4)(d-5)}{(d+1)(d+2)(d+3)(d+4)}$
a_5					$-\frac{(d-1)(d-2)(d-3)(d-4)(d-5)}{(d+1)(d+2)(d+3)(d+4)(d+5)}$

The structural diagram of FPL-PIMR-MRC is presented in Figure 8.

The ideal frequency characteristic of the fractional-order lead compensator is a unity magnitude response and a linear phase response. Assuming the fractional part d is 0.4, the IIR and the FIR fractional-order filters with different M-orders are used to approximate the fractional item z_m^d. The frequency characteristics of these filters are shown in Figure 9. The magnitude response of the IIR is one, while the FIR varies nonlinearly within the entire frequency bandwidth. Furthermore, both the IIR and FIR have excellent linear phase response within the range of 1 kHz. However, the IIR has a better linear phase response than FIR when the frequency exceeds 1 kHz. That is, the IIR has a wider linear phase region within the Nyquist bandwidth range. Moreover, the IIR filter can achieve fractional-order lead compensation with less computational complexity and lower order. However, achieving the same performance requires higher-order FIR filters, resulting in increased input-to-output delay and resource consumption [38]. Therefore, a 3rd-order IIR as the fractional-order phase lead compensator is employed in this paper.

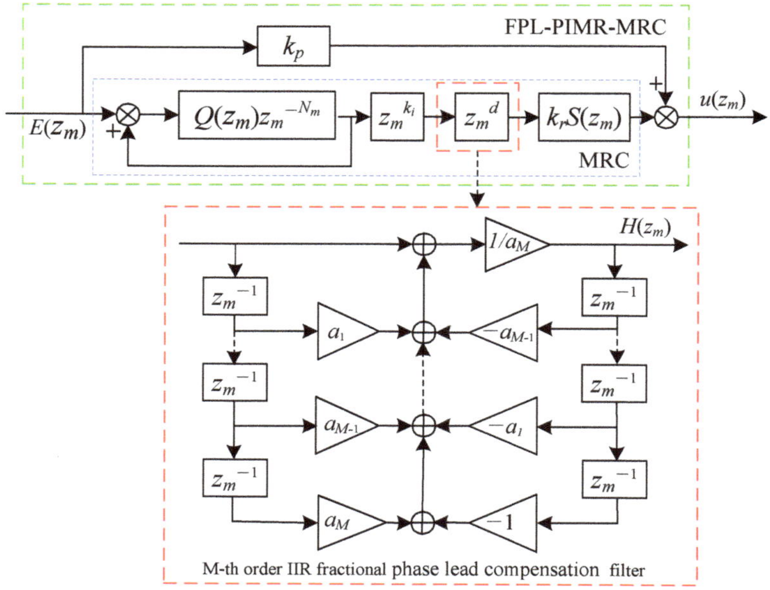

Figure 8. Structural diagram of the FPL-PIMR-MRC.

According to (12), when $M = 3$, a 3rd-order IIR filter can be derived as

$$z_m^d \approx \frac{1 + a_1 z_m^{-1} + a_2 z_m^{-2} + a_3 z_m^{-3}}{a_3 + a_2 z_m^{-1} + a_1 z_m^{-2} + z_m^{-3}}. \quad (14)$$

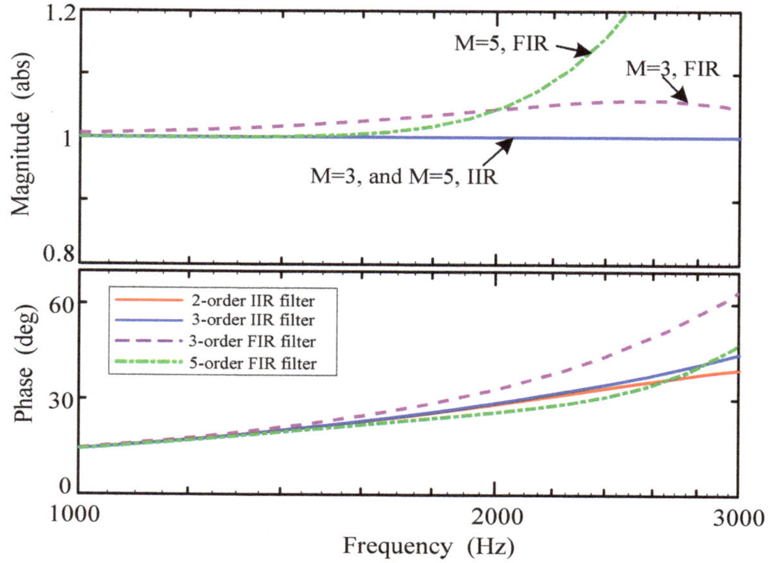

Figure 9. Frequency characteristics of IIR and FIR filters.

To meet the stability of Thrian all-pass filters, the optimal range of the fractional part d of the IIR filter is as follows [39]:

$$M - 0.5 \leq d \leq M + 0.5. \tag{15}$$

For instance, if $k = 3.7, M = 3$, and the sampling ratio $m = 2$, according to (15), a value of d = 2.7 satisfies the stability criteria. Hence, using a 3rd-order IIR-FL filter, the fractional-order lead item $z_m^{3.7}$ can be approximated as:

$$z_m^{3.7} = z_m^1 z_m^{2.7} \approx z_m^1 \frac{z_m^3 + 0.2432z_m^2 - 0.03623z_m + 0.003602}{0.003602z_m^3 - 0.03623z_m^2 + 0.2432z_m + 1}. \tag{16}$$

4.2. Stability Analysis of Fractional-Order Phase Lead Proportional-Integral Multi-Resonant Multirate Repetitive Control

Referring to Figure 6, the tracking error $E(z_m)$ can be derived as

$$\begin{aligned} E(z_m) &= \frac{1}{1 + (G_{mrc}(z_m) + k_p)P(z_m)}(i_{ref}(z_m) - u_g(z_m)) \\ &= \frac{1}{(1 + k_pP(z_m))(1 + G_{mrc}(z_m)P_0(z_m))}(i_{ref}(z_m) - u_g(z_m)), \end{aligned} \tag{17}$$

where $P_0(z_m)$ represents a new equivalent controlled plant of MRC, with its expression given by:

$$P_0(z_m) = \frac{P(z_m)}{1 + k_pP(z_m)}. \tag{18}$$

The FPL-PIMR-MRC system is subject to two stability conditions [11]: ① The roots of the equation $1 + k_pP(z_m) = 0$ lie in a unit circle. ② $|1 + G_{mrc}(z_m)P_0(z_m)| \neq 0$. By selecting a suitable k_p, condition ① can be achieved easily. The condition ② can be met unless the following expression holds:

$$\left|Q(z_m)z_m^{-N_m}(1 - z_m^k k_r S(z_m)P_0(z_m))\right| < 1, \forall z_m = e^{j\omega T_m}, 0 < \omega < \pi/T_m. \tag{19}$$

$$\text{i.e.,} \quad \left|1 - z_m^k k_r S(z_m)P_0(z_m)\right| < \left|z_m^{-N_m}\right|^{-1}|Q(z_m)|^{-1}. \tag{20}$$

When the signal frequency matches the fundamental frequency or its integer multiple within the bandwidth, $\left|z_m^{-N_m}\right|^{-1}$ approaches 1. $Q(z_m)$ is typically chosen as a zero-phase low-pass filter. Thus, $|Q(z_m)|$ also approaches 1. Therefore, condition ② can be derived as

$$\left|1 - z_m^k k_r S(z_m)P_0(z_m))\right| < 1. \tag{21}$$

Thus, the conditions for system stability are the same as for the PIMR-RC system. According to [11], the sufficient conditions for ② are as follows:

$$\left|\theta_s(\omega) + \theta_{P_0}(\omega) + k\omega T_m\right| < 90^\circ, \tag{22}$$

$$0 < k_r < \min_{\omega} \frac{2\cos(\theta_s(\omega) + \theta_{P_0}(\omega) + k\omega T_m)}{N_s(\omega)N_{P_0}(\omega)}, \tag{23}$$

where $N_s(\omega)$ and $\theta_s(\omega)$ represent the magnitude and the phase characteristics of $S(z_m)$, while $N_{P_0}(\omega)$ and $\theta_{P_0}(\omega)$ represent the magnitude and the phase characteristics of $P_0(z_m)$, respectively. Furthermore, the phase lead step k and the MRC gain k_r can be calculated using (22) and (23), respectively.

5. Parameters Design of Fractional-Order Phase Lead Proportional-Integral Multi-Resonant Multirate Repetitive Control

Referring to Figure 8, five parameters need to be designed for the FPL-PIMR-MRC. They are k_p, $Q(z_m)$, $S(z_m)$, k_r, and z_m^k, respectively.

5.1. Proportional Gain k_p

Selecting an appropriate k_p is crucial to expanding the stability region and enhancing the error tracking accuracy [40]. According to condition ①, all the roots of equation $1 + k_pP(z_m) = 0$ should be located within a unit circle at low sampling rates. In Figure 10a, when the value of k_p is varied from 10 to 25, condition ① is satisfied. Figure 10b shows all roots of the equation $1 + k_pP(z_m) = 0$ lie in a unit circle at different sampling rates when k_p = 16. Therefore, to strike a balance between system stability and accuracy of error tracking, the value of k_p has been chosen as 16.

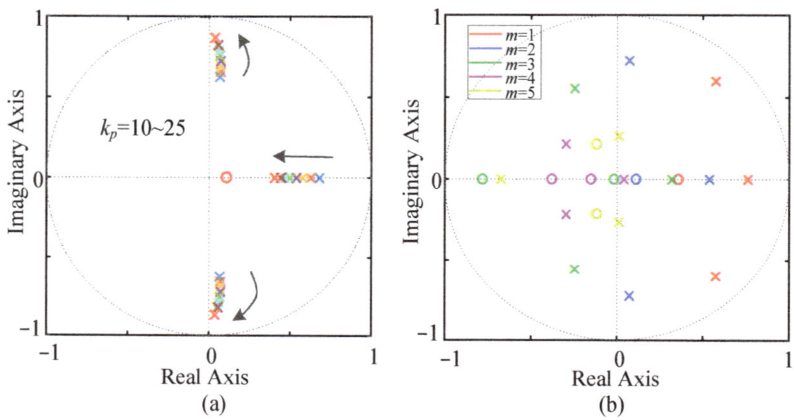

Figure 10. The roots of $1 + P_0(z_m) = 0$. (**a**) With different proportional gain. (**b**) With different sampling rates.

5.2. Internal Mode Filter $Q(z_m)$

To avoid phase delay caused by the filters, $Q(z_m)$ can be selected as a zero-phase low-pass filter to improve the system stability.

$$Q(z_m) = 0.25z_m^{-1} + 0.5 + 0.25z_m. \tag{24}$$

5.3. Low-Pass Filter $S(z_m)$

To effectively suppress high-frequency signals, $S(z_m)$ is set to be a fourth-order Butterworth low-pass filter. At the low sampling rate for different m values (m = 1, 2, 4), and setting the cut-off frequency to be 1000 Hz, $S(z_m)$ can be calculated as follows:

$$S(z_m) = \begin{cases} \frac{0.004824z_m^4+0.0193z_m^3+0.02895z_m^2+0.0193z_m+0.004824}{z_m^4-2.37z_m^3+2.314z_m^2-1.055z_m+0.1874}, m = 1, \\ \frac{0.04658z_m^4+0.1863z_m^3+0.2795z_m^2+0.1863z_m+0.04658}{z_m^4-0.7821z_m^3+0.68z_m^2-0.1827z_m+0.03012}, m = 2, \\ \frac{0.04328z_m^4+1.731z_m^3+2.597z_m^2+1.731z_m+0.4328}{z_m^4+2.37z_m^3+2.314z_m^2+1.055z_m+0.1874}, m = 4. \end{cases} \tag{25}$$

However, the phase lag may be more severe after adding a low-pass filter $S(z_m)$. Phase lag will affect the dynamic response of the system. Thus, a phase lead compensator needs to be introduced.

5.4. RC Gain k_r and Phase Lead Compensator z_m^k

Following expression (23), if the angle $(\theta_s(\omega) + \theta_{P_0}(\omega) + k\omega T_m)$ approaches zero degrees, by selecting an appropriate k, a larger k_r can be obtained. Thereby, the error convergence rate and system stability can be improved significantly. When $m = 2$, the phase-frequency characteristics of $S(z_m)P_0(z_m)z_m^k$ with different k are shown in Figure 11.

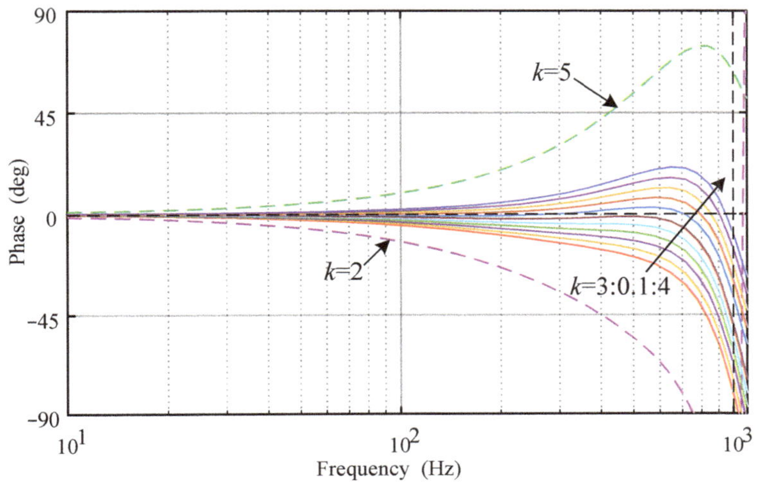

Figure 11. Phase-frequency characteristics of $S(z_m)P_0(z_m)z_m^k$ with different k.

From Figure 11, within the frequency range of 1 kHz, the phase-frequency characteristics of $S(z_m)P_0(z_m)z_m^k$ approach zero degrees when the phase lead step k varies from 3 to 4. Furthermore, the variation range of phase-frequency characteristics is within ±90°, which meets the system stability requirements. However, when k = 2 or k = 5, the $(\theta_s(\omega) + \theta_{P_0}(\omega) + k\omega T_m)$ values deviate from zero degrees significantly or potentially surpass the stable range.

To ascertain the optimal value of phase lead compensation, an IIR fractional-order filter utilizing the Thiran method was employed. When k changes every 0.1 intervals from 3 to 4, the fractional part of the k can be approximated using a 3rd-order IIR-FL filter. Obviously, when k = 3.7, the phase-frequency characteristics can meet the stability requirements and approach zero degrees within 1 kHz.

According to (21), define

$$H(e^{j\omega T_m}) = 1 - k_r e^{j\omega k T_m} S(e^{j\omega T_m}) P_0(e^{j\omega T_m}). \tag{26}$$

To guarantee system stability, the trajectories of $H(e^{j\omega T_m})$ should be inside a unit circle. The closer $H(e^{j\omega T_m})$ is to the unit circle's center, the faster the error convergence of the system [7]. When k = 3.7, and considering system modeling errors, the range of values for k_r is determined as $k_r \in [0 - 22]$.

Figure 12a shows the Nyquist curves $H(e^{j\omega T_m})$ $(\omega \in [0, \pi/T_m])$ within the unit circle when the RC gain k_r varies from 14 to 20. Note that the frequency at the center of the circle is 50 Hz with k_r = 16. Therefore, the Nyquist curve $H(e^{j\omega T_m})$ with k_r = 16 provides a wider stability margin at low frequencies. Figure 12b shows that when m = 1, the Nyquist curve $H(e^{j\omega T_m})$ exceeds the system stable range. When m = 1, the RC rate equals the feedback rate, and both have a high sampling frequency. However, when m = 2, the RC rate is reduced by half, and the low sampling rate repetitive controller can reduce computational delays caused by itself in each control cycle. Therefore, it can be concluded that the stability margins of the MRC system have increased compared to the SRC system. Note that a larger

m will produce imprecise phase compensation, which results in reduced system stability or even makes the system unstable.

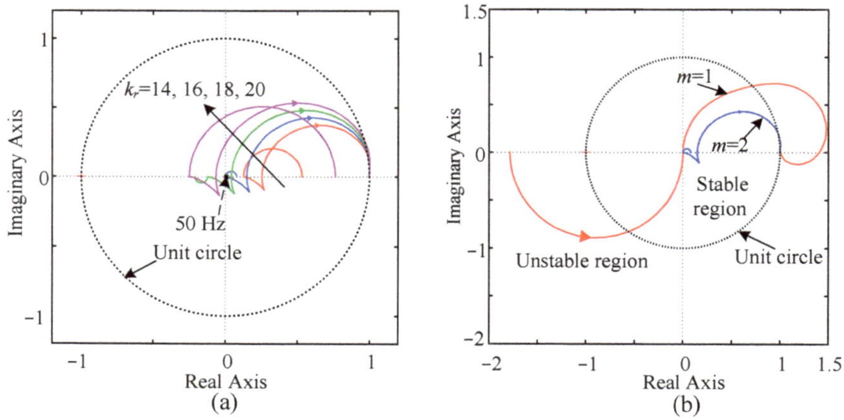

Figure 12. Nyquist curves of $H(e^{j\omega T_m})$. (**a**) With different RC gains. (**b**) With different sampling rates.

In practice, the ideal frequency characteristics of the compensated controlled object are unity gain and zero phase. Figure 13 shows the bode diagrams of $P_0(z_m)$ and $S(z_m)P_0(z_m)k_r z_m^k$ when m = 2. Figure 13a indicates that after compensation, the amplitude characteristic of $S(z_m)P_0(z_m)k_r z_m^k$ with k_r = 16 and k = 3.7 is closer to unity gain compared to $P_0(z_m)$. Furthermore, the gain of $S(z_m)P_0(z_m)k_r z_m^k$ using an IIR-FL filter exceeds that of an FIR-FL filter between 500 Hz and 2000 Hz. A larger gain means smaller tracking errors and lower THD of the system. As shown in Figure 13b, the phase characteristic of $S(z_m)P_0(z_m)k_r z_m^k$ with k_r = 16 and k = 3.7 is closer to the zero-phase line than that with k = 3 or k = 4.

According to Figure 6, the open-loop system transfer function of FPL-PIMR-MRC is

$$G(z_m) = G_{mrc}(z_m)P_0(z_m). \tag{27}$$

The open-loop system transfer function amplitude-frequency characteristic of FPL-PIMR-MRC with different sampling rates is shown in Figure 14.

Figure 14 shows that the system is an SRC system with a high sampling rate when m = 1. The SRC system trades excellent control performance at the cost of huge computation and memory usage. When m = 2, the sampling frequency of the RC controller decreases by half, and diminished computational burden and memory usage will be achieved.

Remark 1. *The FPL-PIMR-MRC system has a large open loop gain at low frequencies and rapid attenuation at high frequencies. Thus, the FPL-PIMR-MRC system can achieve a wider stability region compared to the SRC system. It is crucial to observe that reducing the sampling rate too much may result in poor harmonic suppression or even lead to system instability.*

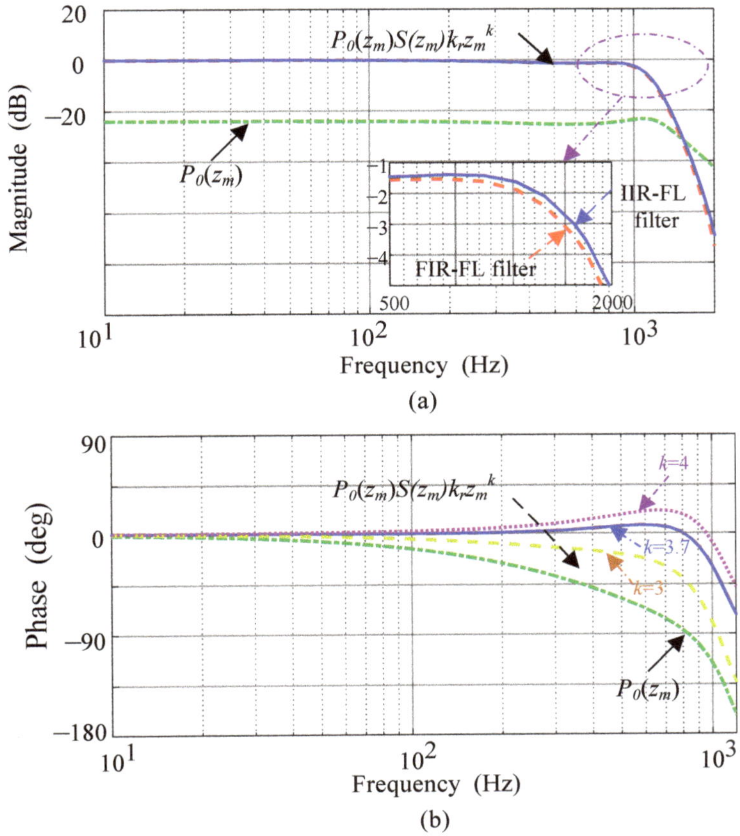

Figure 13. Bode diagrams of $P_0(z_m)$ and $S(z_m)P_0(z_m)k_r z_m^k$. (**a**) Amplitude frequency characteristics. (**b**) Phase frequency characteristics.

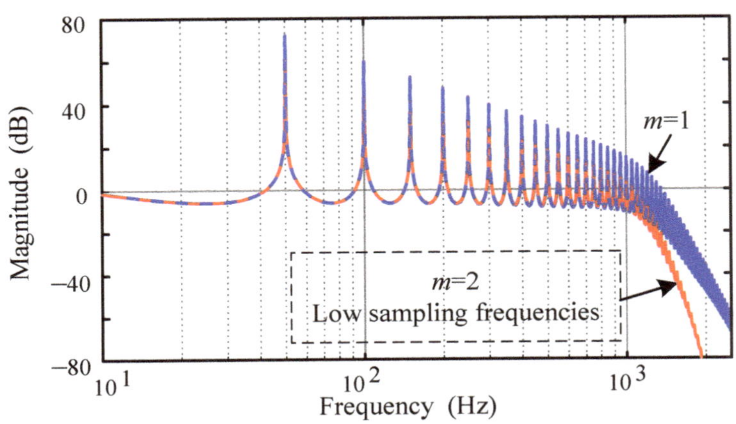

Figure 14. Amplitude-frequency characteristic of FPL-PIMR-MRC with different sampling rates.

6. Simulation

As depicted in Figure 1, a model was constructed for the 2.2 kW LCL-type single-phase grid-tied inverter system. The schemes of PIMR-RC, PIMR-MRC, and FPL-PIMR-MRC for grid-tied inverters were employed, respectively. Then, simulation models for these three control schemes were established using the MATLAB/Simulink environment. The rated value of the i_{ref} is 10 A, and the other parameters are assumed to be configured according to Table 1. Additionally, according to [41], the filters of F_1 and F_2 are selected as follows:

$$F_1 = F_2 = 0.15z^{-1} + 0.7 + 0.15z. \tag{28}$$

6.1. Steady-State Response

Figures 15–21 illustrate the steady-state response performance with different sampling rates. Figures 15 and 16 show the single sampling rate control system; the others are the multirate control systems. The THD of the grid current i_g is 3.14% just by proportional control, as shown in Figure 15. However, when the PIMR-RC system with $k = 9$ and $k_r = 16$ is employed, the THD of i_g decreases to 0.73%, as shown in Figure 16. It can be inferred that the CRC system can achieve excellent harmonic suppression performance with the single sampling rate control. Figure 17 illustrates the steady-state response of the PIMR-MRC system when $m = 2$, $k = 4$, and $k_r = 16$. The THD value is 0.95% because the sampling rate is reduced by half, which leads to the output current being distorted.

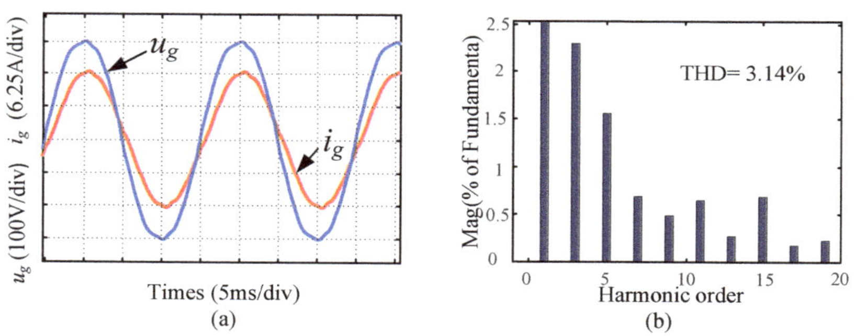

Figure 15. Steady-state response without RC with $m = 1$. (**a**) Waveforms of u_g and i_g. (**b**) Spectrum analysis of i_g.

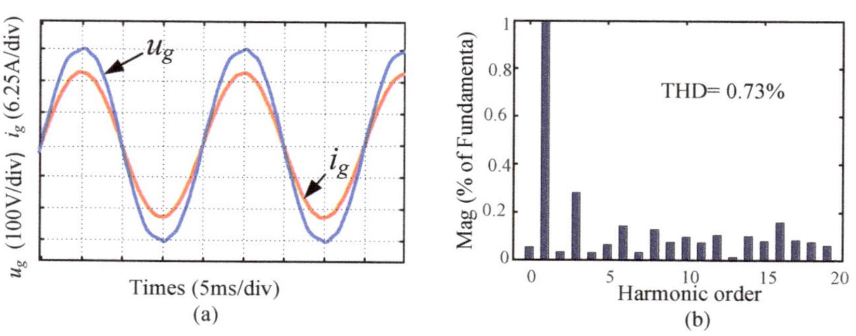

Figure 16. Steady-state response of the PIMR-RC system with $m = 1$, $k = 9$, and $k_r = 16$. (**a**) Waveforms of u_g and i_g. (**b**) Spectrum analysis of i_g.

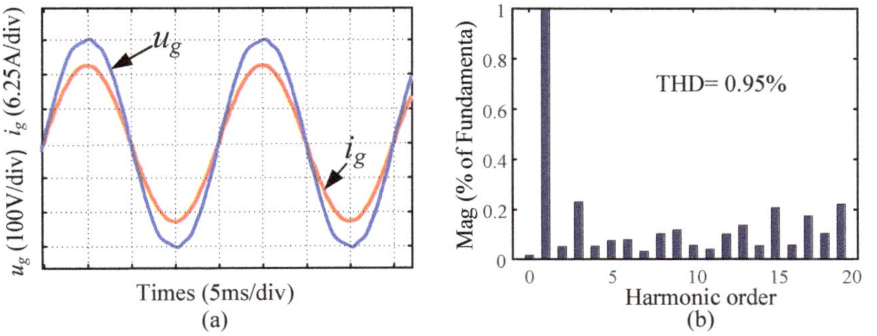

Figure 17. Steady-state response of the PIMR-MRC system with $m = 2$, $k = 4$, and $k_r = 16$. (**a**) Waveforms of u_g and i_g. (**b**) Spectrum analysis of i_g.

Figures 18 and 19 show that the THD values of the FPL-PIMR-MRC system based on a 3rd-order FIR-FL filter or a 3rd-order IIR-FL filter with $m = 2$, $k = 3.7$, and $k_r = 16$ are 0.66% and 0.51%, respectively. Obviously, due to the introduction of fractional-order phase lead compensation, the waveform quality of the grid current is improved. Additionally, because the IIR-FL filter exhibits a higher magnitude gain than the FIR-FL filter in the 500 Hz–2000 Hz range, it enhances the harmonic suppression performance for FPL-PIMR-MRC with the IIR-FL filter.

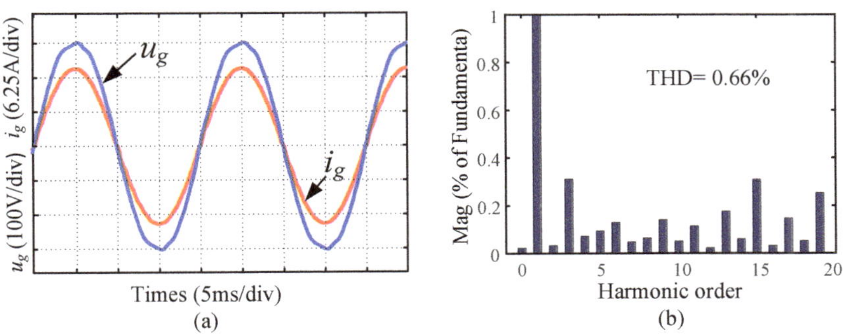

Figure 18. Steady-state response of the FPL-PIMR-MRC system based on FIR-FL with $m = 2$, $k = 3.7$, and $k_r = 16$. (**a**) Waveforms of u_g and i_g. (**b**) Spectrum analysis of i_g.

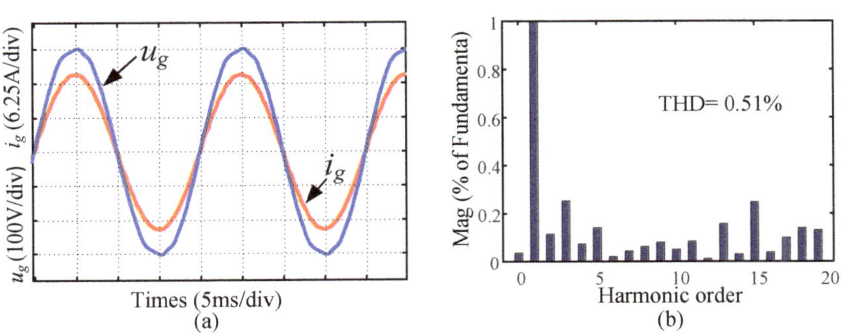

Figure 19. Steady-state response of the FPL-PIMR-MRC system based on IIR-FL with $m = 2$, $k = 3.7$, and $k_r = 16$. (**a**) Waveforms of u_g and i_g. (**b**) Spectrum analysis of i_g.

However, when the sampling rate of the RC controller is further reduced in the FPL-PIMR-MRC system, such as in Figures 20 and 21, the sampling rate is decreased to one quarter of its original when $m = 4$. The THD values of the i_g using the two fractional-order filters are increased to 1.6% and 1.32%, respectively. It is worth noting that as the m increases, the high-order harmonics increase significantly due to the open-loop gain decrease at high frequencies.

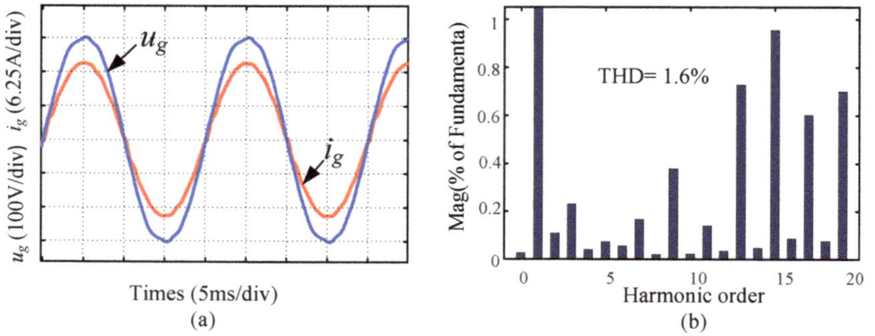

Figure 20. Steady-state response of the FPL-PIMR-MRC system based on FIR-FL with $m = 4$, $k = 3.7$, and $k_r = 16$. (**a**) Waveforms of u_g and i_g. (**b**) Spectrum analysis of i_g.

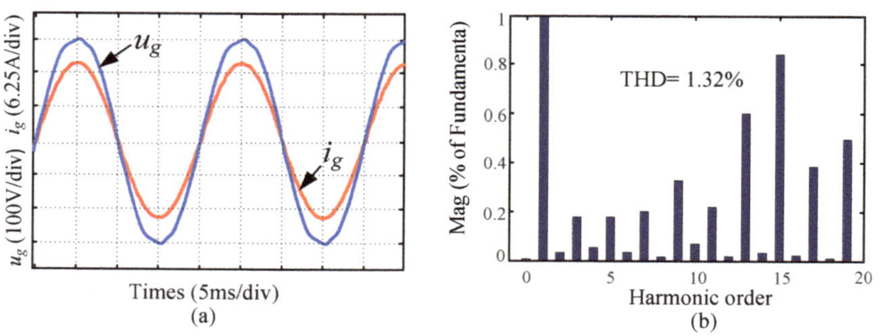

Figure 21. Steady-state response of the FPL-PIMR-MRC system based on IIR-FL with $m = 4$, $k = 3.7$, and $k_r = 16$. (**a**) Waveforms of u_g and i_g. (**b**) Spectrum analysis of i_g.

6.2. Transient Response

Figure 22 illustrates the tracking error of the PIMR-MRC system at different sampling rates. i_g, i_{error}, and i_{ref} represent the output current of the grid, the current error, and the reference current, respectively. When the rated value of the i_{ref} is adjusted from 5 A to 10 A, it is observed that, whether it is an SRC system ($m = 1$) or an MRC system ($m = 2$), the convergence time of the i_{ref} remains consistent at approximately 62 ms.

Figure 23 illustrates the tracking error of the FPL-PIMR-MRC system at different sampling rates. When the rated value of the i_{ref} is changed from 5 A to 10 A, the convergence time of the FPL-PIMR-MRC system, whether utilizing the IIR-FL filter or the FIR-FL filter, is approximately 42 ms, which is faster than for the PIMR-MRC system. The main reason is that the value of the RC gain k_r has increased from 16 to 18, which improves the error convergence rate. Accordingly, fractional-order phase lead compensation can extend the RC gain, thereby improving the control performance of the system.

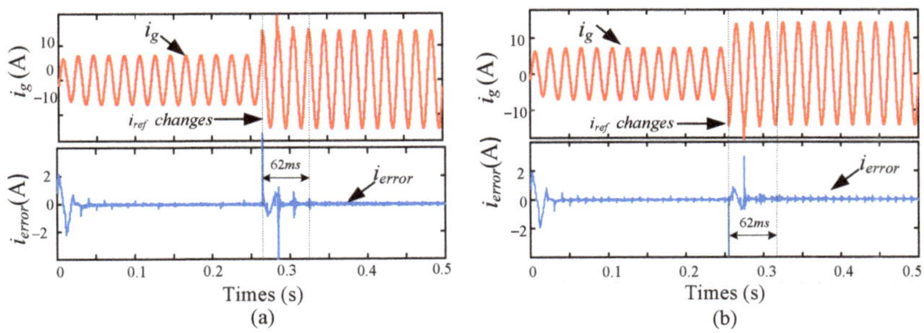

Figure 22. Transient responses of the PIMR-MRC system at different sampling rates. (**a**) $m = 1$, $k = 9$, and $k_r = 16$. (**b**) $m = 2$, $k = 4$, and $k_r = 16$.

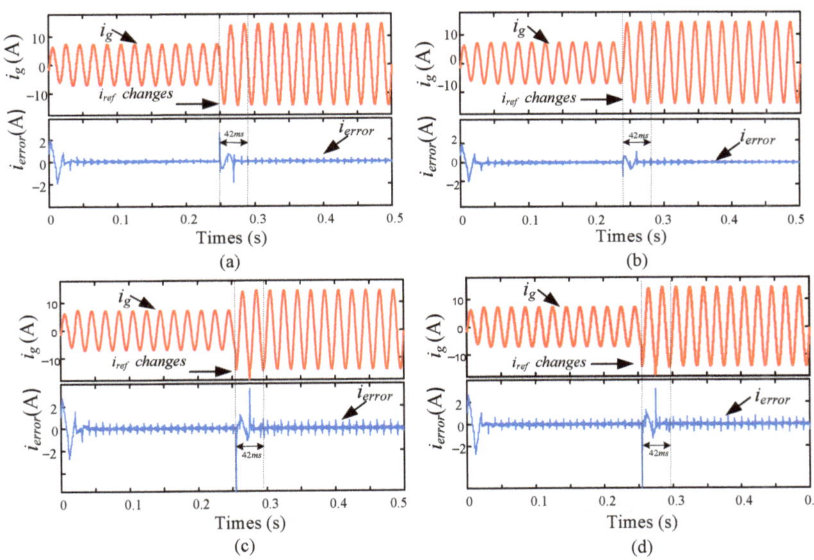

Figure 23. Transient responses of the FPL-PIMR-MRC system at different sampling rates. (**a**) Using FIR-FL, $m = 2$, $k = 3.7$, and $k_r = 18$. (**b**) Using IIR-FL, $m = 2$, $k = 3.7$, and $k_r = 18$. (**c**) Using FIR-FL, $m = 4$, $k = 3.7$, and $k_r = 18$. (**d**) Using IIR-FL, $m = 4$, $k = 3.7$, and $k_r = 18$.

From Figures 22 and 23, a larger i_{error} is produced if the sampling rates increase. Furthermore, the higher the value of m, the greater the peak value of i_{error}. For instance, the peak value of i_{error} is approximately ±0.2 A when $m = 1$. Nevertheless, when m increases to 2, the peak value of i_{error} is enlarged to approximately ±0.4 A, or even more to about ±0.7 A when $m = 4$, as shown in Figure 23c. In addition, for the proposed FPL-PIMR-MRC system with fractional-order phase ($k = 3.7$) lead compensation, such as in Figure 23b, the peak value of the error is ±0.3 A, which is smaller than that for the PIMR-MRC system without fractional-order compensation. Furthermore, as the IIR-FL filter exhibits a larger magnitude gain in some frequency ranges, the proposed FPL-PIMR-MRC system utilizing the IIR-FL filter has a smaller tracking error compared to systems employing the IIR-FL filter with identical parameters and sampling rates.

Remark 2. *Based on the Simulink experiments of the steady-state and transient response, it can be proved that when the fractional-order phase lead compensation is introduced, the proposed FPL-PIMR-MRC scheme can provide an accurate fractional phase lead step and a high gain, enlarging the stability margin of the system, and can then achieve better control performance in improving the quality of the grid-injected current.*

7. Conclusions

This article introduces a fractional-order FPL-PIMR-MRC scheme utilizing an IIR-FL filter to achieve fractional-order compensation for grid-tied inverters. This design is both flexible and efficient, and is characterized by diminished computation and lower memory loss due to the RC loop having a reduced sampling rate. The sampling frequency f_m and the down-sampling ratio m can be selected flexibly in practical applications. Therefore, it represents a universal design method for single-rate or multirate control systems. Furthermore, the proposed approach can achieve precise linear phase lead compensation by using a 3rd-order IIR-FL filter to approximate the fractional phase lead step k. The simulation results demonstrate the effectiveness of the proposed FPL-PIMR-MRC, which not only reduces hardware consumption but also has a wide stability region, low THD, and small tracking errors by selecting a suitable down-sampling ratio.

Additionally, the introduction of the fractional-order system based on an IIR-FL filter provides a more accurate phase lead compensation step for grid-tied inverters, improving the compensation accuracy so as to improve the control performance of the system. Moreover, due to the advantages of high control accuracy and excellent stability of fractional-order control systems, their application value has been highlighted in some fields requiring high control accuracy and performance. The proposed approach can also be promoted for application in APF or other power conversion units.

Author Contributions: Conceptualization, F.L., H.-J.L. and H.Z.; methodology, F.L. and H.-J.L.; software, F.L.; validation, F.L.; formal analysis, F.L. and H.-J.L.; writing—original draft preparation, F.L.; writing—review and editing, F.L. and H.Z.; supervision, H.-J.L.; project administration, H.-J.L.; funding acquisition, H.-J.L. All authors have read and agreed to the published version of the manuscript.

Funding: This work was supported in part by the National Research Foundation of Korea, funded by the Korean government (MSIT) (NRF-RS-2022-00143178); in part by the Basic Science Research Program through the National Research Foundation of Korea, funded by the Ministry of Education (NRF-2020R1I1A3069569); in part by the Incubation Program for Young Master Supervisors under the grant of China (No. D202213); in part by the Foundation for University Youth Key Teachers by Henan province of China (No. 2018GGJS-213); and in part by the Graduate Student Research Innovation Program of Zhongyuan University of Technology under the grant of China (No. YKY2023ZK24).

Data Availability Statement: The authors confirm that the data supporting the findings of this study are available within the article.

Conflicts of Interest: The authors declare no conflict of interest.

Abbreviations

The following abbreviations are used in this manuscript:

MRC	Multirate repetitive control
PIMR-RC	Proportional–integral multi-resonant repetitive control
PIMR-MRC	Proportional–integral multi-resonant multirate repetitive control
IIR	Infinite impulse response
FPL-PIMR-MRC	Fractional-order phase lead proportional–integral multi-resonant multirate repetitive control
PWM	Pulse-width modulated
THD	Total harmonic distortion
FOC	Fractional-order control
RC	Repetitive control
PI	Proportional integral

PR	Proportional resonant
SRC	Single-rate repetitive control
CRC	Conventional repetitive control
APF	Active power filters
FIR-FL	Finite impulse response fractional-order lead
IIR-FL	Infinite impulse response fractional-order lead
PLL	Phase-locked loop
PCC	Point of common coupling

References

1. Chen, D.; Zhang, J.; Qian, Z. An Improved Repetitive Control Scheme for Grid-Connected Inverter with Frequency-Adaptive Capability. *IEEE Trans. Power Electron.* **2013**, *60*, 814–823. [CrossRef]
2. Prasad, B.S.; Jain, S.; Agarwal, V. Universal Single-Stage Grid-Connected Inverter. *IEEE Trans. Energy Convers.* **2008**, *23*, 128–137. [CrossRef]
3. Wu, Y.; Ye, Y. Internal Model-Based Disturbance Observer with Application to CVCF PWM Inverter. *IEEE Trans. Ind. Electron.* **2018**, *65*, 5743–5753. [CrossRef]
4. di Benedetto, M.; Faro, A.; Bigarelli, L.; Lidozzi, A.; Solero, L. Constant Delay-Line Frequency Adaptive Repetitive-Resonant Control for Grid-Tied and Intentional Islanding Operations. *IEEE Trans. Ind. Appl.* **2023**, *59*, 1944–1955. [CrossRef]
5. Zhou, K.; Wang, D. Digital Repetitive Learning Controller for Three-Phase CVCF PWM Inverter. *IEEE Trans. Ind. Electron.* **2001**, *48*, 820–830. [CrossRef]
6. Ye, J.; Liu, L.; Xu, J.; Shen, A. Frequency Adaptive Proportional-Repetitive Control for Grid-Connected Inverters. *IEEE Trans. Ind. Electron.* **2021**, *68*, 7965–7974. [CrossRef]
7. Zou, Z.; Zhou, K.; Wang, Z.; Cheng, M. Frequency-Adaptive Fractional-Order Repetitive Control of Shunt Active Power Filters. *IEEE Trans. Ind. Electron.* **2015**, *62*, 1659–1668. [CrossRef]
8. Xu, F.; Zhu, M.; Ye, Y. The Stability of LCL-Type Grid-Tied Inverter Based on Repetitive Control and Grid Voltage Feed-Forward. *IEEE J. Emerg. Sel. Top. Power Electron.* **2023**, *11*, 1496–1506. [CrossRef]
9. Zhang, K.; Kang, Y.; Xiong, J.; Chen, J. Direct Repetitive Control of SPWM Inverter for UPS Purpose. *IEEE Trans. Power Electron.* **2003**, *18*, 784–792. [CrossRef]
10. Yang, Y.; Zhou, K.; Blaabjerg, F. Enhancing the Frequency Adaptability of Periodic Current Controllers with a Fixed Sampling Rate for Grid-Connected Power Converters. *IEEE Trans. Power Electron.* **2016**, *31*, 7273–7285. [CrossRef]
11. Zhao, Q.; Ye, Y. A PIMR-Type Repetitive Control for a Grid-Tied Inverter: Structure, Analysis, and Design. *IEEE Trans. Power Electron.* **2018**, *33*, 2730–2739. [CrossRef]
12. Wu, W.; He, Y.; Blaabjerg, F. An LLCL Power Filter for Single-Phase Grid-Tied Inverter. *IEEE Trans. Power Electron.* **2012**, *27*, 782–789. [CrossRef]
13. Zundert, J.; Oomen, T.; Verhaegh, J.; Aangenent, W.; Antunes, D.J.; Heemels, W.P.M.H. Beyond Performance/Cost Tradeoffs in Motion Control: A Multirate Feedforward Design with Application to a Dual-Stage Wafer System. *IEEE Trans. Control Syst. Technol.* **2020**, *28*, 448–461. [CrossRef]
14. Fujimoto, H.; Kawakami, F.; Kondo, S. Multirate Repetitive Control and Applications. In Proceedings of the American Control Conference, Denver, CO, USA, 4–6 June 2003; pp. 2875–2880.
15. Xie, C.; Zhao, X.; Savaghebi, M.; Meng, L.; Guerrero, J.M.; Vasquez, J.C. Multirate Fractional-Order Repetitive Control of Shunt Active Power Filter Suitable for Microgrid Applications. *IEEE J. Emerg. Sel. Top. Power Electron.* **2017**, *5*, 809–819. [CrossRef]
16. Zhang, B.; Zhou, K.; Wang, D. Multirate Repetitive Control for PWM DC/AC Converters. *IEEE Trans. Ind. Electron.* **2014**, *61*, 2883–2890. [CrossRef]
17. Zeng, C.; Li, S.; Wang, H.; Miao, H. A Fractional Phase Compensation Scheme of PRMRC for LCL Inverter Connected to Weak Grid. *IEEE Access* **2021**, *9*, 167027–167038. [CrossRef]
18. Ohnishi, W.; Strijbosch, N.; Oomen, T. Multirate State Tracking for Improving Inters ample Behavior in Iterative Learning Control. In Proceedings of the IEEE International Conference on Mechatronics (ICM), Kashiwa, Japan, 7–9 March 2021; pp. 1–6.
19. Sadegh, N.; Hu, A.; James, C. Synthesis, Stability Analysis, and Experimental Implementation of a Multirate Repetitive Learning Controller. *J. Dyn. Syst. Meas. Control* **2002**, *124*, 668–674. [CrossRef]
20. Chen, S.; Zhao, Q.; Ye, Y.; Qu, B. Using IIR Filter in Fractional Order Phase Lead Compensation PIMR-RC for Grid-Tied Inverters. *IEEE Trans. Ind. Electron.* **2023**, *70*, 9399–9409. [CrossRef]
21. Zhao, Q.; Ye, Y. Fractional Phase Lead Compensation RC for an Inverter: Analysis, Design, and Verification. *IEEE Trans. Ind. Electron.* **2017**, *64*, 3127–3136. [CrossRef]
22. Han, B.; Lai, J.; Kim, M. Down-Sampled Repetitive Controller for Grid-Connected Ćuk CCM Inverter. *IEEE Trans. Emerg. Sel. Top. Power Electron.* **2022**, *10*, 1125–1137. [CrossRef]
23. Kurniawan, E.; Cao, Z.; Man, Z. Design of Robust Repetitive Control with Time-varying Sampling Periods. *IEEE Trans. Ind. Electron.* **2014**, *61*, 2834–2841. [CrossRef]
24. Liu, Z.; Zhang, B.; Zhou, K. Universal Fractional-Order Design of Linear Phase Lead Compensation Multirate Repetitive Control for PWM Inverters. *IEEE Trans. Ind. Electron.* **2017**, *64*, 7132–7140. [CrossRef]

25. Zou, Z.; Wang, Z.; Cheng, M. Modeling, Analysis, and Design of Multifunction Grid-Interfaced Inverters with Output LCL Filter. *IEEE Trans. Power Electron.* **2014**, *29*, 3830–3839. [CrossRef]
26. Albalawi, H.; Bakeer, A.; Zaid, S.A.; Aggoune, E.-H.; Ayaz, M.; Bensenouci, A.; Eisa, A. Fractional-Order Model-Free Predictive Control for Voltage Source Inverters. *Fractal Fract.* **2023**, *7*, 433. [CrossRef]
27. Azghandi, M.A.; Barakati, S.M.; Yazdani, A. Passivity-Based Design of a Fractional-Order Virtual Capacitor for Active Damping of Multiparalleled Grid-Connected Current-Source Inverters. *IEEE Trans. Power Electron.* **2022**, *37*, 7809–7818. [CrossRef]
28. Roohi, M.; Zhang, C.; Taheri, M.; Basse-O'Connor, A. Synchronization of Fractional-Order Delayed Neural Networks Using Dynamic-Free Adaptive Sliding Mode Control. *Fractal Fract.* **2023**, *7*, 682. [CrossRef]
29. Fečkan, M.; Wang, J.; Pospíšil, M. *Fractional-Order Equations and Inclusions*; Walter de Gruyter GmbH & Co KG: Berlin, Germany; Boston, MA, USA, 2017; pp. 37–90.
30. Micula, S. An iterative numerical method for fractional integral equations of the second kind. *J. Comput. Appl. Math.* **2018**, *339*, 124–133. [CrossRef]
31. Jafari, H.; Tuan, N.A.; Ganji, R.M. A new numerical scheme for solving pantograph type nonlinear fractional integro-differential equations. *J. King Saud Univ. Sci.* **2021**, *33*, 101185. [CrossRef]
32. Moumen, A.; Mennouni, A.; Bouye, M. A Novel Vieta–Fibonacci Projection Method for Solving a System of Fractional Integrodifferential Equations. *Mathematics* **2023**, *11*, 3985. [CrossRef]
33. Xiong, Y.; Ye, Y.; Zhu, M. UDE-Based Current Controller with Enhanced Grid Frequency Fluctuation Adaptability for LCL-Type Grid-Tied Inverters. *IEEE Trans. Power Electron.* **2023**, *38*, 1683–1691. [CrossRef]
34. Solatialkaran, D.; Khajeh, K.G.; Zare, F. A Novel Filter Design Method for Grid-Tied Inverters. *IEEE Trans. Power Electron.* **2021**, *36*, 5473–5485. [CrossRef]
35. Zhang, Z.; Chu, B.; Liu, Y.; Ren, H.; Li, Z.; Owens, D.H. Multiperiodic Repetitive Control for Functional Electrical Stimulation-Based Wrist Tremor Suppression. *IEEE Trans. Control Syst. Technol.* **2022**, *30*, 1494–1509. [CrossRef]
36. Guo, Q.; Wang, J.; Ma, H. Frequency Adaptive Repetitive Controller for Grid-Connected Inverter with an All-pass Infinite Impulse Response (IIR) filter. In Proceedings of the IEEE 23rd International Symposium on Industrial Electronics (ISIE), Istanbul, Turkey, 1–4 June 2014; pp. 491–496.
37. Shukl, P.; Singh, B. Combined IIR and FIR Filter for Improved Power Quality of PV Interfaced Utility Grid. *IEEE Trans. Ind. Appl.* **2021**, *57*, 774–783. [CrossRef]
38. Huang, L.; Yang, J.; Wei, K.; Chen, Z.; Zhang, X. Peak Cancellation Method with IIR Filters and Its FPGA Implementation. *IEEE Trans. Circuits Syst. II Express Briefs* **2023**, *70*, 3704–3708. [CrossRef]
39. Valimaki, V.; Laakso, T.I. Principles of Fractional Delay Filters. In Proceedings of the IEEE International Conference on Acoustics, Speech, and Signal Processing, Istanbul, Turkey, 5–9 June 2000; Volume 6, pp. 3870–3873.
40. Zhao, Q.; Zhang, H.; Gao, Y.; Chen, S.; Wang, Y. Novel Fractional-Order Repetitive Controller Based on Thiran IIR Filter for Grid-Connected Inverters. *IEEE Access* **2022**, *10*, 82015–82024. [CrossRef]
41. Zhao, Q.; Liu, K.; Li, H. A Fractional-Order Multi-Rate Repetitive Controller for Single-Phase Grid-Connected Inverters. *Electronics* **2023**, *12*, 1021. [CrossRef]

fractal and fractional

MDPI

Article

Optimal Fractional-Order Controller for the Voltage Stability of a DC Microgrid Feeding an Electric Vehicle Charging Station

Sherif A. Zaid [1,*], Abualkasim Bakeer [2], Hani Albalawi [1,3], Aadel M. Alatwi [1,4], Hossam AbdelMeguid [5,6] and Ahmed M. Kassem [7]

1 Electrical Engineering Department, Faculty of Engineering, University of Tabuk, Tabuk 47913, Saudi Arabia
2 Department of Electrical Engineering, Faculty of Engineering, Aswan University, Aswan 81542, Egypt; abualkasim.bakeer@aswu.edu.eg
3 Renewable Energy & Energy Efficiency Centre (REEEC), University of Tabuk, Tabuk 47913, Saudi Arabia
4 Industrial Innovation and Robotic Center (IIRC), University of Tabuk, Tabuk 47731, Saudi Arabia
5 Department of Mechanical Engineering, Faculty of Engineering, University of Tabuk, Tabuk 47913, Saudi Arabia
6 Mechanical Power Engineering Department, Faculty of Engineering, Mansoura University, El-Mansoura 35516, Egypt
7 Electrical Engineering Department, Faculty of Engineering, Sohag University, Sohag 82524, Egypt
* Correspondence: shfaraj@ut.edu.sa

Abstract: Charging stations are regarded as the cornerstone of electric vehicle (EV) development and utilization. Electric vehicle charging stations (EVCSs) are now energized via standalone microgrids that utilize renewable energy sources and reduce the stress on the utility grid. However, the control and energy management of EVCSs are challenging tasks because they are nonlinear and time-varying. This study suggests a fractional-order proportional integral (FOPI) controller to improve the performance and energy management of a standalone EVCS microgrid. The microgrid is supplied mainly by photovoltaic (PV) energy and utilizes a battery as an energy storage system (ESS). The FOPI's settings are best created utilizing the grey wolf optimization (GWO) method to attain the highest performance possible. The grey wolf is run for 100 iterations using 20 wolves. In addition, after 80 iterations for the specified goal function, the GWO algorithm almost discovers the ideal values. For changes in solar insolation, the performance of the proposed FOPI controller is compared with that of a traditional PI controller. The Matlab/Simulink platform models and simulates the EVCS's microgrid. The results demonstrate that the suggested FOPI controller significantly improves the transient responsiveness of the EVCS performance compared to the standard PI controller. Despite all PV insolation disruptions, the EV battery continues to charge while the ESS battery precisely stores and balances PV energy changes. The results support the suggested FOPI control's robustness to parameter mismatches. The microgrid's efficiency fluctuations with the insolation level and state of charge of the EV battery are discussed.

Keywords: fractional-order proportional integral (FOPI); electric vehicle charging station (EVCS); grey wolf optimization (GWO)

1. Introduction

Electric vehicles (EVs) are currently replacing conventional internal combustion engine (ICE) cars [1,2]. Actually, ICEs have several flaws that EVs can help overcome. Compared to ICEs, EVs produce little pollution, are more energy efficient, make less noise, and require less maintenance. The infrastructure of the charging stations, the duration of the charging process, and the impact of these stations on the current electrical grid are just a few of the challenges that have yet to be overcome. Rapid charging techniques may drastically decrease the charging time to a few minutes [3,4]. However, these methods place heavy electrical demands and negatively affect the electricity grid. Numerous issues would be created, such as severe overload, voltage fluctuation, and voltage instability, particularly when many charging stations are connected concurrently to the utility grid [5,6]. The improvement of the electrical system is one remedy for these issues, but it will be expensive. Utilizing an energy storage system (ESS) that may serve as a buffer between the utility and the EV charging station (EVCS) is another superior approach [7]. The utility grid will feel less strain with the deployment of an ESS, but there will still be difficulties due to the anticipated high number of EVCSs in the future.

If most of the grid electricity used at the charging stations comes from fossil fuels, the claim that electric vehicles are ecologically beneficial may not hold up. Therefore, it is necessary to use renewable energy sources in EVCSs to emphasize the effects of EVs on the environment. It is well known that renewable sources are typically intermittent rather than continuous. In order to resolve the discontinuity in these sources, ESSs are used.

Biogas, PV, and wind energy systems are the most often employed renewable energy sources for EV charging stations [8,9]. Compared to wind energy systems, PV energy systems are easier to use and more efficient. PV electricity is, therefore, more desirable for EV charging stations. PV-based charging stations have been the subject of several research articles [10]. The authors of one study put forth a concept for an EV charging station powered by solar energy [11]. Additionally, the authors provide a mathematical model of the charging station and employ simulation tools to evaluate the system's performance under varying conditions. The simulation findings demonstrate that the suggested charging station can operate independently of the power grid and satisfy an EV's charging needs. The authors of [12] suggest a creative method for developing an EV charging station that combines vehicle-to-grid (V2G) technology with solar and wind energy. The authors' detailed system design includes a solar panel array, a wind turbine, a battery energy storage system, an EV charging station, and a V2G interface. The technology is meant to charge EVs and feed extra electricity generated during high demand back into the grid. According to the scientists, the suggested system can deliver dependable and sustainable energy for grid support and EV charging. The authors of [13] propose a solar PV-powered EV charging station. To increase the efficiency of the charging process, the study discusses the design of a PV array, a DC–DC converter, and using a perturb and observe method. According to the simulation results, electric car charging at the station may be efficient and dependable. A control method that guarantees a charging station's steady functioning while optimizing the use of renewable energy sources is suggested by [14]. A secondary control loop optimizes the power flow between the various energy sources based on the available energy and the demand for EV charging. By contrast, a primary control loop smooths the voltage and current of the converter as part of the control strategy. Research on the viability of a hybrid system that combines solar and wind power to provide energy for a grid-connected EV charging station is presented in [15]. The research uses HOMER software (V4.9) to simulate the system, which optimizes the system's architecture by identifying the ideal mix of solar and wind resources, energy storage capability, and EV charging demand. According to the findings, the hybrid system is technically possible and has a respectable payback period. The report also analyzes the hybrid system's potential advantages, such as promoting local economic growth by generating green employment, lowering emissions, and strengthening energy security. In [16], a hybrid energy power system that uses the utility grid as a backup is utilized to charge EVs using the electric railway power system. Ref. [17] integrated PV electricity into an EV quick charging station using PV panels and a bank of batteries using a three-level boost converter. Although output capacitor voltage balancing must be considered, this relieves some pressure on switching devices.

Several renewable energy systems have used fractional-order control (FOC), a relatively novel control method [18]. FOC is a development of conventional integer-order control and has several benefits over traditional control methods, including higher performance, better resilience, and more flexibility. Despite these benefits, FOC has certain drawbacks as well. Its intricacy, difficulty of implementation, and computationally demanding nature are the key drawbacks. Ref. [19] presents a survey of recent advances in FOC autotuning techniques. A method for controlling fractional-order semilinear systems that have limitations given by linear equations was introduced by [20]. The suggested approach is used in a drug delivery system to regulate the drug concentration, and the results of extensive simulations are used to evaluate its performance. However, the technique has a complex procedure. Ref. [21] introduces a new adaptive FOPID compensator that self-adjusts fractional instructions to get the most power possible out of a standalone PV system as the environment changes. However, the numerical computations are very tedious.

This study presents an implementation of the FOPI controller to manage and control a photovoltaic-powered autonomous EV charging station. The ESS's battery is charged and discharged using a bidirectional converter, while the EV is charged using a unidirectional converter. The energy management and control of the proposed microgrid are based on optimal fractional-order controllers. The system's main objective is to maintain DC bus voltage regulation, control the storage battery charging, control the EV charging, and manage the system energy. The gains of the fractional-order controller are optimally chosen based on the GWO optimization technique. Moreover, performance comparisons were carried out between the proposed FOPI controller and the traditional PI controller. The proposed EVCS microgrid was modeled and simulated using the Matlab/Simulink platform. The following are this study's objectives:

- To enhance the functionality of the suggested EV charging station, FOPI controllers were incorporated.
- The suggested FOPI controller's ideal gains were determined using the metaheuristic optimization technique GWO.
- The suggested system's performance using the FOPI controller and the traditional PI were compared. The performance of the controller was evaluated under various solar isolation disturbances.
- Modeling the suggested system using Matlab. In order to determine the impact of fluctuations of the solar insolation on the microgrid's response, the system's performance was evaluated.
- The suggested control system's durability was explored in the face of parameter uncertainty.

This paper is organized as follows: the suggested charging station is shown in Section 2; Section 3 describes the microgrid modeling and design considerations of the charging station; the proposed control structure is presented in Section 4; the outcomes of the simulation are covered in Section 5; and, finally, the conclusions are presented in Section 6.

2. The Studied EVCS Description

Figure 1 depicts the proposed EVCS system's design. The off-grid EVCS system uses solar energy as its source of electricity. It is connected to a PV array, producing EVCS electricity. The primary energy source for the EVCS is the PV panel. However, the generated energy varies depending on various environmental factors, like the amount of solar insolation. Therefore, ESS batteries are typically utilized to address the issue of intermittent energy supply. A boost converter is attached to the PV's output terminals. Its purpose is to use the PV panel's maximum power point tracking (MPPT) condition by matching the PV voltage level to that of the DC link. The energy storage and EV charging converters are connected to the DC bus. Typically, these converters are DC/DC converters. A unidirectional buck converter is used to manage the charging process of the EV battery. The energy storage converter, on the other hand, is a bidirectional DC/DC converter. Its function is to regulate how the storage battery is charged and discharged. Additionally, it plays a role in controlling the DC bus voltage to counteract changes in EV load and insolation level. The following paragraphs will discuss the modeling and operating principles of these converters.

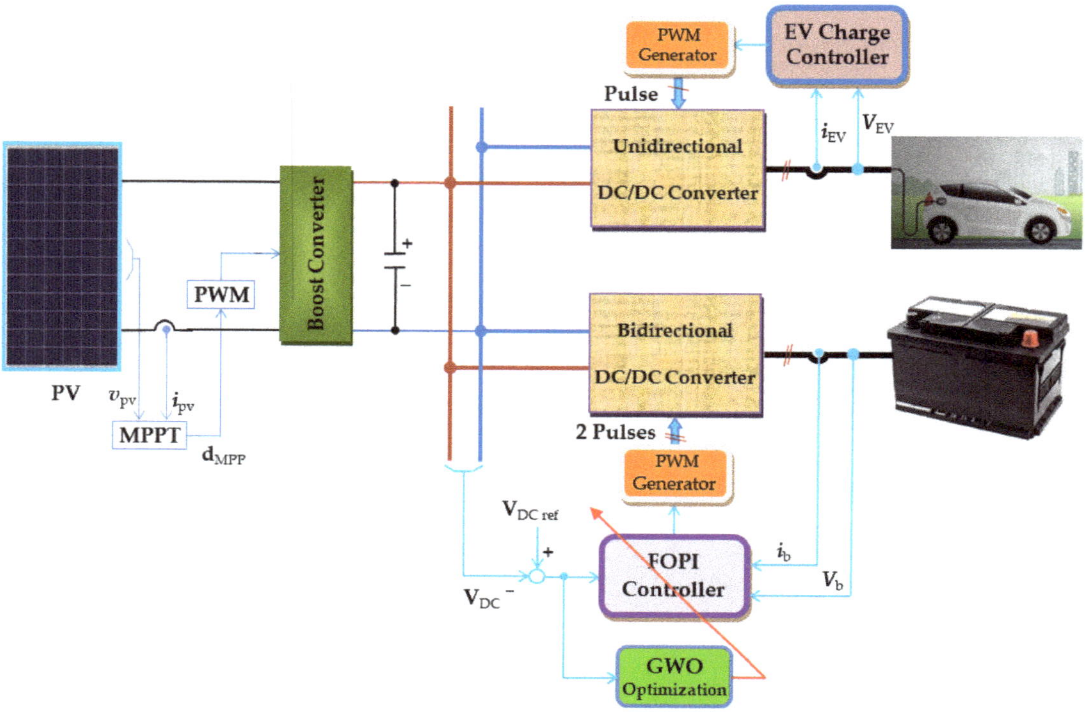

Figure 1. The proposed microgrid of the charging station.

3. Modeling of the EVCS Microgrid

For the purpose of designing the control system of the proposed EVCS microgrid, the microgrid model must be identified carefully. Hence, the model of the EVCS microgrid will be discussed in the following subsections. Specifically, the model of the power, energy, and DC/DC converters will be explained. The PV array model is common in the literature, so it will not be repeated here.

3.1. Power and Energy Model

The energy and power relationships of the EVCS microgrid are presented in this section. It is assumed that the EV station (P_{EV}) is constant and instantaneous PV-panel power (P_{pv}) is given by [11]:

$$P_{pv} = \begin{cases} P_{max}\left(1 - \frac{t^2}{36}\right) & -6 \le t \le 6 \\ 0 & 6 \le t \le 18 \end{cases} \tag{1}$$

where (t) is the time in hours starting at noon and (P_{max}) is the maximum power generated by the PV. It is assumed that the origin of the time axis is fixed at noon when insolation is at its highest. It is believed that solar energy lasts for 12 h, beginning at 6:00 a.m.

The instantaneous power of the proposed microgrid is governed by:

$$P_{EV} = P_{pv} - P_b - P_{loss} \tag{2}$$

where (P_{loss}) is the microgrid power loss and (P_b) is the ESS battery power. A useful equation for the design purpose is given by:

$$P_{max} = 3P_{EV} - \frac{\int_0^T P_{loss}dt}{T} \tag{3}$$

where (T) is the time period. The state of charge (SOC_b) of the ESS battery may be determined using:

$$SOC_b = \frac{\int P_b dt}{E_{rated}} \tag{4}$$

where (E_{rated}) is the ESS battery's rated energy.

The ESS battery has a maximum stored energy ($\Delta E_b|_{max}$) determined using [11]:

$$\Delta E_b|_{max} = 12.53P_{EV} - 2\int_0^T P_{loss}dt \tag{5}$$

3.2. EVCS Converter Model

There are two charging converters in the suggested system. A straightforward unidirectional buck converter serves as the EV charging converter. The energy storage battery, however, is a bidirectional DC/DC converter. Figure 2 displays the circuit diagram for the converters, where $L_1 = L_2 = L$. The unidirectional converter is included in the operation, modeling, and analysis of the bidirectional converter. Continuous mode is necessary for the converter's operation. The converter was made of a filter, two antiparallel diodes, and two transistors (Q_1, Q_2). The storage battery and the DC bus are connected to the converter terminals. (E_b, r_b) denote the internal voltage and resistance of the ESS battery. The filter inductance is believed to be sufficiently large to retain enough energy to charge/discharge the ESS battery. Therefore, the continuous conduction mode of operation is ensured. Buck mode and boost mode are the converter's two operating modes. The bidirectional converter is used to discharge the battery while switch Q_1 acts as a diode and switch Q_2 is in boost mode. However, it works in the buck mode, the battery charging mode, when switch Q_1 is turned on and switch Q_2 functions as a diode. The state-space model of the converter is given as follows:

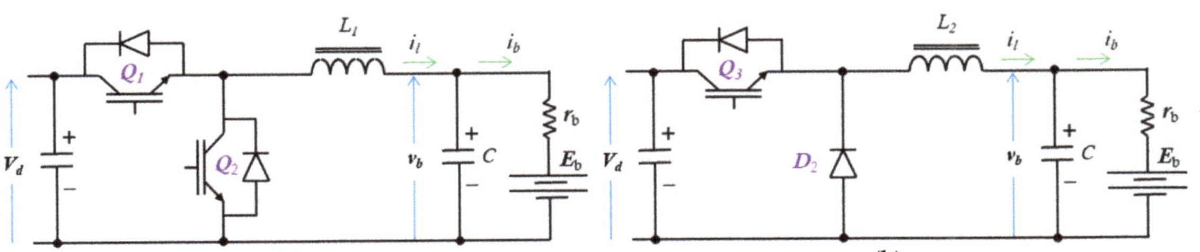

Figure 2. The circuit diagram of the (**a**) bidirectional converter and (**b**) unidirectional converter.

Charging mode:

$$\dot{x} = Ax + Bu_1 + D_1 \tag{6}$$

$$x = \begin{bmatrix} v_c \\ i_l \end{bmatrix},\ A = \begin{bmatrix} \frac{1}{C} & \frac{-1}{Cr_b} \\ 0 & \frac{-1}{L} \end{bmatrix},\ B = \begin{bmatrix} 0 \\ \frac{V_d}{L} \end{bmatrix},\ u_1 = \begin{cases} 1 & Q_1\ is\ on \\ 0 & Q_1\ is\ off \end{cases},\ D_1 = \begin{bmatrix} \frac{E_b}{Cr_b} \\ 0 \end{bmatrix} \tag{7}$$

where (L, C) are the filter's inductance and capacitance, (i_l, v_c) are the inductor's current and capacitor's voltage, and the DC link voltage is (V_d).

Discharging mode:

$$\dot{x} = Ax - Bu_2 + D_2 \tag{8}$$

$$u_2 = \begin{cases} 1 & Q_2 \text{ is on} \\ 0 & Q_2 \text{ is off} \end{cases}, \quad D_2 = \begin{bmatrix} \frac{E_b}{Cr_b} \\ \frac{V_{dc}}{L} \end{bmatrix} \tag{9}$$

4. Proposed Control Structure

The three controllers responsible for the stable operation of the proposed microgrid, as depicted in Figure 3, include the MPPT controller, the EV converter controller, and the DC link regulator using the DC–DC converter of the battery energy storage. The primary task of the MPPT regulator is to continually capture the maximum power output from the photovoltaic (PV) array. It achieves this by calculating and providing the appropriate duty cycle to the boost converter, ensuring the MPPT is effectively maintained. On the other hand, the EV converter controller plays a crucial role in ensuring a well-regulated charging process for the electric vehicle. Additionally, the converter's controller of the battery energy storage regulates both the DC bus voltage and the charge/discharge operations of the energy storage system (ESS). It comprises two loops: the first loop (i.e., outer loop) controls the DC bus voltage, while the second (i.e., inner loop) regulates the battery current. In this section, we will not delve into the details of the MPPT algorithm, as it is a relatively simple approach based on the perturb and observe (P&O) technique. However, maintaining the DC link voltage stable across the terminals of the EV charger is paramount for ensuring the electric vehicle's smooth functioning during the charging process. It is crucial to understand that the EV acts as a power-consuming load, and it relies on either the photovoltaic (PV) system or the battery to provide the necessary power, based on their availability.

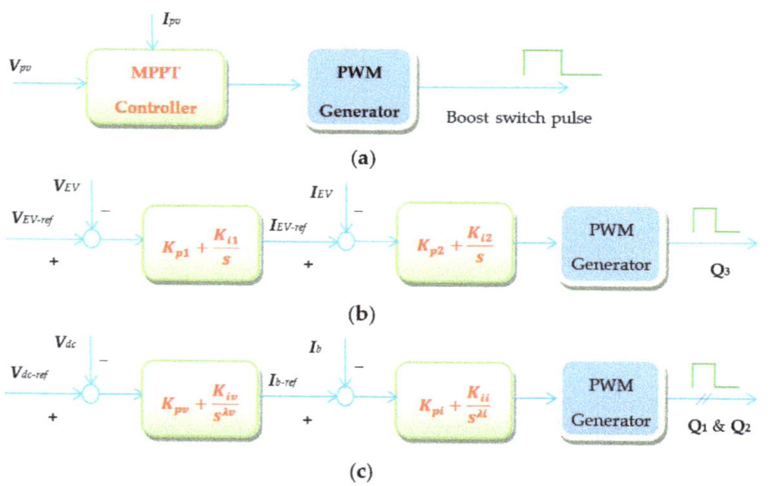

Figure 3. Structure of each control stage for the proposed standalone EV charging station: (**a**) MPPT regulator, (**b**) EV converter controller, and (**c**) DC bus based on the BESS converter.

4.1. Proposed Fractional-Order Controller

The objective of this regulator is to maintain a steady DC link voltage (V_{dc}) at a specific set value ($V_{d\text{-}ref}$). To achieve this, the controller adjusts the charging and discharging processes of the battery energy storage accordingly. The proposed controller is constructed using the fractional-order proportional integral (FOPI), which is employed to design both the inner loop, which maintains the DC link stability, and the outer loop, which regulates the battery current.

The utilization of fractional operators in the controller enables the representation of any real number through a sophisticated and versatile integral or differential notation [22]. The fundamental mathematical correlation between the FO differential and integral operator, for a certain order q, can be expressed in the following function:

$$D^q_{lb,ub}f(t) = \begin{cases} \frac{d^q}{dt^q} f(t) & q > 0 \\ f(t) & q = 0 \\ \int_{lb}^{ub} f(t) d\tau^{-q} & q < 0 \end{cases} \tag{10}$$

where l_b denotes the lower band and u_b denotes the upper band; when the value of the order q is positive (that is, $q > 0$), it is classified as a FO differential transfer function; conversely, when the value of the order q is negative (that is, $q < 0$), it is classified as a first-order integral.

Scholars have proposed various definitions in order to make the concept of fractional order (FO) more accessible, as it can be challenging to comprehend its physical implications. One such definition, known as the Riemann–Liouville (R-L) approach, offers a method to calculate the function's order derivative, aiding our understanding of the underlying principles of FO [23]:

$$D^q_{lb,ub}f(t) = \frac{1}{\Gamma(n-q)} \left(\frac{d}{dt}\right)^n \int_{lb}^{ub} \frac{f(\tau)}{(t-\tau)^{q-n+1}} d\tau \tag{11}$$

where $n \in \mathrm{N}$, $n - 1 < q < n$, and the Gamma function $\Gamma(w)$ is defined as,

$$\Gamma(w) = \int_0^\infty t^{w-1} e^{-t} dt \tag{12}$$

Equation (11)'s fractional derivative of R-L may be transformed using the Laplace method to provide Equation (13)'s answer [22]. Caputo's definition, a second definition related to the concept of FO, is used to express the time domain representation of the q order of the function *f(t)*, as in Equation (14) [24].

$$\mathcal{L}\left\{D^q_0 f(t)\right\} = s^q F(s) - \sum_{z=0}^{n-1} s^z \left(D_0^{q-z-1} f(t)\right)\Big|_{t=0} \tag{13}$$

$$D^q_{lb,ub}f(t) = \begin{cases} \frac{1}{\Gamma(n-q)} \left(\int_{lb}^{ub} \frac{f^n(\tau)}{(t-\tau)^{1-n+q}} d\tau \right) & n-1 < q < n \\ \left(\frac{d}{dt}\right)^n f(t) & q = n \end{cases} \tag{14}$$

When the Laplace transformation is applied to (14), the integral order of the equation is accompanied by an initial condition. This initial condition holds significant physical meaning and can be explained through Equation (15), where s is the Laplace operator.

$$\mathcal{L}\left\{D^q_0 f(t)\right\} = s^q F(s) - \sum_{z=0}^{n-1} s^{q-z-1} f^{(z)}(0) \tag{15}$$

Using the FO operators in the time domain involves performing complicated math calculations. To implement FO operators, we often use the recursive approximation method [25,26]. The Laplace transformation of the qth derivative is a way to represent it using a different mathematical formula, as follows:

$$s^q \approx K \prod_{k=-N}^{N} \frac{s + \omega'_k}{s + \omega_k} \tag{16}$$

where

$$K = \omega_h^q,$$
$$\omega_k' = \omega_b \left(\frac{\omega_h}{\omega_b}\right)^{\frac{k+N+(1-q)/2}{2N+1}},$$
$$\omega_k = \omega_b \left(\frac{\omega_h}{\omega_b}\right)^{\frac{k+N+(1+q)/2}{2N+1}},$$

and (N) is the approximation order of the Oustaloup method in the effective frequency range $[\omega_b, \omega_h]$ that can be chosen as $[-1000, 1000]$ rad/s. In the present study, we chose a value of N equivalent to 5.

The present study utilizes the FO proportional integral controller consisting of three tuning parameters: proportional gain (K_p), integral gain (K_i), and integral fractional order λ. Controllers constructed with these particular parameters have been discovered to exhibit enhanced stability, transient time, and overall precision in comparison to conventional PI regulators [27]. Furthermore, this controller offers greater adaptability and robustness in the face of system disruptions, enabling it to manage various disturbances effectively. Moreover, Equation (17) presents the comprehensive expression for the transfer function of the FOPI in Laplace form, denoted as $G_c(s)$, where λ is frequently in the range of [0, 1]. At the same time, Figure 4 illustrates the basic configuration of the control structure.

$$G_c(s) = K_p + K_i\left(\frac{1}{s}\right)^{\lambda} \tag{17}$$

The voltage across the DC link is monitored and compared to a set reference voltage. The FOPI controller is responsible for regulating the voltage difference by producing the desired battery current reference value. This reference value helps ensure proper control and management of the DC link voltage. Hence, the desired ESS's battery current, as determined by the reference, is then compared to the actual battery current. The FOPI controller utilizes this comparison to compute and modify the duty cycle of the bidirectional DC/DC converter. The dual loop control system ensures that the current drawn from the battery remains at a safe level, thus providing protection.

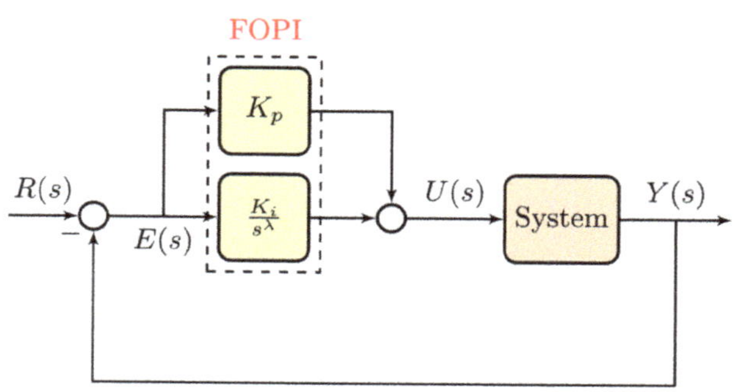

Figure 4. The FOPI controller's fundamental architecture.

4.2. Utilized Optimization Approach

Determining the FOPI parameters by trial and error can be complex, relying greatly on the practitioner's skill and knowledge. It can be quite a daunting challenge to identify suitable values for the proposed FOPI parameters. However, it is of utmost importance to carefully carry out this process to enhance the system's performance and ensure its stability, preventing disruptions. To achieve this, a metaheuristic optimization technique, GWO, is employed to identify the most optimal values for the FOPI controller's parameters.

Recently, there has been a surge in the popularity of a particular optimization technique that draws inspiration from the behaviors exhibited by gray wolves. This technique, known as gray wolf optimization (GWO), is highly regarded as a meta-heuristic approach [28]. In a wolf pack, a well-defined social structure based on dominance exists. These packs typically consist of members ranging from 5 to 12 individuals. Leading this pack is the alpha wolf (α), who holds the highest position of authority. Assisting the alpha is the beta wolf (β), a trusted ally who aids in making important decisions for the group. On the other end of the hierarchy, we find the omega wolf (ω) occupying the lowest rank. This wolf serves as a scapegoat and is often subjected to the blame for any mishaps or conflicts within the pack. The remaining wolves in the group are known as delta wolves (δ), and they dutifully follow the leadership of the alphas and betas, embodying a sense of loyalty and obedience. The GWO method involves several roles, as depicted in Figure 5. To successfully capture prey, the method consists of three primary steps: firstly, the wolves engage in a search to locate the prey and closely approach it; secondly, they encircle the prey, restricting its movement; and finally, they initiate an attack to capture and ultimately bring down the prey.

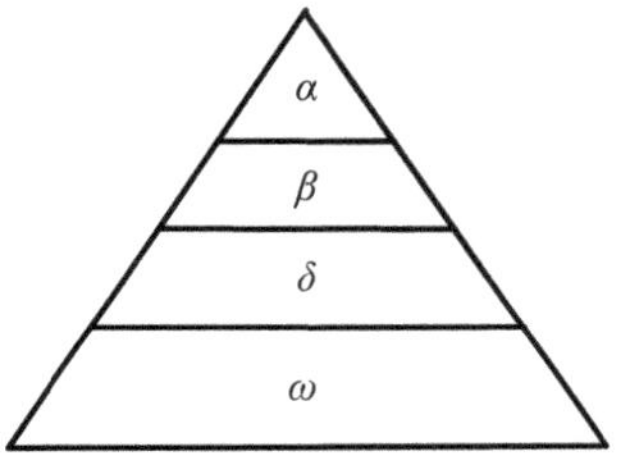

Figure 5. The general hierarchy structure of the grey wolves.

The environment in which the prey exists can be represented using Equation (18), where $\vec{P}_{pi}$ is the position of the prey, $\vec{P}_i$ is the position of the grey wolf, $\vec{S}$ is the distance between them, $\vec{A}$ and $\vec{C}$ are vectors computed from Equations (20) and (21), respectively.

$$\vec{S} = \left|\vec{C}\vec{P}_{pi} - \vec{P}_i\right| \tag{18}$$

$$\vec{P}_{i+1} = \vec{P}_{pi} - \vec{A}\vec{S} \tag{19}$$

$$\vec{A} = 2\vec{a}\,\vec{n}_1 - \vec{a} \tag{20}$$

$$\vec{C} = 2\vec{n}_2 \tag{21}$$

where

$$\vec{a} = 2 - \frac{2t}{Max\_iter} \tag{22}$$

where $\vec{n}_1$ and $\vec{n}_2$ are random numbers in the range of zero to one. The value of the factor $\vec{a}$ gradually decreases from two to zero as the number of iterations increases. The divergence technique is employed when looking for a prey location with $|A| > 1$, whereas the convergence technique is utilized to obtain prey in locations with $|A| < 1$. The hunting process involves the utilization of α, then β, and δ as presented in Equations (23) to (25).

$$\begin{cases} \vec{S}_\alpha = \left|\vec{C}_1\vec{P}_{\alpha i} - \vec{P}_i\right| \\ \vec{S}_\beta = \left|\vec{C}_2\vec{P}_{\beta i} - \vec{P}_i\right| \\ \vec{S}_\delta = \left|\vec{C}_3\vec{P}_{\delta i} - \vec{P}_i\right| \end{cases} \tag{23}$$

$$\begin{cases} \vec{P}_1 = \vec{P}_{\alpha i} - \vec{A}_1 \vec{S}_\alpha \\ \vec{P}_2 = \vec{P}_{\beta i} - \vec{A}_2 \vec{S}_\beta \\ \vec{P}_3 = \vec{P}_{\delta i} - \vec{A}_3 \vec{S}_\delta \end{cases} \tag{24}$$

$$\vec{P}_{i+1} = \frac{1}{3}\left(\vec{P}_1 + \vec{P}_2 + \vec{P}_3\right) \tag{25}$$

4.3. Objective Function Definition

The tuning process of the FOPI gains for the dual loop of the BESS controller is presented in Figure 6a. The following points summarize the GWO routine to find the optimal parameters of the proposed FOPI:

- Initialize the population of wolves, which are considered the candidate solutions for the FOPI parameters (i.e., six values).
- Simulate the proposed EVCS using the parameters generated from GWO.
- Calculate the objective function based on the integral square error (ISE) to quantify the control system's performance.
- Identify the population's alpha, beta, and delta wolves based on their fitness values. Alpha represents the best solution, beta the second-best, and delta the third-best.
- Update the positions of the remaining wolves in the population.
- Check the new updated positions of the grey wolves that remain within the constraints.
- Repeat the process until the termination criteria are achieved.

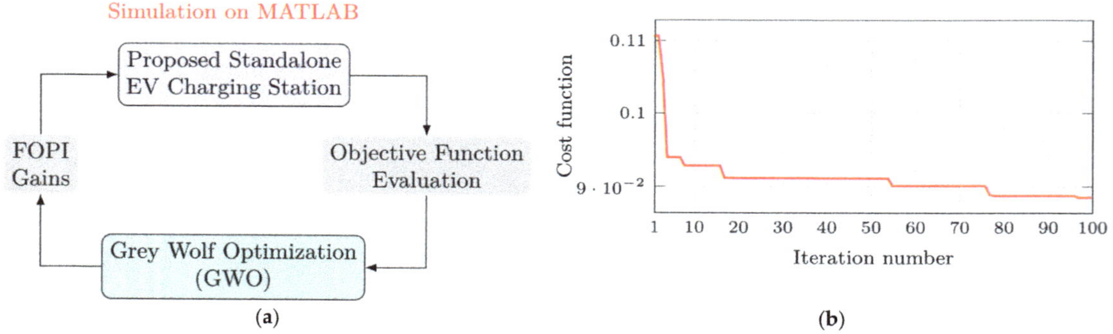

Figure 6. (**a**) The tuning procedure of the proposed FOPI gains to stabilize the EV charging station; (**b**) the convergence curve of the employed GWO to tune the FOPI gains.

Table 1 summarizes the associated parameters of the GWO. In Equation (26), the integral square error (ISE) is used as the fitness function for the GWO, in which t_{sim} is the simulation time. The GWO algorithm was executed on a personal computer with an Intel CoreTM i5-8265U CPU running at 1.60 GHz and 16 GB of RAM. The applied GWO's convergence curve is shown in Figure 6b, and Table 2 lists the best FOPI values. The GWO algorithm that has been proposed successfully attains a remarkably low fitness function value of around 0.0886. Furthermore, it is worth mentioning that the GWO algorithm can approach the optimal parameters in a relatively short period, typically within 80 iterations, when considering the *ISE* objective function.

$$ISE = \int_0^{tsim} \left(V_{dc-ref} - V_{dc}\right)^2 dt \tag{26}$$

Table 1. Parameters of the GWO for tuning the gains of the proposed FOPI.

Description	Value
Number of wolves	20
Number of iterations	100
Minimum range	[0,0,0,0,0,0]
Maximum range	[50,500,1,10,50,1]

Table 2. The optimal parameters of the FOPI using GWO.

Control Loop	Parameter	Value
Inner loop (DC link voltage stabilization)	K_{pv}	8.809
	K_{iv}	341.984
	λ_v	0.753
Outer loop (BESS current regulation)	K_{pi}	3.797
	K_{ii}	48.138
	λ_i	0.874

5. Simulation Results and Discussion

The suggested EVCS's microgrid, represented in Figure 3, is simulated using the Matlab/Simulink platform to validate the paper's hypothesis. The microgrid's parameters are listed in Table 3, including the PV array, ESS battery, EV battery, and converters. The converters' switching frequency utilized for the PWM carrier is 4 kHz.

Table 3. The parameters of the EVCS's microgrid.

Item	Item	Value
EV Battery	I_n, V_n	6.5 Ah, 3.7 V
	I_{ch}	3 A
	I_{disch}	5 A
ESS Battery	I_n, V_n	65 Ah, 12 V
	I_{ch}	13 A
PV Array (Copex-P120)	Type	Polycrystalline
	MPPT	120 W
	V_{oc}	19.2 V
	I_{sc}	8.82 A
Filter	C	1000 μF
	C_{dc}	2200 μF
	L	560 μH

Figure 7 displays the DC link voltage performances for FOPI and traditional PI controllers in response to step changes in solar irradiation. It may be seen in Figure 7a that the DC bus voltage closely follows its reference value precisely for both controllers. There is no steady-state error for both the FOPI and traditional PI controller responses. Nevertheless, the FOPI controller response has the lowest overshoot ($\leq$3.2%). On the other hand, the traditional PI controller response has an overshoot of $\leq$10.4%. Hence, the reduction in the system overshoot is (~1/3). Also, the settling time of the step response of DC link voltage using the FOPI controller has a lower value of $\leq$0.04 s. However, the settling time with the traditional PI controller is $\leq$ 0.1 s. Hence, the DC link voltage settling time reduction is (~2/5). These issues indicate a great improvement in the system response using the proposed FOPI controller. It is also noted that the DC link voltage response with the traditional PI controller has contaminated ripples that are not present in the response with the FOPI controller. The disturbance of solar radiation is presented in Figure 7b.

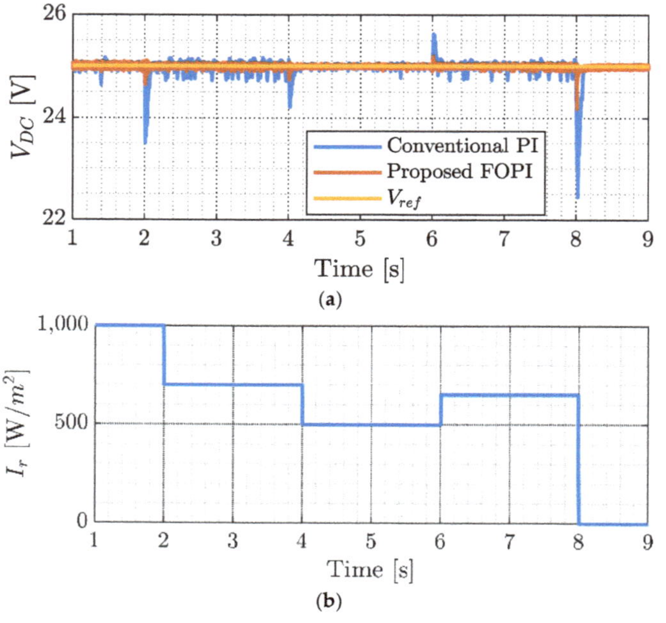

Figure 7. (**a**) Comparison of the DC bus voltage for both controllers under the variation of (**b**) PV insolation.

For the same disturbance profile in solar radiation, the charging station response with the proposed FOPI controller compared to the conventional PI controller is presented in Figure 8. The responses of the PV current corresponding to the step change in the solar radiation for both controllers are shown in Figure 8a,b. The PV current values match the MPPT circumstances. Figure 8c,d present the response of the PV voltage for both controllers. During the period [8 s to 9 s], the solar insolation drops to zero; therefore, the PV output voltage and current are zero. Figure 8e,f display the ESS battery's current response for both controllers. For both controllers, it tracks its reference fairly well. The DC bus voltage controller produces the reference value for the ESS battery's current. The charging and discharging procedures also keep track of their references and account for the radiation changes. Figure 8g,h display the ESS battery's voltage for the proposed FOPI and traditional PI controllers. When charged, its voltage rises, and when discharged, it falls. The performance of the ESS battery's SOC for the proposed FOPI and traditional PI controllers is shown in Figure 8i,j. The response of the SOC is nearly identical for both controllers. However, the charging and discharging processes are indicated. The insolation is about 70% for the first four seconds. Therefore, The EV battery may be charged, and the reserve may be stored in the ESS battery using the generated PV power. However, the insolation during the next two seconds, [4s to 6s], is just 50%, insufficient to provide the EV with energy. In order to make up for the decrease in solar energy, the storage battery drains. The insolation is 65% in the subsequent interval [6s to 8s], barely sufficient to charge the EV battery and keep the reserve in the ESS battery. Consequently, the SOC has a modest and positive slope. The sun insolation finally disappears during the period [8s to 9s]. As a result, no energy is created, and the ESS battery empties to make up for the solar energy.

Figure 9 shows the EV charging response with the proposed FOPI controller compared to the conventional PI controller. For both controllers in Figure 9a,b, the current of the EV closely matches its reference produced by the EV's converter controller. However, its response with the suggested FOPI controller is the best and has no overshoots or ripples. The voltage of the EV battery is shown in Figure 9c,d for both controllers. The replies remain the same as the EV battery charges continually. Figure 9e,f display the SOC of the EV battery for the two controllers, respectively. The responses remain the same as the EV battery charges continuously.

Figure 10 compares the proposed FOPI controller to the conventional PI controller for PV, ESS, and EV battery power. Both controllers' replies and how they track the PV's MPPT level remain constant. Figure 10a,b display the PV power responses, respectively. The irradiation in the first four seconds is around 70%. As a result, the PV energy produced is sufficient to charge the EV battery and store the remaining energy in the ESS battery. The following two seconds' insolation, from [4 s to 6 s], is 50%, insufficient to power the EV. In the ensuing period [6 s to 8 s], the insolation is 65%, barely sufficient to charge the EV battery and maintain the reserve in the ESS battery.

The sun's insolation completely fades between [8 s and 9 s]. As a result, no energy is produced, and to make up for the decrease in solar output, the ESS battery discharges. As seen in Figure 10c,d, the ESS battery drains to make up for the decrease in solar energy. Also, the procedures of charging/discharging account for the radiation changes. Figure 10e,f, which depict the two controllers, illustrate that the EV power is consistent under all circumstances. However, the proposed FOPI controller has a better response, free of disturbances and overshoots.

Table 4 summarizes the comparison of the maximum overshoot during each interval of the results using the proposed FOPI and the conventional PI controller.

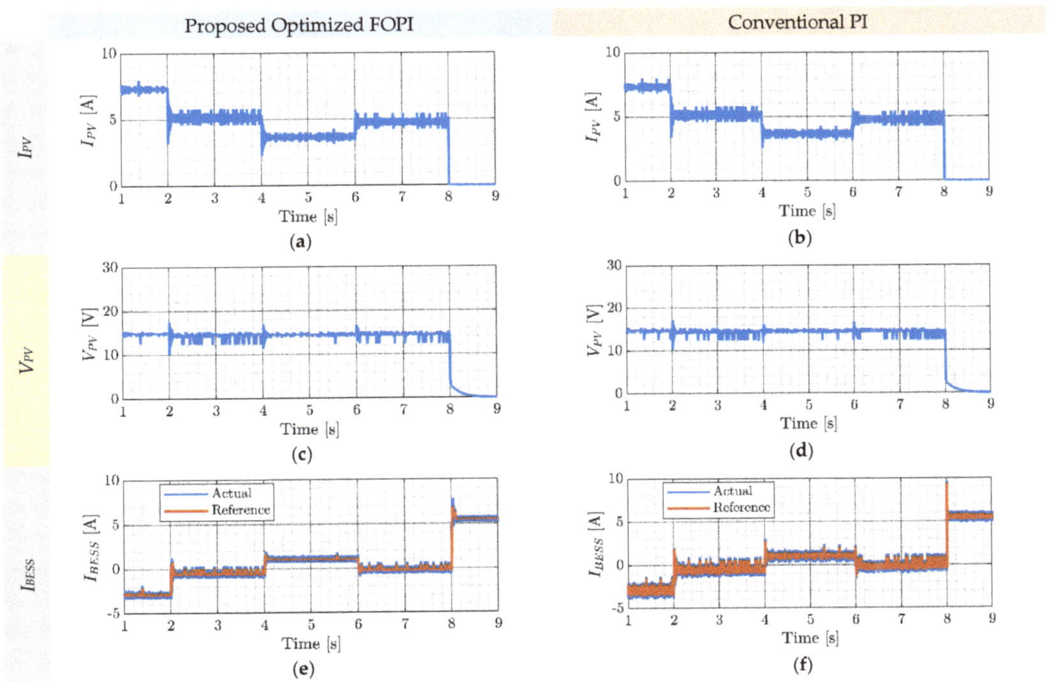

Figure 8. *Cont.*

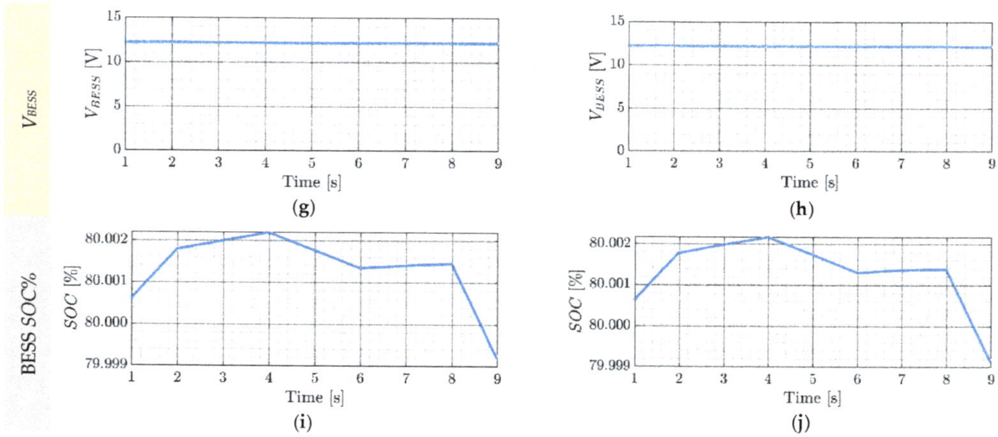

Figure 8. The charging station response with the proposed FOPI controller (**a**,**c**,**e**,**g**,**i**) compared to the conventional PI controller (**b**,**d**,**f**,**h**,**j**).

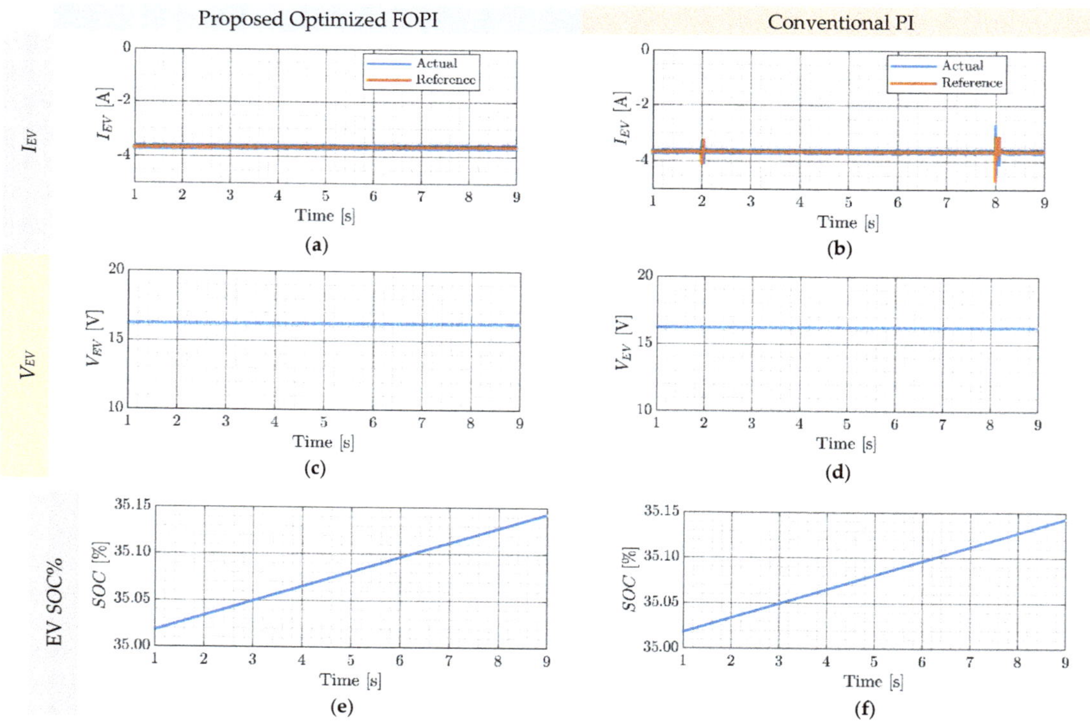

Figure 9. The EV charging response with the proposed FOPI controller (**a**,**c**,**e**) compared to the conventional PI controller (**b**,**d**,**f**).

Table 4. Maximum overshoot of the DC link voltage with the proposed FOPI and the conventional PI controller.

Time Interval (s)	Solar Irradiation Level (%)	Proposed FOPI	Conventional PI
1–2	100	1.6%	6%
2–4	70	1.2%	3.2%
4–6	50	0.8%	2.4%
6–8	65	3.2%	10.4%
8–9	0	0	0

Figure 11 depicts the FOPI controller's system efficiency fluctuation with insolation level and EV battery SOC. It should be noticed that both controllers' values for system efficiency are the same. As the EV_{SOC} increases, the system efficiency increases. However, that makes sense given that as the EV_{SOC} increases, the charging current decreases, the power losses decrease, and the EV's power decreases. Nevertheless, the power losses vary with the square of the current, but the EV's power is proportional to the current. Hence, microgrid efficiency increases. On the other hand, the efficiency decreases as the solar insolation level increases. That happened due to increased losses while charging the ESS battery of the EVCS's microgrid.

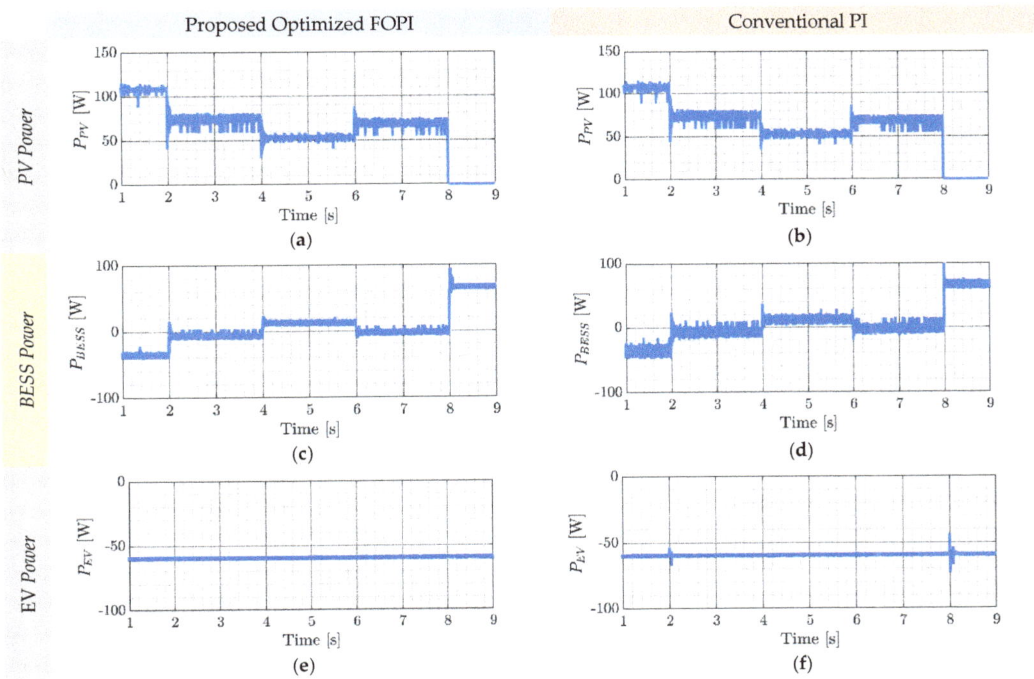

Figure 10. The power response of the charging station with the proposed FOPI controller (**a**,**c**,**e**) compared to the conventional PI controller (**b**,**d**,**f**).

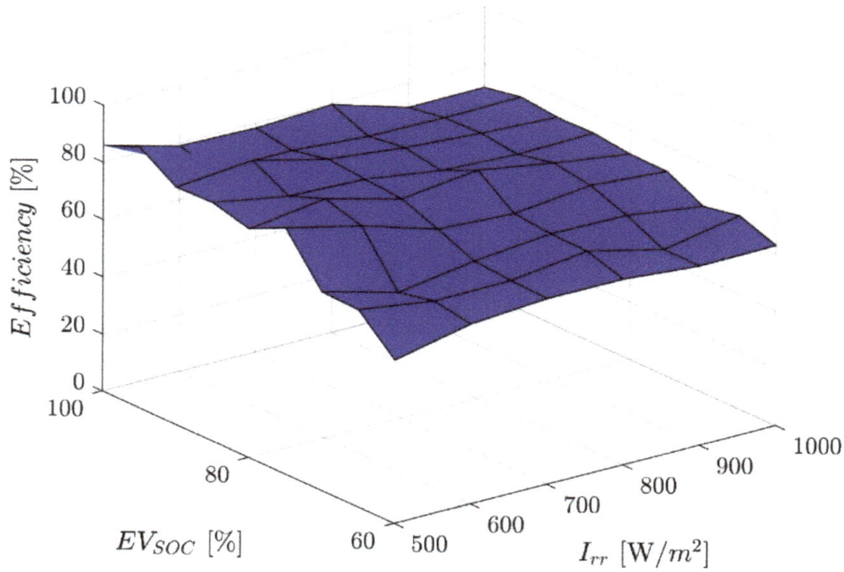

Figure 11. The system efficiency variations against the PV insolation and the EV battery SOC.

The impact of a ±10% variation in the value of the filter inductors (L_1, L_2) on the performance of the DC link voltage with the suggested controller is shown in Figure 12 to test the resilience of the proposed FOPI controller under parameter mismatches. It should be mentioned that the controller tracks the reference signal effectively. This demonstrates the control system's resistance to parameter mismatches. On the other hand, the overshoots are somewhat decreased, but the ripples are slightly enhanced.

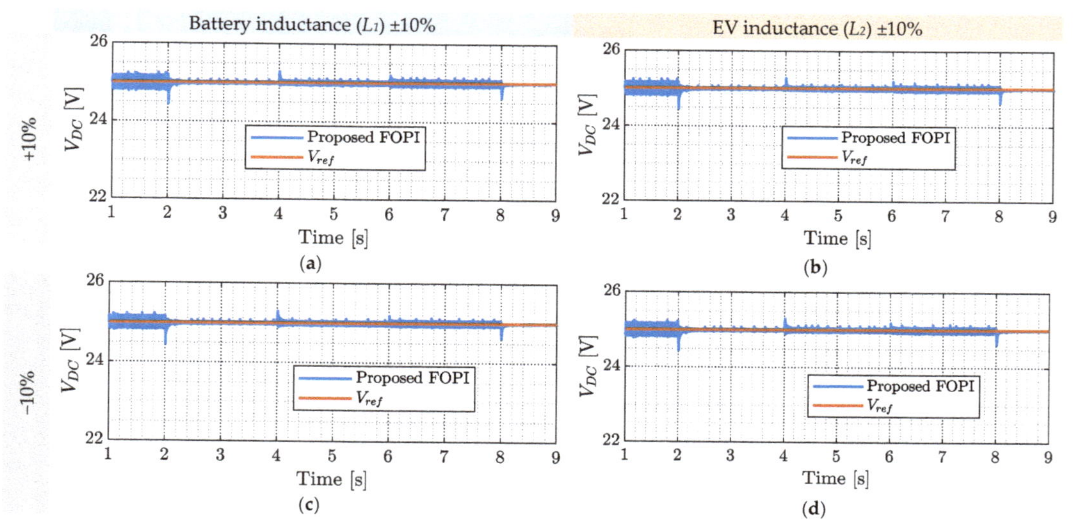

Figure 12. Effect of the (**a**,**c**) battery inductance (L_1) mismatch ±10% and (**b**,**d**) EV inductance (L_2) mismatch ±10% on the DC link voltage using the proposed FOPI controller.

Despite the merits of FOPI controllers compared to traditional PI controllers, they also come with certain limitations, such as complexity, tuning difficulty, limited industrial adoption, modeling challenges, and performance trade-offs. However, the last limitation is considered the main limitation of the proposed controller. Even though they can offer more flexibility, FOPI controllers may not necessarily perform better than conventional controllers in all circumstances. If not correctly tuned, the extra degrees of freedom in FOPI controllers might result in overfitting or inadequate performance. The recommended future work is the real-world implementation of the introduced system. The proposed power system has the advantage of simple scalability to the standard EVCS rating. Though many control software and hardware platforms are built around integer-order controllers like PID, FOPI controllers might not be readily supported in these systems. Hence, their integration and implementation are more challenging tasks.

6. Conclusions

The objective of this research paper is to introduce a new electric vehicle (EV) charging station that operates independently using a photovoltaic (PV) energy source. The proposed system is composed of several components, including a PV panel, boost converter, battery energy storage system (BESS), two DC/DC charging converters, and an EV battery. To ensure efficient operation, the control system encompasses three controllers: a maximum power point tracking (MPPT) controller, an EV charger controller, and a BESS controller. The battery controller utilizes the fractional-order technique in its dual loop control system to ensure a stable DC link voltage for the EV charger, even in the face of varying levels of sunlight. The grey wolf optimization (GWO) effectively achieved the desired optimal gains for the proposed FOPI controller. In order to test the effectiveness of the proposed EV charging station, a simulation using MATLAB/Simulink and the proposed optimal FOPI was conducted. The charging process of the EV battery remains stable regardless of the amount of sunlight, while the BESS effectively stores and compensates for variations in PV energy using the proposed optimal FOPI controller. Compared to the traditional PI controller, the proposed optimal FOPI controller reduces the voltage deviation of the DC link during changes in sunlight by approximately 8%. The converters' current and voltage controllers perform well and accurately follow their desired values. Moreover, the MPPT controller effectively monitors and adjusts to the optimal conditions of the PV system. In terms of future work, there is potential to further enhance the system by incorporating additional energy storage technologies. This would result in the development of hybrid distributed energy systems that facilitate the rapid charging of electric vehicles.

Author Contributions: Formal analysis and conceptualization were performed by S.A.Z.; modeling and controller tuning were assisted by A.B.; the article was evaluated and reviewed by H.A. (Hani Albalawi) and A.M.A.; and funding acquisition was aided by A.M.K. and H.A. (Hossam AbdelMeguid). All authors have read and agreed to the published version of the manuscript.

Funding: This research was funded by the Deputyship for Research and Innovation, Ministry of Education, Saudi Arabia through the University of Tabuk, grant number S-1443-0007.

Institutional Review Board Statement: Not applicable.

Informed Consent Statement: Not applicable.

Data Availability Statement: Not applicable.

Acknowledgments: The authors extend their appreciation to the Deputyship for Research and Innovation, Ministry of Education, Saudi Arabia for funding this research work through the project number (S-1443-0007).

Conflicts of Interest: The authors declare no conflict of interest.

References

1. Sun, X.; Li, Z.; Wang, X.; Li, C. Technology Development of Electric Vehicles: A Review. *Energies* **2020**, *13*, 90. [CrossRef]
2. Irle, R. Global EV Sales for the 1st Half of 2019. EV Volumes. 2019. Available online: http://www.ev-volumes.com/country/total-world-plug-in-vehicle-volumes/ (accessed on 20 November 2019).
3. Chakraborty, S.; Vu, H.-N.; Hasan, M.M.; Tran, D.-D.; Baghdadi, M.E.; Hegazy, O. DC-DC Converter Topologies for Electric Vehicles, Plug-in Hybrid Electric Vehicles and Fast Charging Stations: State of the Art and Future Trends. *Energies* **2019**, *12*, 1569. [CrossRef]
4. Luc, Vehicles & Charging Tips. Fastned. 2019. Available online: https://www.greencars.com/greencars-101/electric-car-charging-tips (accessed on 12 July 2023).
5. Minh, P.V.; Le Quang, S.; Pham, M.-H. Technical Economic Analysis of Photovoltaic-Powered Electric Vehicle Charging Stations under Different Solar Irradiation Conditions in Vietnam. *Sustainability* **2021**, *13*, 3528. [CrossRef]
6. Liu, Y.; Dong, H.; Wang, S.; Lan, M.; Zeng, M.; Zhang, S.; Yang, M.; Yin, S. An Optimization Approach Considering User Utility for the PV-Storage Charging Station Planning Process. *Processes* **2020**, *8*, 83. [CrossRef]
7. Francfort, J.; Salisbury, S.; Smart, J.; Garetson, T.; Karner, D. *Considerations for Corridor and Community DC Fast Charging Complex System Design*; Idaho National Lab: Idaho Falls, ID, USA, 2017.
8. Rafi, M.A.H.; Bauman, J. A Comprehensive Review of DC Fast Charging Stations with Energy Storage: Architectures, Power Converters, and Analysis. *IEEE Trans. Transp. Electrif.* **2021**, *7*, 345–368. [CrossRef]
9. Nicholas, M.; Hall, D. *Lessons Learned on Early Fast Electric Vehicle Charging Systems*; The National Academies of Sciences, Engineering, and Medicine: Washington, DC, USA, 2018.
10. Mouli, C.G.R.; Schijffelen, J.; Heuvel, M.; Kardolus, M.; Bauer, P. A 10 kW Solar-Powered Bidirectional EV Charger Compatible With Chademo and COMBO. *IEEE Trans. Power Electron.* **2019**, *34*, 1082–1098. [CrossRef]
11. Atawi, I.E.; Hendawi, E.; Zaid, S.A. Analysis and Design of a Standalone Electric Vehicle Charging Station Supplied by Photovoltaic Energy. *Processes* **2021**, *9*, 1246. [CrossRef]
12. Fathabadi, H. Novel grid-connected solar/wind powered electric vehicle charging station with vehicle-to-grid technology. *Energy* **2017**, *132*, 1–11. [CrossRef]
13. Awad, M.; Ibrahim, A.M.; Alaas, Z.M.; El-Shahat, A.; Omar, A.I. Design and analysis of an efficient photovoltaic energy-powered electric vehicle charging station using perturb and observe MPPT algorithm. *Front. Energy Res.* **2022**, *10*, 969482. [CrossRef]
14. Zhang, Y.; He, J.; Ionel, D.M. Modeling and control of a multiport converter based EV charging station with PV and battery. In Proceedings of the 2019 IEEE Transportation electrification conference and EXPO (ITEC), Detroit, MI, USA, 19–21 June 2019; pp. 1–5.
15. Singh, S.; Chauhan, P.; Singh, N.J. Feasibility of grid-connected solar-wind hybrid system with electric vehicle charging station. *J. Mod. Power Syst. Clean Energy* **2020**, *9*, 295–306. [CrossRef]
16. Ahmadi, M.; Kaleybar, H.J.; Brenna, M.; Castelli-Dezza, F.; Carmeli, M.S. DC Railway Micro Grid Adopting Renewable Energy and EV Fast Charging Station. In Proceedings of the 2021 IEEE International Conference on Environment and Electrical Engineering and 2021 IEEE Industrial and Commercial Power Systems Europe (EEEIC/I&CPS Europe), Bari, Italy, 7–10 September 2021; pp. 1–6.
17. Oulad-Abbou, D.; Doubabi, S.; Rachid, A.; García-Triviño, P.; Fernández-Ramírez, L.M.; Fernández-Ramírez, C.A.; Sarrias-Mena, R. Combined control of MPPT, output voltage regulation and capacitors voltage balance for three-level DC/DC boost converter in PV-EV charging stations. In Proceedings of the 2018 International Symposium on Power Electronics, Electrical Drives, Automation and Motion (SPEEDAM), Amalfi, Italy, 20–22 June 2018; pp. 372–376.
18. Traver, J.E.; Nuevo-Gallardo, C.; Tejado, I.; Fernández-Portales, J.; Ortega-Morán, J.F.; Pagador, J.B.; Vinagre, B.M. Cardiovascular Circulatory System and Left Carotid Model: A Fractional Approach to Disease Modeling. *Fractal Fract.* **2022**, *6*, 64. [CrossRef]
19. Muresan, C.I.; Birs, I.; Ionescu, C.; Dulf, E.H.; De Keyser, R. A Review of Recent Developments in Autotuning Methods for Fractional-Order Controllers. *Fractal Fract.* **2022**, *6*, 37. [CrossRef]
20. Yaghooti, B.; Hosseinzadeh, M.; Sinopoli, B. Constrained Control of Semilinear Fractional-Order Systems: Application in Drug Delivery Systems. In Proceedings of the IEEE Conference on Control Technology and Applications (CCTA), Montreal, QC, Canada, 24–26 August 2020; pp. 833–838. [CrossRef]
21. Saleem, O.; Ali, S.; Iqbal, J. Robust MPPT Control of Stand-Alone Photovoltaic Systems via Adaptive Self-Adjusting Fractional Order PID Controller. *Energies* **2023**, *16*, 5039. [CrossRef]
22. Zaid, S.A.; Bakeer, A.; Magdy, G.; Albalawi, H.; Kassem, A.M.; El-Shimy, M.E.; AbdelMeguid, H.; Manqarah, B. A New Intelligent Fractional-Order Load Frequency Control for Interconnected Modern Power Systems with Virtual Inertia Control. *Fractal Fract.* **2023**, *7*, 62. [CrossRef]
23. Bingi, K.; Prusty, R.; Singh, A. A Review on Fractional-Order Modelling and Control of Robotic Manipulators. *Fractal Fract.* **2023**, *7*, 77. [CrossRef]
24. Almasoudi, F.M.; Magdy, G.; Bakeer, A.; Alatawi, K.S.S.; Rihan, M. A New Load Frequency Control Technique for Hybrid Maritime Microgrids: Sophisticated Structure of Fractional-Order PIDA Controller. *Fractal Fract.* **2023**, *7*, 435. [CrossRef]
25. Oustaloup, A.; Levron, F.; Mathieu, B.; Nanot, F.M. Frequency-band complex noninteger differentiator: Characterization and synthesis. *IEEE Trans. Circuits Syst. I Fundam. Theory Appl.* **2000**, *47*, 25–39. [CrossRef]

26. Morsali, J.; Zare, K.; Hagh, M.T. Applying fractional order PID to design TCSC-based damping controller in coordination with automatic generation control of interconnected multi-source power system. *Eng. Sci. Technol. Int. J.* **2017**, *20*, 1–17. [CrossRef]
27. Mohamed, E.A.; Ahmed, E.M.; Elmelegi, A.; Aly, M.; Elbaksawi, O.; Mohamed, A.-A.A. An Optimized Hybrid Fractional Order Controller for Frequency Regulation in Multi-Area Power Systems. *IEEE Access* **2020**, *8*, 213899–213915. [CrossRef]
28. Salama, H.S.; Magdy, G.; Bakeer, A.; Vokony, I. Adaptive coordination control strategy of renewable energy sources, hydrogen production unit, and fuel cell for frequency regulation of a hybrid distributed power system. *Prot. Control. Mod. Power Syst.* **2022**, *7*, 34. [CrossRef]

fractal and fractional

MDPI

Article

Automatic Voltage Regulator Betterment Based on a New Fuzzy FOPI+FOPD Tuned by TLBO

Mokhtar Shouran [1,2] and Mohammed Alenezi [3,*]

[1] Libyan Centre for Engineering Research and Information Technology (LCERIT), Bani Walid P.O. Box 38645, Libya; moktar4520@gmail.com
[2] Control Engineering Department, College of Electronics Technology, Bani Walid P.O. Box 38645, Libya
[3] Wolfson Centre for Magnetics, School of Engineering, Cardiff University, Cardiff CF24 3AA, UK
* Correspondence: alenezim1@cardiff.ac.uk

Abstract: This paper presents a novel Fuzzy Logic Controller (FLC) framework aimed at enhancing the performance and stability of Automatic Voltage Regulators (AVRs) in power systems. The proposed system combines fuzzy control theory with the Fractional Order Proportional Integral Derivative (FOPID) technique and employs cascading control theory to significantly improve reliability and robustness. The unique control architecture, termed Fuzzy Fractional Order Proportional Integral (PI) plus Fractional Order Proportional Derivative (PD) plus Integral (Fuzzy FOPI+FOPD+I), integrates advanced control methodologies to achieve superior performance. To optimize the controller parameters, the Teaching–Learning-Based Optimization (TLBO) algorithm is utilized in conjunction with the Integral Time Absolute Error (ITAE) objective function, ensuring precise tuning for optimal control behavior. The methodology is validated through comparative analyses with controllers reported in prior studies, highlighting substantial improvements in performance metrics. Key findings demonstrate significant reductions in peak overshoot, peak undershoot, and settling time, emphasizing the proposed controller's effectiveness. Additionally, the robustness of the controller is extensively evaluated under challenging scenarios, including parameter uncertainties and load disturbances. Results confirm its ability to maintain stability and performance across a wide range of conditions, outperforming existing methods. This study presents a notable contribution by introducing an innovative control structure that addresses critical challenges in AVR systems, paving the way for more resilient and efficient power system operations.

Keywords: AVR; TLBO; fuzzy FOPI+FOPD+I; ITAE; power system stability

1. Introduction

The Automatic Voltage Regulator (AVR) is a critical component in modern power systems, tasked with maintaining the terminal voltage of synchronous generators at a stable and nominal level [1]. By dynamically adjusting the excitation voltage, the AVR ensures consistent power delivery and enhances overall system reliability. Its role becomes particularly vital in environments characterized by varying load demands and dynamic operational conditions [2].

Voltage regulation is a complex challenge due to the high inductance of generator field windings and the inherent variability of load conditions. These factors can lead to instability, slower response times, and deviations in terminal voltage. Thus, achieving a stable and rapid response from the AVR is crucial for ensuring power system stability and maintaining electrical power quality [3]. In scenarios such as no load, partial load, or full load, the AVR continuously adjusts the excitation current of the synchronous generator's exciter to counteract fluctuations and stabilize the terminal voltage [4].

While AVRs play a fundamental role in power systems, their performance is often constrained by system nonlinearity and the challenges associated with traditional control methods. Load variations, time delays, and the high inductance of alternator windings further exacerbate these challenges, necessitating advanced control strategies. A well-designed control system is essential to achieve the following goals [5,6]:

Improved Stability: Maintain terminal voltage consistency under all operating conditions.

Faster Response: Reduce transient times and ensure quick recovery from disturbances.

Reduced Overshoot: Minimize deviations during voltage adjustments.

From a control engineering perspective, ensuring the stable and reliable operation of modern power systems remains a significant challenge. This complexity arises from the continuous expansion of power system size, which increases structural and operational intricacies. Additionally, power systems are subject to a range of unpredictable internal and external disturbances, further complicating their dynamic behavior [7]. To address these challenges, researchers have extensively explored various control strategies for analyzing and improving the performance of AVR systems, aiming to achieve enhanced dynamic response and system stability.

Notably, the classical Proportional, Integral, and Derivative (PID) controller was the most commonly proposed technique to function as an AVR system [8]. This controller was implemented in the literature based on different techniques. However, there has been considerable consideration for using optimization algorithms to tune its parameters in order to achieve the optimal values of its parameters. In [9], a memory-based smoothed functional algorithm (MSFA) is used to efficiently tune PID controllers in AVR systems, enhancing performance while reducing computational costs. An AVR system based on a PID controller optimized by a sine–cosine algorithm (SCA) is proposed in [10]. This design has outperformed PID tuned by Artificial Bee Colony (ABC) and Differential Evolution (DE). A Symbiotic Organisms Search (SOS) algorithm is utilized in [11] to efficiently determine the optimum values of PID used for the AVR system. AVR based on a PID controller tuned by the Naked Mole Rat Algorithm (NMRA) is proposed in [12]; this controller has bettered the performance of PID tuned by other optimization tools in the literature. Moreover, recent studies have explored more advanced optimization algorithms for enhancing PID controllers in AVR systems, demonstrating significant improvements in performance and reliability. The Tree Seed Algorithm (TSA) [13] was shown to enhance stability and dynamic response, while the Grasshopper Optimization Algorithm (GOA) [14] effectively balances exploration and exploitation to improve transient and steady-state performance. The World Cup Optimization (WCO) algorithm [15], inspired by competitive dynamics, has achieved superior control performance with reduced overshoot and faster response times. Similarly, the Ant Lion Optimizer (ALO) [16] and Particle Swarm Optimization (PSO) [17] have demonstrated their ability to optimize PID parameters efficiently, achieving robust and accurate tuning with improved system stability. The Local Unimodal Sampling (LUS) algorithm [18] has provided efficient local search capabilities, resulting in better transient response. Moreover, the Water Wave Optimization (WWO) algorithm [19] has shown excellent performance in handling complex, nonlinear search spaces for AVR systems. Additionally, the hybrid Equilibrium Optimizer–Evaporation Rate Water Cycle Algorithm (EO–ERWCA) [20] has combined exploration and refinement capabilities to achieve optimal PID parameters, leading to faster settling times, reduced overshoot, and enhanced stability under various conditions. These advancements highlight the effectiveness of modern optimization techniques in addressing the challenges of PID tuning for AVR systems.

Although the aforementioned designs have demonstrated good performance, more advanced designs were proposed to achieve superior results. Consequently, to improve the performance of PID controllers for AVRs, enhanced versions were developed, leveraging various theories and optimization techniques. PID Acceleration (PIDA) is another version of PID but with three more gains; K_a, which is the acceleration gain, in addition to $\propto$ *and* β, which are the filter coefficients to improve stability. This controller was implemented in [21] for an AVR based on the whale optimization algorithm (WOA) and outperformed the traditional PID tuned by the same algorithm. PID and PIDA optimized by a novel hybrid technique between the Harmony Search (HS) and Dwarf Mongoose Optimization (DMO) algorithms was proposed in [22] as an AVR controller; the introduced designs illustrated robust performance under different conditions. A PID controller with filtered derivative action (PID-F) incorporates four independent parameters that are precisely optimized through the application of the Symbiotic Organism Search (SOS) algorithm is presented in [23]. Another unique design of PID is proposed in [24] for AVR applications. This design benefits from double derivative actions; PID plus second-order derivative (PIDD^2). The results showed that PIDD^2 provided a better dynamic response than the classical PID controller.

Fractional Order PID (FOPID or PI$^\lambda$D$^\mu$) was proposed as an effective solution for AVRs. The FOPID controller offers two additional degrees of flexibility compared to the traditional PID controller, enhancing its performance by introducing fractional orders (λ, which is the order of the integration, and μ, which is the order of the differentiator) alongside the standard three parameters [25]. FOPID optimized by an improved version of the Marine Predators Algorithm (MPA) within an AVR system is demonstrated in [26]. A modified smoothed function algorithm (MSFA)-based approach for tuning the FOPID controller of an AVR system is proposed in [27]; the algorithm is approved to minimize the number of function evaluations required per iteration and the controller provided a good performance as it outperformed other controllers in the literature. The study in [28] presents a FOPID controller for AVR systems, tuned with the Dandelion Optimizer (DO), achieving superior robustness and transient response compared to traditional PID controllers. It emphasizes its practical application for reliable voltage regulation. In [29], the Chaotic Yellow Saddle Goatfish Algorithm (CYSGA) is introduced for FOPID tuning, offering enhanced dynamic response and stability margins while effectively handling system uncertainties, making it a robust solution for AVR systems.

Furthermore, extended improvements to the FOPID were proposed in the literature to further enhance the performance of AVRs. The authors in [30] proposed a cascaded Real PID with Double Derivative (RPIDD^2) and Fractional Order Proportional Integral (FOPI) controller for AVR systems. This advanced controller enhances precision, stability, and adaptability in voltage regulation. The RPIDD^2 controller improves phase margin and stability, while the FOPI controller adds flexibility through fractional order terms. This was fine-tuned using the Quadratic Wavelet-Enhanced Gradient-Based Optimization (QWGBO) algorithm, where the cascaded RPIDD^2-FOPI controller outperforms conventional methods by minimizing overshoot, rise time, and settling time. The study in [31] presents a comparative study of AVR control systems, introducing a novel Optimized Proportional Integral Derivative Model Reference Fractional Adaptive (OPIDMR-FA) controller. This controller integrates model reference adaptive control and fractional order design principles to achieve superior dynamic performance. Compared to traditional controllers such as PID, PIDA, and FAOPID, OPIDMR-FA demonstrated smoother transient responses with zero steady-state error. The research in [32] introduces a modified Tilt-Integral-Derivative

(TID) controller, named $I^{\lambda}DND^2N^2 - T$, for AVR systems. This fractional order controller integrates advanced tuning flexibility and enhanced robustness through second-order derivative terms and low-pass filters. The proposed $I^{\lambda}DND^2N^2 - T$ tuned using an Equilibrium Optimizer (EO) demonstrated superior performance over traditional PID, $PIDD^2$, and FOPID controllers, showcasing its efficacy in improving AVR system reliability.

While traditional controllers like PID are prevalent in AVR applications, advanced control strategies such as Sliding Mode Control (SMC), Model Predictive Control (MPC), H-infinity (H_∞) control, adaptive control, and Linear Quadratic Regulator (LQR) have seen comparatively limited implementation. These methodologies offer robust performance in the presence of system uncertainties and disturbances.

SMC is recognized for its robustness against matched uncertainties and was applied in various domains, including power electronics. It ensures system trajectories reach and maintain a sliding manifold, providing insensitivity to certain disturbances. However, its practical application can be hindered by chattering phenomena and implementation complexity [33,34].

MPC handles multi-variable control and constraints by predicting future system behavior and optimizing control actions accordingly. This predictive capability makes it suitable for complex systems, but its reliance on accurate models and computational intensity can limit real-time applications [35–37].

H_∞ control techniques are designed to manage systems with nonlinearities and uncertainties, offering robust performance under varying conditions. They aim to achieve stabilization with guaranteed performance by formulating the control problem as a mathematical optimization problem. The complexity of designing H∞ controllers and the need for precise system models can restrict their practical use [38,39].

Adaptive control adjusts controller parameters in real-time to cope with system changes, enhancing performance in uncertain environments. It was applied in various fields, including aerospace, to handle systems with significant uncertainties. The design and analysis of adaptive controllers can be complex, which may limit their widespread adoption [40].

LQR is an optimal control strategy that minimizes a quadratic cost function, balancing state error and control effort. It provides a systematic approach to designing controllers for linear systems but may not directly address robustness to model uncertainties [41,42].

Despite their theoretical advantages, the practical application of these advanced control strategies in AVR systems remains less common, possibly due to implementation complexity and the need for precise system models.

Recent advancements in AVR systems have explored the integration of Fuzzy Logic Controllers (FLCs) to enhance performance under varying operational conditions. FLCs, known for their ability to handle system nonlinearities and uncertainties, were effectively applied to AVR systems to improve voltage regulation. A robust fuzzy PI and Type-2 Fuzzy PI controllers optimized using metaheuristic algorithms like Artificial Bee Colony (ABC) and PSO, which demonstrate improved robustness and transient response under varying parameters and disturbances, are presented in [43]. The study presented in [44] implements an FLC for AVR systems, showing significant enhancements in dynamic performance and voltage stability; the FLC-based AVR system outperformed traditional methods in mitigating load variations and achieving faster stabilization. Similarly, the integration of a fuzzy PID controller with Programmable Logic Controllers (PLCs), tuned through genetic algorithms and radial-basis function networks, achieves efficient voltage regulation and reduces oscillations, combining the adaptability of fuzzy logic with the reliability of PLCs, as studied in [45]. Additionally, the study in [46] presents an FLC designed with triangular membership functions that outperforms traditional PID controllers, achieving better performance, and ensuring superior transient and steady-state performance. Table 1 provides an overview of the controllers and optimization algorithms utilized for tuning AVR control systems. It highlights the diverse range of controllers and algorithms implemented in the literature.

Table 1. An outline of the control methods and optimization tools of AVR systems in the literature.

Reference	Control Method	Optimization Tool
Ref. [9]	PID	Memory-based smoothed functional algorithm (MSFA)
Ref. [10]	PID	Sine–Cosine Algorithm (SCA)
Ref. [11]	PID	Symbiotic Organisms Search (SOS) algorithm
Ref. [12]	PID	Naked Mole Rat Algorithm (NMRA)
Ref. [13]	PID	Tree Seed Algorithm (TSA)
Ref. [14]	PID	Grasshopper Optimization Algorithm (GOA)
Ref. [15]	PID	World Cup Optimization (WCO) algorithm
Ref. [16]	PID	Ant Lion Optimizer (ALO)
Ref. [17]	PID	Particle Swarm Optimization (PSO)
Ref. [18]	PID	Local Unimodal Sampling (LUS) algorithm
Ref. [19]	PID	Water Wave Optimization (WWO)
Ref. [20]	PID	Equilibrium Optimizer–Evaporation Rate Water Cycle Algorithm (EO–ERWCA)
Ref. [21]	PIDA	Whale optimization algorithm (WOA)
Ref. [22]	PIDA	Hybrid Harmony Search (HS) and Dwarf Mongoose Optimization (DMO) algorithms
Ref. [23]	PIDF	Symbiotic Organism Search (SOS) algorithm
Ref. [24]	$PIDD^2$	Particle Swarm Optimization (PSO)
Ref. [26]	FOPID	Marine Predators Algorithm (MPA)
Ref. [27]	FOPID	A Modified Smoothed Function Algorithm (MSFA)
Ref. [28]	FOPID	Dandelion Optimizer (DO)
Ref. [29]	FOPID	Chaotic Yellow Saddle Goatfish Algorithm (CYSGA)
Ref. [30]	$RPIDD^2$-FOPI	Quadratic Wavelet-Enhanced Gradient-Based Optimization (QWGBO) algorithm
Ref. [31]	FOPID-MRAC	Manually tuned
Ref. [32]	$I^{\lambda}DND^2N^2-T$	Equilibrium Optimizer (EO)
Ref. [33]	SMC	Sine–Cosine Algorithm and Grey Wolf Optimization
Ref. [34]	SMC	Sine–Cosine Algorithm (SCA)
Ref. [35]	MPC	Mathematically calculated
Ref. [36]	MPC	Marine Predators Algorithm (MPA)
Ref. [38]	H_2/H_∞	Fuzzy logic
Ref. [39]	H_∞ and μ-analysis	Mathematically calculated
Ref. [40]	Adaptive	Mathematically calculated
Ref. [41]	LQR	Mathematically calculated
Ref. [42]	LQR	Biogeography-Based Optimization (BBO)
Ref. [43]	Fuzzy PI	Artificial Bee Colony (ABC) and PSO
Ref. [44]	Fuzzy Control	Mathematically tuned
Ref. [45]	Fuzzy and PLC	Genetic algorithm
Ref. [46]	Fuzzy Control	Mathematically tuned
Proposed	Fuzzy FOPI+FOPD + I	Teaching–Learning-Based Optimization (TLBO) algorithm

In AVR systems, achieving an optimal balance between response speed and voltage stability is crucial. A fast response with overshoot enables quick voltage stabilization, which is essential in dynamic environments with frequent load changes. However, this can temporarily exceed safe voltage limits, potentially harming sensitive equipment. Conversely, a slower response without overshoot maintains voltage within desired thresholds, safeguarding equipment, but may result in temporary undervoltage or overvoltage during load transitions. Therefore, the ideal controller design depends on specific application requirements and system constraints, aiming to harmonize rapid response with minimal overshoot to ensure both performance and safety [47].

While the aforementioned techniques demonstrate distinct merits and have substantially addressed the challenges associated with Automatic Voltage Regulator (AVR) systems, their implementation in power systems is not without limitations. Traditional controllers, for instance, often fail to deliver adequate performance under nonlinear operating conditions or high system sensitivity. Despite their low cost being a significant advantage, these controllers lack the adaptability required for modern AVR applications. On the other hand, advanced methods such as adaptive control and SMC provide effective solutions for AVR systems but are hindered by the complexity of their design and the high computational overhead they entail. Similarly, while various fuzzy control approaches were proposed for voltage regulation, most fail to incorporate reliability considerations into their design.

To overcome these challenges, this study proposes an innovative FLC configuration designed to ensure a high degree of reliability for AVR systems. The proposed controller, termed Fuzzy cascade FOPI plus FOPD plus I (Fuzzy FOPI+FOPD + I), integrates a two-input-one-output fuzzy controller with a cascaded FOPI and FOPD controller at its output terminal in addition to an integral controller to further enhance the reliability of the AVR system. To achieve optimal performance, the controller parameters are fine-tuned using the Teaching–Learning-Based Optimization (TLBO) algorithm, a robust and efficient optimization technique. Concisely, the key contributions of the study are as follows:

1. Novel Controller Design: Introduces a unique Fuzzy Logic Controller configuration for AVR systems, featuring a two-input-one-output fuzzy logic design combined with a cascade FOPI-FOPD controller, forming the Fuzzy FOPI-FOPD+I controller.
2. Performance Validation: Demonstrates the superiority of the proposed design through comparative analysis with traditional PID controllers optimized using multiple state-of-the-art tools.
3. Robustness Assessment: Examines the resilience of the proposed controller under parametric uncertainties and load disturbance, validating its robustness and reliability in diverse operating conditions.

The rest of the paper is categorized as follows: Section 2 describes the AVR system and its mathematical model with dynamic response without a controller. Section 3 details the proposed controller and provides a brief idea about the mechanism of the proposed algorithm and the used objective function. Section 4 presents the main results and illustrates a detailed discussion of the findings. Section 5 investigates the robustness of the testbed system with the proposed controller. Finally, Section 6 provides a concise conclusion, mentions the limitations of this study, and proposes different directions for possible future works.

2. AVR System

Figure 1 illustrates a simplified general model of the AVR system developed for this study, which regulates the generator's output voltage by controlling its excitation through an error signal generated by comparing the sensed terminal voltage of the generator with a predefined reference voltage. During a voltage drop, typically caused by increased loading conditions, the error signal increases positively, prompting the AVR to enhance the generator's excitation until the sensed voltage aligns with the reference value, after which the excitation stabilizes to maintain a constant supply voltage. Conversely, when the load decreases, a voltage rise occurs, generating a negative error signal that causes the AVR to reduce the generator's excitation until equilibrium is achieved, ensuring voltage stability. The transfer function of the AVR system is modeled using Laplace transform. Each component of the AVR—

namely the amplifier, exciter, generator, and sensor—is linearized for analytical convenience. This system has been widely used in the literature for AVR applications.

The closed-loop transfer function of the model presented in Figure 2 based on the parameters given in Table 2 is illustrated in Equation (1). This model was an extensively investigated system in the literature to study the dynamic behavior of different control concepts for AVR systems [13]. The step response of the system is demonstrated in Figure 3 with the following characteristics: peak amplitude = 1.5066 pu; peak time = 0.7522 s; settling time = 6.9865 s; rise time = 0.2607 s; and steady-state error = 0.091 pu. The model has one zero at −100, two real poles at −99.9712 + 0i, and −12.4892 + 0i, and two complex poles at −0.5198 + 4.6642i and −0.5198 − 4.6642i, as shown in the root locus plot presented in Figure 4.

$$\mathrm{TF_{AVR}} = \frac{0.1\mathrm{s} + 10}{0.0004\,\mathrm{s}^4 + 0.045\,\mathrm{s}^3 + 0.555\,\mathrm{s}^2 + 1.51\,\mathrm{s} + 11}$$

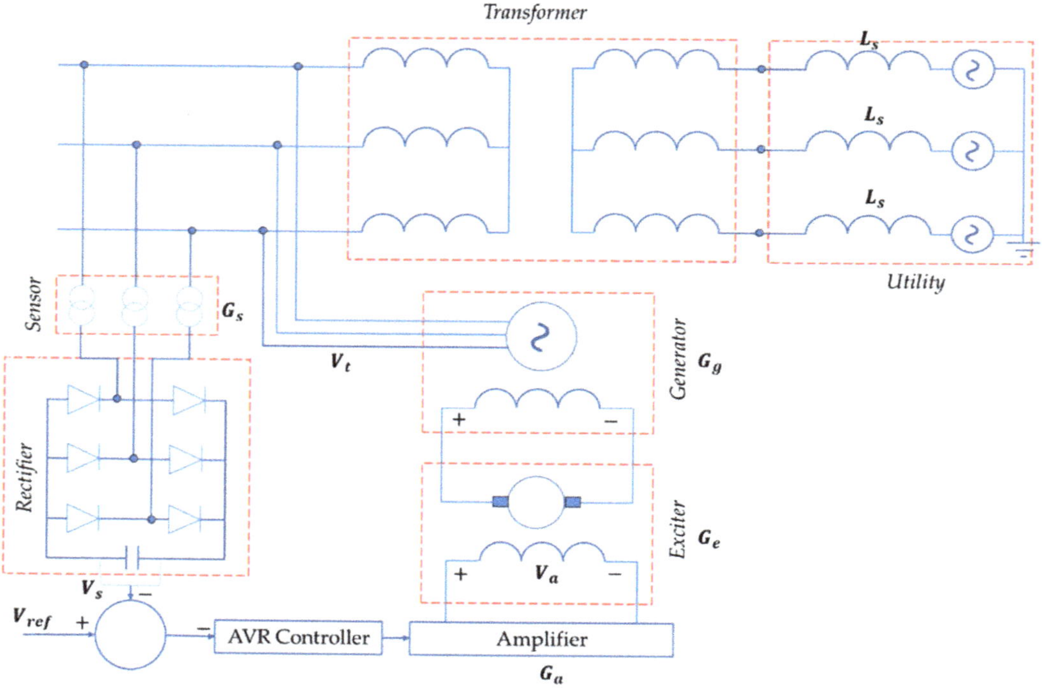

Figure 1. Schematic representation of generalized AVR components.

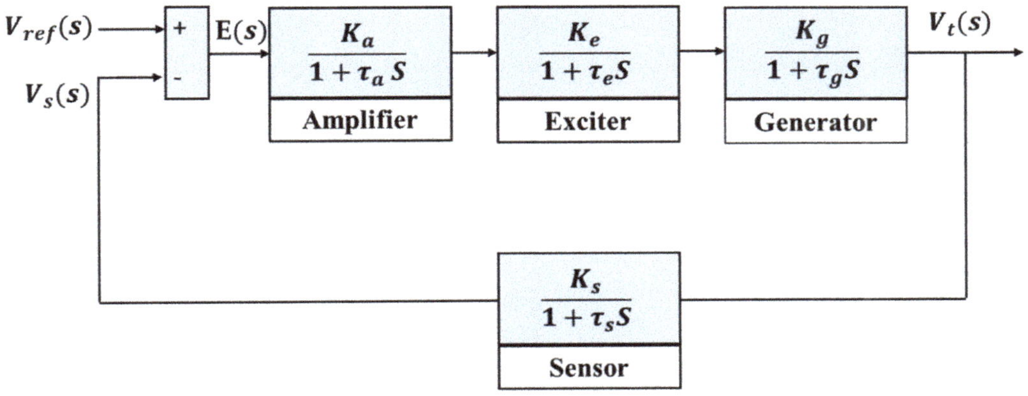

Figure 2. The traditional AVR model without a controller.

Table 2. The range and implemented values of the AVR model.

AVR Component	Transfer Function	Value Range	Implemented Value
Generator	$\frac{K_g}{1+\tau_g}$	$0.7 \leq K_g \leq 1$ $1 \leq \tau_g \leq 2$	$K_g = 1, \tau_g = 1$
Excitor	$\frac{K_e}{1+\tau_e}$	$1 \leq K_e \leq 10$ $0.4 \leq \tau_e \leq 1$	$K_e = 1, \tau_e = 0.4$
Sensor	$\frac{K_s}{1+\tau_s}$	$1 \leq K_s \leq 2$ $0.001 \leq \tau_s \leq 0.06$	$K_s = 1, \tau_s = 0.01$
Amplifier	$\frac{K_a}{1+\tau_a}$	$10 \leq K_a \leq 400$ $0.02 \leq \tau_a \leq 1$	$K_g = 10, \tau_g = 0.1$

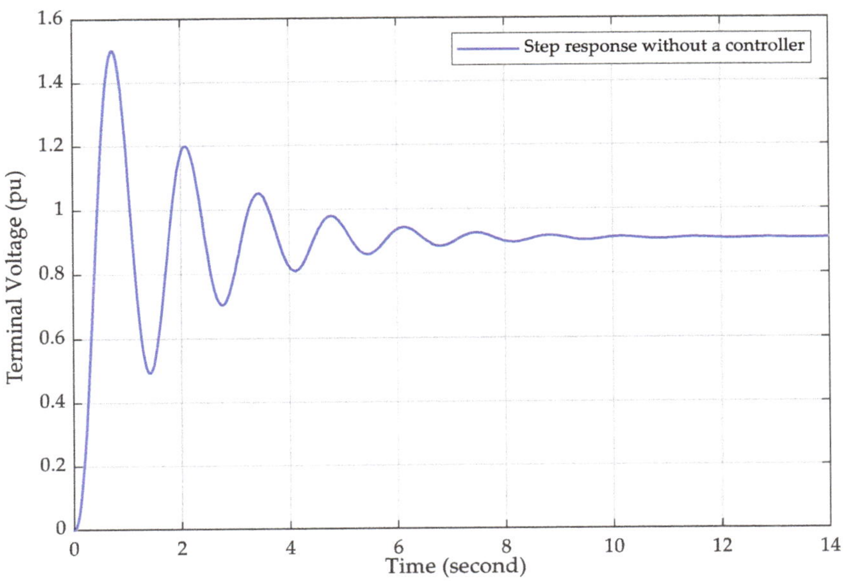

Figure 3. Step response of AVR system without controller.

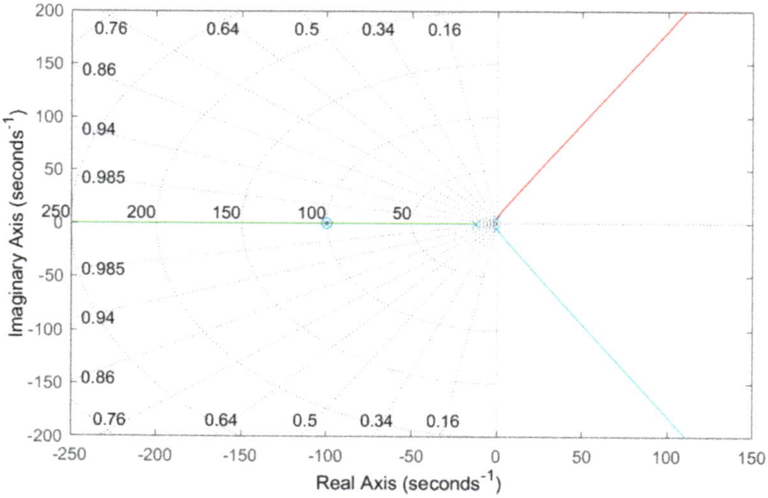

Figure 4. A root locus diagram of the AVR system without a controller.

From the characteristics presented above, it is clear that a proper control design is needed to improve the overall performance of the system in its transient and steady-state responses. The design of the controller is investigated in Section 3.

3. The Proposed Controller and Implemented Optimization Tool and Cost Function

3.1. The Proposed Fuzzy FOPI Plus FOPD Plus I Controller

Figure 5 illustrates the structural diagram of the proposed controller, which integrates two main components: a Fuzzy FOPI plus FOPID controller and a classical I controller. The fuzzy controller is designed with two inputs: the error and the derivative of the error. These inputs are scaled by two input gain factors, denoted as K_1 *and* K_2, respectively. The controller produces a single output, which is subsequently connected to the FOPI controller. To maintain simplicity and minimize computational time, five triangular membership functions are employed for both the input and output variables. These membership functions, as shown in Figure 6, are defined as Negative Big, Negative Small, Zero, Positive Small, and Positive Big.

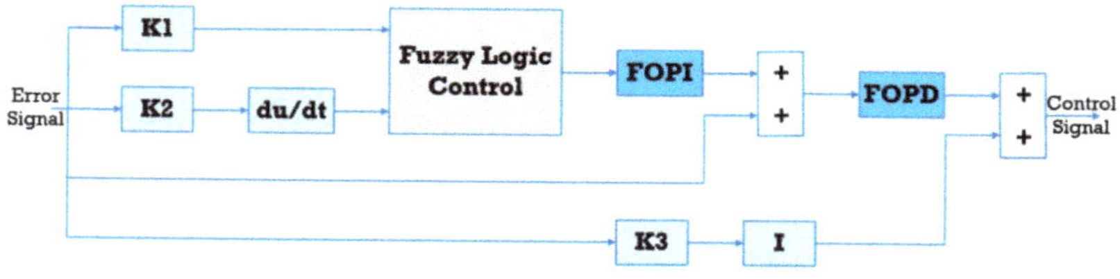

Figure 5. The proposed Fuzzy FOPI+FOPD+I AVR system.

The controller utilizes 25 fuzzy rules to generate its output, with the complete rule base provided in Table 3. These rules were meticulously developed based on a detailed analysis of the dynamic behavior of the testbed system. The Mamdani inference system is used for the fuzzification process, ensuring efficient mapping of crisp inputs to fuzzy values. For the defuzzification stage, the centroid method is applied, converting the fuzzy output into a crisp real-valued control signal. This design ensures an effective and computationally efficient implementation of the controller.

Table 3. The rule base of the fuzzy controller.

Error	Change in Error				
	NB	**NS**	**Z**	**PS**	**PB**
NB	NB	NB	NB	NS	Z
NS	NB	NB	NS	Z	PS
Z	NB	NS	Z	BS	PB
PS	NS	Z	PS	PB	PB
PB	Z	PS	PB	PB	PB

The output of the fuzzy controller serves as an input of the FOPI controller. The FOPI controller's transfer function is given in Equation (1), where λ is the order of integration, K_{p1} is the proportional gain, and K_i is the integral gain.

$$\text{FOPI Controller}_c(S) = K_{p1} + \frac{K_i}{S^{\lambda}} \tag{1}$$

The FOPD controller, in turn, receives its input from two signals: the output of the FOPI controller and the error signal. The FOPD controller's transfer function is given in Equation (2), where μ is the order of differentiator, K_{p2} is the proportional gain, and K_d is the derivative gain.

$$\text{FOPD Controller}_c(S) = K_{p2} + K_d S^{\mu} \tag{2}$$

The final control signal is generated by summing the output of the FOPD controller with the output of the classical I controller with a gain named K_3. This comprehensive approach enhances the performance and robustness of the proposed control strategy.

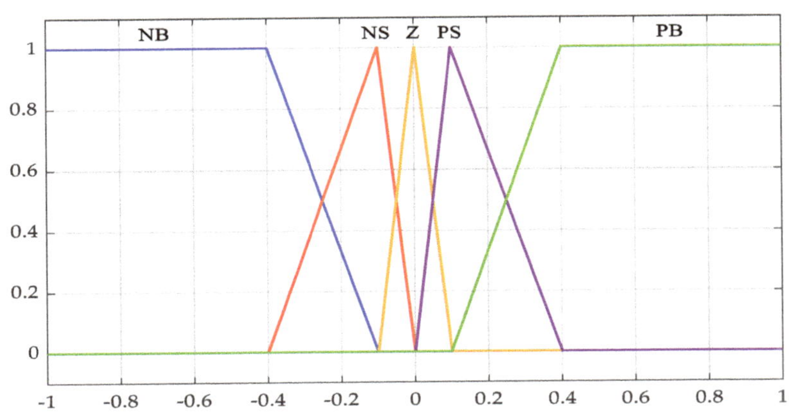

Figure 6. The membership functions of the fuzzy controller.

3.2. The Implemented TLBO Algorithm and the Utilized Cost Function

To enhance the performance of a controller, it is critical to thoroughly understand the behavior of the controlled plant and design the controller to achieve the desired system dynamics. A pivotal step in the control design process is the determination of the optimal parameters for the controller. However, accurately estimating these parameters to ensure optimal performance often proves challenging. To address these limitations, soft computing based on optimization algorithms were introduced, offering a robust solution to this challenge and other complex optimization problems.

Therefore, in order to achieve most of the proposed fuzzy controller for AVR, a powerful and modern heuristic algorithm, the Teaching–Learning-Based Optimization (TLBO) algorithm, is used to find the optimal values of the controller's parameters: K_1, K_2, K_3, K_{P1}, K_I, λ, K_{P2}, K_D, and μ, as explained in Figure 7.

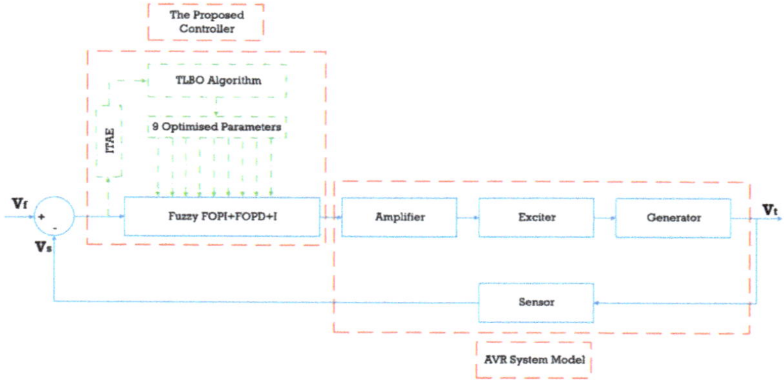

Figure 7. The TLBO-tuned Fuzzy FOPI+FOPD+I for AVR.

TLBO is a heuristic optimization technique introduced in [48]. It is inspired by the teaching and learning processes that occur in a classroom setting, particularly the interactions between a teacher and students. This approach models the influence of a teacher on students' academic performance. In this analogy, the teacher represents the most knowledgeable and skilled individual, providing quality education to enhance students' learning. The improvement in students' performance depends not only on the teacher's effectiveness but also on the students' own efforts and their collaborative knowledge sharing with classmates. The performance of the students is ultimately assessed based on their grades or outcomes in the class.

TLBO is a nature-inspired, parameter-free algorithm that employs a population of solutions to converge toward the optimal result. For TLBO, the population is likened to students in a class, while the control variables represent the subjects they study. The algorithm operates in two main phases, the teaching phase and the learning phase, as detailed below [49]. The flowchart shows the TLBO methodology, as presented in Figure 8

Teaching Phase: This is the first part of the algorithm in which the teacher tries the level best according to their potential to improve the performance of students in a class. The average result of the classroom is improved by the influence of the teacher to some extent, i.e., μ_j^k. If the new average grade of jth subject at kth iteration is μ_{newj}^k, the difference between the existing mean and the new mean of the jth subject at the kth iteration may be given as follows:

$$\mu_{\text{diffj}}^{\text{k}} = \text{rand}\left(\mu_{\text{newj}}^{\text{k}} - (\text{TF})\ \mu_{\text{j}}^{\text{k}}\right) \tag{3}$$

where TF is the teaching factor, which is evaluated randomly by the following equation:

$$\text{TF} = \text{round}\ (1+\text{rand}(0,1)) \tag{4}$$

The grade of the *j*th subject of the *i*th student at *k* + 1*th* iteration is updated as follows:

$$x_{ij}^{k+1} = x_{ij}^{k} + \mu_{diffj}^{k} \tag{5}$$

Learning Phase: It is the last part of the algorithm where students upgrade their results by mutual interactions among themselves. Any two students such as x_i and x_j are randomly selected from the class and their grades are updated based on the better student. The learning process may be expressed mathematically as follows:

$$x_{ij}^{k+1} = x_{ij}^{k} + \text{rand} \times \left(x_{ij}^{k} - x_{lj}^{k}\right) \text{ if } f(x_i) < f(x_l) \tag{6}$$

Otherwise,

$$x_{ij}^{k+1} = x_{ij}^{k} + \text{rand} \times \left(x_{lj}^{k} - x_{ij}^{k}\right) \text{ if } f(x_l) < f(x_i) \tag{7}$$

where x_{ij}^{k+1} and x_{ij}^{k} are the grade point of the *j*th subject of the *i*th student at the *k*th and $k+1th$ iteration; x_{ij}^{k} is the grade point of the *j*th subject of the lth student (randomly selected) at *k*th iteration; and $f(x_i)$ is the overall grade point of the *i*th student.

$$X_i = \left[x_{i1}x_{i2}\ldots,\ x_{ij}\ldots, x_{iND}\right] \tag{8}$$

In control theory, achieving both fast response and stability for a controlled system is highly desirable but often presents a significant challenge. Balancing these two requirements necessitates a careful design, as optimizing one typically impacts the other. A proper control design aims to find an equilibrium between quick response and adequate stability by carefully selecting an appropriate controller and designing it to minimize a well-defined objective function and using optimization techniques.

In control system design, four commonly considered performance criteria include the Integral of Time-Weighted Absolute Error (ITAE), Integral of Squared Error (ISE), Integral of Time-Weighted Squared Error (ITSE), and Integral of Absolute Error (IAE). Among these, the ISE and ITAE criteria are frequently utilized in the literature due to their superior performance compared to the IAE and ITSE criteria.

The ISE criterion evaluates the integral of the squared error over time, giving greater emphasis to larger errors since the square of a large error significantly outweighs smaller ones. Consequently, control systems designed to minimize ISE tend to eliminate substantial errors rapidly but may allow smaller errors to persist over extended periods. This often results in systems with fast response times accompanied by low-amplitude oscillations.

On the other hand, the ITAE criterion integrates the absolute error multiplied by time over the response duration, placing greater weight on errors that persist over longer periods compared to those occurring at the beginning. As a result, ITAE tuning methods generally produce systems that settle more quickly than those tuned using ISE [50]. Accordingly, in this study, the parameters of the Fuzzy FOPI+FOPD+I proposed for AVR applications are optimized using the TLBO algorithm by minimizing the ITAE objective function, which is mathematically expressed as follows:

$$\text{Objective function} = \text{ITAE} = \int_0^t |e|.t.dt \tag{9}$$

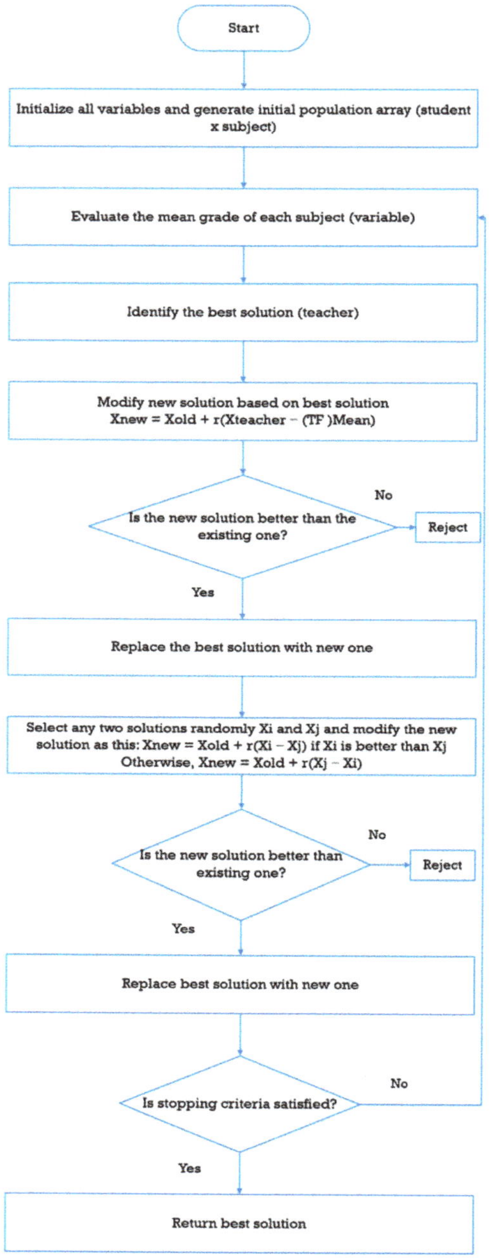

Figure 8. The flowchart of the TLBO algorithm.

For a reasonable computational time needed by the algorithm to tune the parameters, the number of population and number of iterations are set to 50. The convergence curve of the TLBO is presented in Figure 9. The tuning process is performed only prior to the deployment of the actual controller, making it a one-time procedure for each specific application. The process can be further improved by employing alternative optimization techniques with less computational time needed for tuning. However, it is common to encounter extended computational times when tuning controllers that involve fuzzy logic and a relatively large number of parameters. The optimum values of the proposed AVR controller obtained by TLBO by minimizing ITAE are presented in Table 4.

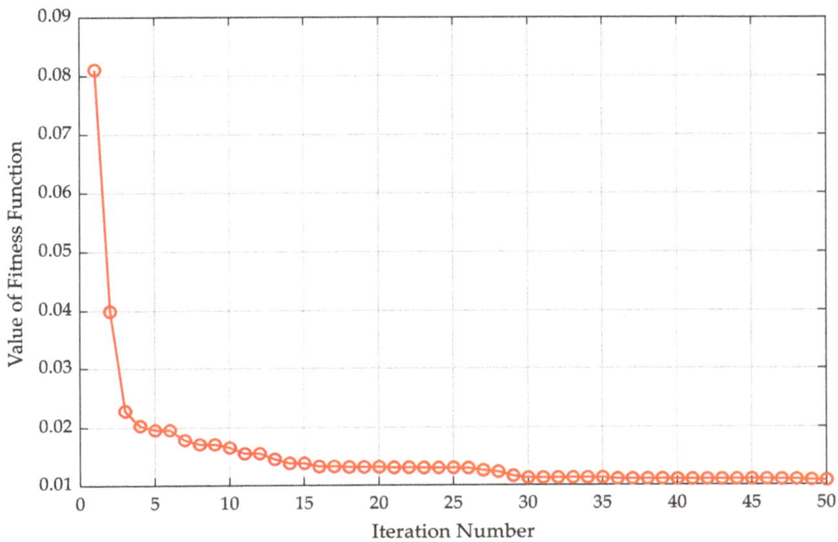

Figure 9. The convergence curve of TLBO.

Table 4. The optimum values of the Fuzzy FOPI+FOPD+I.

Controller	Parameters								
	K_1	K_2	K_3	K_{P1}	K_I	λ	K_{P2}	K_D	μ
Fuzzy FOPI+FOPD + I	0.284	0.7304	1.9193	0.5200	0.4816	0.6971	1.6179	0.7885	0.1499

4. Results and Discussion

This study was implemented on MATLAB 2022a; the TLBO code was programmed in .m file while the AVR model and the controller were implemented in MATLAB Simulink environment. A step response of 1 pu amplitude was set as a reference point. To investigate the supremacy of the proposed technique, the obtained results are compared with other studies presented in the literature based on PID tuned by IWOA [51] ($K_P = 0.8167$; $K_I = 0.6898$; $K_D = 0.2799$), ABC [52] ($K_P = 0.6352$; $K_I = 0.4235$; $K_D = 0.2241$), Many Optimizing Liaisons (MOL) [53] ($K_P = 0.5857$; $K_I = 0.4189$; $K_D = 0.1772$), and TSA [13] ($K_P = 1.1281$; $K_I = 0.9567$; $K_D = 0.5671$). The characteristics of the response represented by the peak overshoot (POs) in pu, peak undershoot (PUs) in pu, settling time (Ts) in seconds, rise time (Tr) in seconds, and the value of the ITAE are presented in Table 5. Figure 10 illustrates the response of the AVR system based on the proposed fuzzy configuration and other PID controllers tuned by different optimization tools previously presented in the literature.

Table 5. The characteristics of the AVR system based on different techniques.

Controller	Characteristics				
	OS (pu)	US (pu)	Ts	Tr	ITAE
Proposed Fuzzy	0.055239	**0.9915**	**0.1612**	**0.1094**	**0.01092**
PID IWOA	0.069178	0.9912	0.3276	0.2258	0.07078
PID ABC	**0.0002**	0.9689	0.4559	0.2951	0.06165
PID MOL	**0.019606**	**0.9967**	0.5153	0.3429	**0.05131**
PID TSA	0.15593	0.90455	**0.1897**	**0.1311**	0.08791

Figures in **black bold** represent the **best results** and in **blue bold** represent the **second-best.**

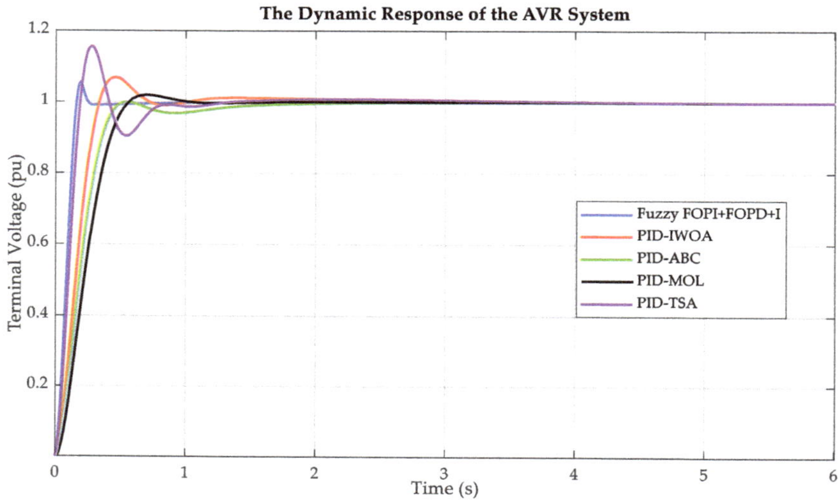

Figure 10. The dynamic response of the AVR system based on different controllers.

As mentioned above the proposed controller is designed to compromise between quick response and robust stability. From the presented findings of Figure 10 and Table 5, it is obvious that the proposed fuzzy structure provides the best overall response. It secured the best settling time, rise time, and objective function with values of 0.1612 s, 0.1094 s, and 0.01092, respectively. Furthermore, it offers the second-best peak overshoot and undershoot. Also, it is clear that the PID-TSA offers the worst overall response.

For unit suitability, Figures 11–13 present a comparative analysis of the characteristics of AVR control systems designed using various approaches, including the proposed Fuzzy FOPI+FOPD+I method. Specifically, Figure 11 illustrates the settling time and rise time, while Figure 12 provides bar chart comparisons of the ITAE cost function. Figure 13 highlights the maximum overshoot (in pu) and maximum undershoot (in pu).

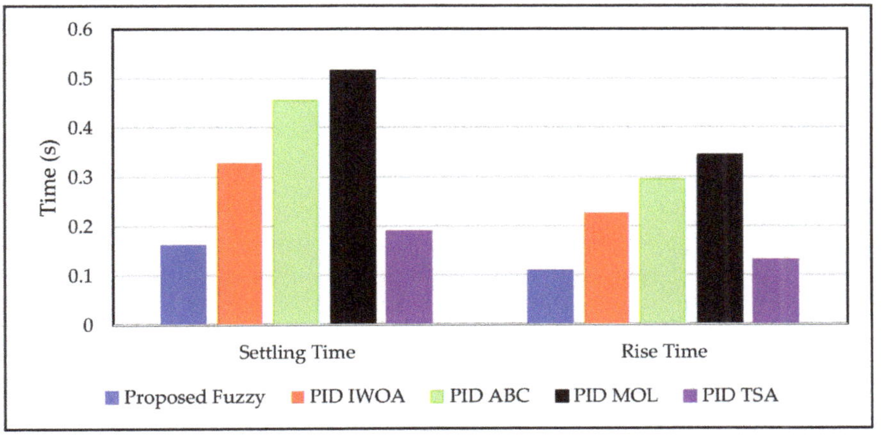

Figure 11. Settling and rise times of different controllers.

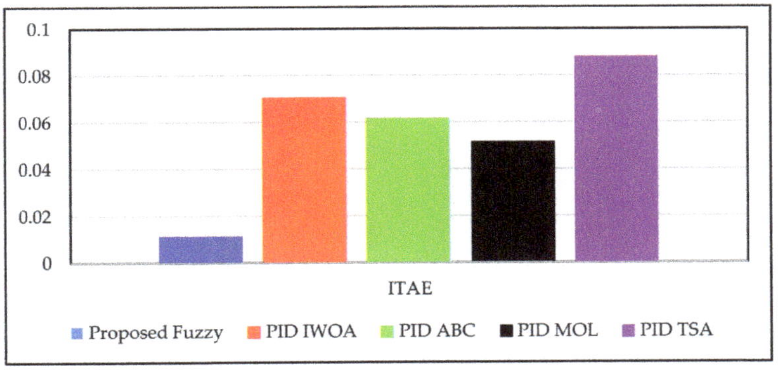

Figure 12. ITAE of different controllers.

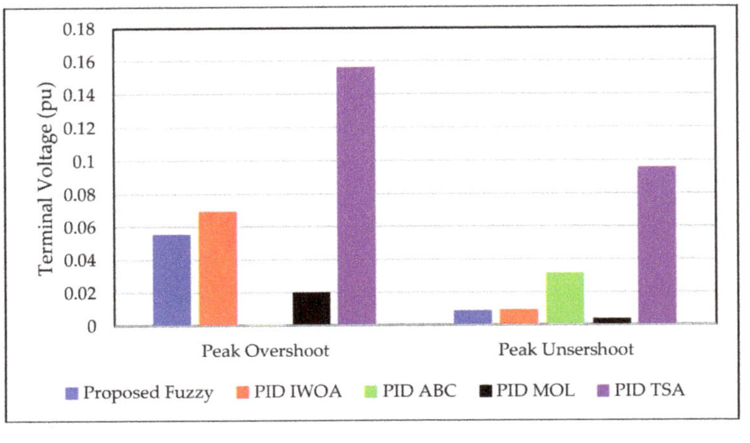

Figure 13. Peak overshoot and undershoot of different controllers.

From Figures 10–13 and Table 5, it is evidenced that the proposed Fuzzy FOPI+FOPD+I demonstrated excellent performance as an AVR controller and has outperformed several techniques proposed in the literature. This controller offered a stable and quick response, which gives clear evidence that this controller can be implemented in real-time applications.

5. Robustness Analysis

The system parameters such as the gains and time constants are inherently prone to constant fluctuations. These variations can significantly deteriorate the performance of closed-loop control systems. Despite their critical impact, this issue has received comparatively less attention in the context of AVR systems within existing research. The influence of parameter variations on the overall performance of the AVR system is thoroughly analyzed. To this end, a detailed investigation is conducted to assess the effects of parametric uncertainties within the system. Each parameter varies by ±50% from its nominal value to examine the impact of this variation on the system's stability, as shown in Table 6. Figure 14 illustrates the parametric uncertainty impact on the AVR system without a controller.

Table 6. Different considered scenarios for parametric uncertainty analysis.

Case	Nominal Value	Variation Range	New Value
Case 1: Generator Coefficient	$K_g = 1$	+50% and − 50%	1.5 and 0.5
Case 2: Generator Time Constant	$\tau_g = 1$	+50% and − 50%	1.5 and 0.5
Case 3: Excitor Coefficient	$K_e = 1$	+50% and − 50%	1.5 and 0.5
Case 4: Exciter Time Constant	$\tau_e = 0.4$	+50% and − 50%	0.6 and 0.2
Case 5: Sensor Coefficient	$K_s = 1$	+50% and − 50%	1.5 and 0.5
Case 6: Sensor Time Constant	$\tau_s = 0.01$	+50% and − 50%	0.015 and 0.005
Case 7: Amplifier Coefficient	$K_a = 10$	+50% and − 50%	15 and 5
Case 8: Amplifier Time Constant	$\tau_a = 0.1$	+50% and − 50%	0.15 and 0.05
Random scenario	$K_g = 1$	−25%	0.75
	$\tau_g = 1$	−50%	0.5
	$K_e = 1$	−50%	0.5
	$\tau_e = 0.4$	+50%	0.6
	$K_a = 10$	+50%	15
	$\tau_a = 0.1$	−50%	0.05
	$\tau_s = 0.01$	−50%	0.005

It is obvious from the subfigures shown in Figure 14 that the variation in the parameters leads to a significant deterioration of the overall performance of the model. Moreover, in some scenarios, the system was slightly oscillated. Accordingly, a random severe scenario of parametric uncertainty of the AVR with six parameters deviated by +50% or −50% from their nominal values is assumed, as shown in Table 6. The proposed Fuzzy FOPI+FOPD+I will be equipped in the investigated system to examine its robustness against a critical scenario of parametric uncertainty and evaluate the overall response and the controller performance. Importantly, the optimal gains determined under nominal conditions remain unchanged when the model is subjected to these parameter variations, highlighting the controller's ability to adapt to dynamic system changes without re-tuning.

Figure 15 illustrates the step response of the testbed system under conditions of parametric uncertainty. Despite the significant parametric variations imposed on the system, it continues to operate within acceptable performance limits. The response under uncertainty is slightly slower compared to nominal conditions and exhibits a minimal steady-state error, which is negligible. This is attributed to the cumulative impact of the deviations across all parameters. The obtained result from the robustness test demonstrates the remarkable robustness and reliability of the proposed AVR fuzzy control system.

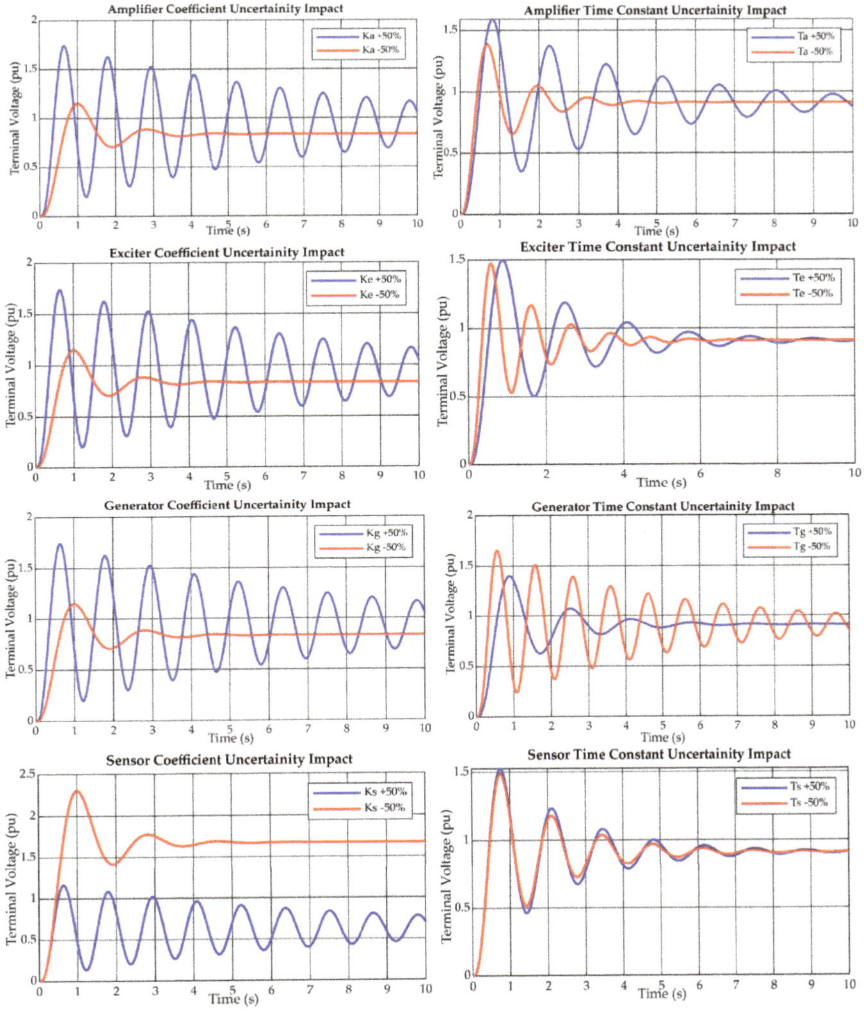

Figure 14. Step responses of AVR systems without controller under different parametric uncertainty conditions.

To thoroughly evaluate the overall performance of the proposed controller, the system will be tested under varying load conditions while being subjected to parametric uncertainty. This scenario represents the most challenging situation the system may encounter during real-time operation.

The rejection of load disturbance effects on AVR output is another critical capability of the controller, demonstrating its robustness. According to [54], disturbances of up to 5% of the generator output voltage are permissible; however, controllers are expected to suppress disturbances exceeding this threshold. In alignment with prior studies, this investigation subjected the system to disturbances rated at 10% of the reference voltage, which were injected into the generator output.

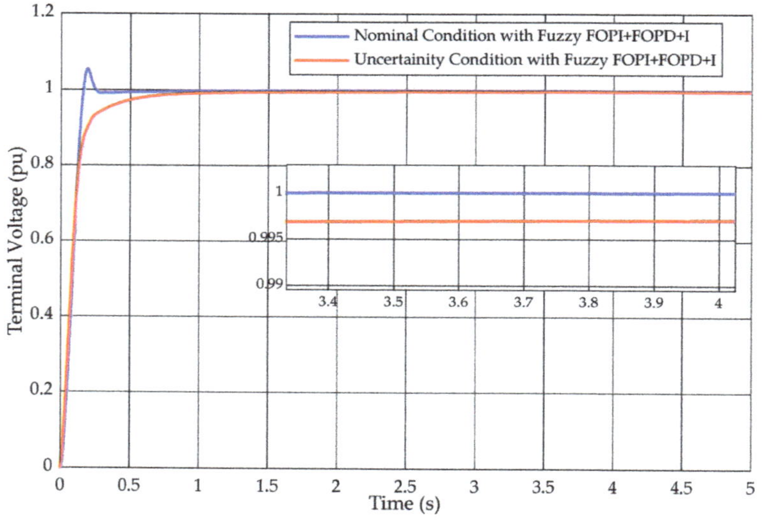

Figure 15. Step responses of AVR systems when system is subjected to parametric uncertainties.

Figure 16 illustrates the injected disturbances, AVR outputs, and corresponding time-domain specifications. The proposed controller effectively suppressed these disturbances, delivering a stable and reliable response across various operating conditions. This highlights the controller's strong ability to maintain performance and robustness under challenging scenarios.

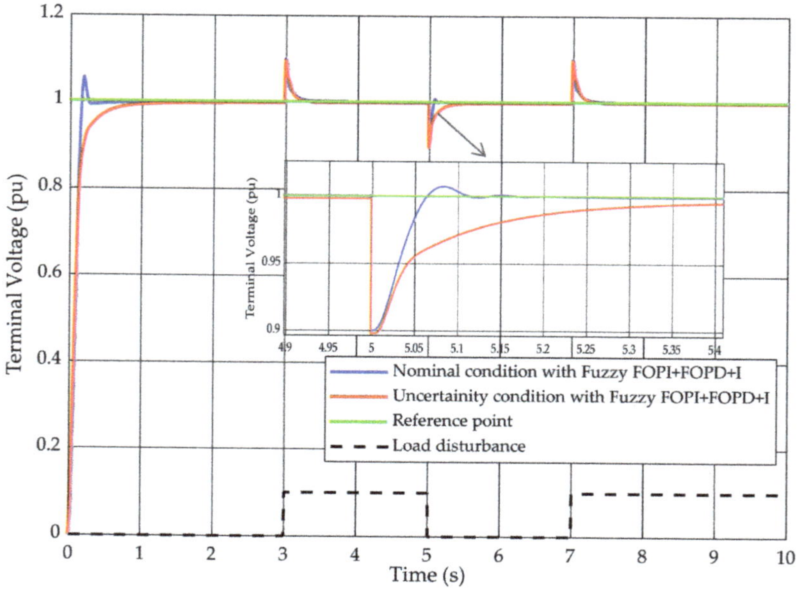

Figure 16. Step responses of AVR systems when system is subjected to parametric uncertainties with load disturbance.

6. Conclusions

This study successfully introduced and implemented a novel fuzzy configuration for AVR systems. The proposed Fuzzy FOPI+FOPD+I controller was optimized using the TLBO algorithm by minimizing the ITAE cost function. This controller was designed to benefit from different control theory advantages such as fuzzy control theory, cascading control, and fractional calculus. The proposed AVR control system demonstrated exceptional performance and outperformed other controllers reported in the literature. These results highlight the controller's readiness for real-time application. The simulation results demonstrate that the Fuzzy FOPI+FOPD+I controller, tuned using the TLBO algorithm, outperforms the other reported methods. Specifically, the peak overshoot was reduced from 0.15593 pu to 0.055239 pu, the peak undershoot was enhanced from 0.90455 pu to 0.9915 pu, and the settling time was significantly improved from 0.5153 s to 0.1612 s. Additionally, the robustness of the proposed controller was thoroughly evaluated under parametric uncertainties, where it exhibited excellent resilience. The controller also maintained superior performance under varying load conditions, further confirming its reliability and effectiveness.

One minor limitation of this study is the computational time required by the TLBO algorithm to tune the controller parameters, which could be improved by adopting a more computationally efficient optimization technique. Furthermore, the slight overshoot observed in the results highlights an area for potential improvement, which could be addressed through advanced tuning strategies or minor adjustments to the control design.

Future work could explore alternative control designs to further enhance the system's performance. For instance, integrating a hybrid fuzzy controller with LQR control could potentially yield even greater improvements in the overall performance of the AVR system. Also, it is important to study the performance of the system considering aspects of nonlinearities. Furthermore, the stability analysis of systems employing fuzzy control remains an underexplored area in the existing literature. As a prospective avenue for future research, examining the zero and pole dynamics of the system under the influence of the proposed fuzzy controller within the AVR model presents a compelling opportunity to enhance understanding in this domain.

Author Contributions: M.S. and M.A. contributed the same in producing this manuscript. All authors have read and agreed to the published version of the manuscript.

Funding: No funds were received to conduct this research.

Data Availability Statement: The data used and produced in this manuscript is available upon request from the corresponding author.

Acknowledgments: The authors thank Cardiff University for paying the APC towards publishing this manuscript.

Conflicts of Interest: The authors declare no conflicts of interest.

References

1. Gopi, P.; Reddy, S.V.; Bajaj, M.; Zaitsev, I.; Prokop, L. Performance and robustness analysis of V-Tiger PID controller for automatic voltage regulator. *Sci. Rep.* **2024**, *14*, 7867. [CrossRef] [PubMed]
2. Vani, E.; Ilambirai, R.C.; Ganga, M.; Gnanamalar, J. Jellyfish Search Optimized Gated Recurrent Unit Based FOPID/PID for Enhanced Automatic Voltage Regulator. *Electr. Power Components Syst.* **2024**, *2322004*, 1–22. [CrossRef]
3. Gopi, P.; Varma, P.S.; Kalyan, C.N.S.; Ravikumar, C.V.; Srinivasulu, A.; Bohara, B.; Rajesh, A.; Ab Wahab, M.N.; Sathish, K. Dynamic Behavior and Stability Analysis of Automatic Voltage Regulator with Parameter Uncertainty. *Int. Trans. Electr. Energy Syst.* **2023**, *2023*, 6662355. [CrossRef]

4. Mohapatra, S.K.; Singh, A.K.B.; Ray, N.K. *PSO with Moderate Random Search Strategy (MRSPSO)-Based PI/PID Controller in AVR System*; Springer Nature: Singapore, 2021; pp. 225–237. [CrossRef]
5. Li, Y.; Ni, L.; Wang, G.; Aphale, S.S.; Zhang, L. Q-Learning-Based Dumbo Octopus Algorithm for Parameter Tuning of Fractional-Order PID Controller for AVR Systems. *Mathematics* **2024**, *12*, 3098. [CrossRef]
6. Amin, M.S.; Attia, M.A.; Khamees, A.K.; Mekhamer, S.F.; Kotb, H.; AboRas, K.M.; Yousef, A. Development of AVR controller performance using exponential distribution and transit search optimization techniques. *Front. Energy Res.* **2024**, *12*, 1356978. [CrossRef]
7. Shouran, M.; Anayi, F. TLBO Tuned a Novel Robust Fuzzy Control Structure for LFC of a Hybrid Power System with Photovoltaic Source. *Eng. Proc.* **2022**, *19*, 1. [CrossRef]
8. Ekinci, S.; Hekimoglu, B. Improved Kidney-Inspired Algorithm Approach for Tuning of PID Controller in AVR System. *IEEE Access* **2019**, *7*, 39935–39947. [CrossRef]
9. Islam, M.S.; Ahmad, M.A.; Hao, M.R.; Suid, M.H.; Tumari, M.Z.M. Fast PID Tuning of AVR System Using Memory-Based Smoothed Functional Algorithm. In Proceedings of the 2024 IEEE Symposium on Industrial Electronics & Applications (ISIEA), Kuala Lumpur, Malaysia, 6–7 July 2024; IEEE: New York, NY, USA, 2024; pp. 1–5. [CrossRef]
10. Hekimoğlu, B. Sine-cosine algorithm-based optimization for automatic voltage regulator system. *Trans. Inst. Meas. Control.* **2019**, *41*, 1761–1771. [CrossRef]
11. Çelik, E.; Durgut, R. Performance enhancement of automatic voltage regulator by modified cost function and symbiotic organisms search algorithm. *Eng. Sci. Technol. Int. J.* **2018**, *21*, 1104–1111. [CrossRef]
12. Gopi, P.; Mahdavi, M.; Alhelou, H.H. Robustness and Stability Analysis of Automatic Voltage Regulator Using Disk-Based Stability Analysis. *IEEE Open Access J. Power Energy* **2023**, *10*, 689–700. [CrossRef]
13. Kose, E. Optimal Control of AVR System With Tree Seed Algorithm-Based PID Controller. *IEEE Access* **2020**, *8*, 89457–89467. [CrossRef]
14. Hekimoglu, B.; Ekinci, S. Grasshopper optimization algorithm for automatic voltage regulator system. In Proceedings of the 2018 5th International Conference on Electrical and Electronic Engineering (ICEEE), Istanbul, Turkey, 3–5 May 2018; IEEE: New York, NY, USA, 2018; pp. 152–156. [CrossRef]
15. Razmjooy, N.; Khalilpour, M.; Ramezani, M. A New Meta-Heuristic Optimization Algorithm Inspired by FIFA World Cup Competitions: Theory and Its Application in PID Designing for AVR System. *J. Control Autom. Electr. Syst.* **2016**, *27*, 419–440. [CrossRef]
16. Pradhan, R.; Majhi, S.K.; Pati, B.B. Design of PID controller for automatic voltage regulator system using Ant Lion Optimizer. *World J. Eng.* **2018**, *15*, 373–387. [CrossRef]
17. Gaing, Z.-L. A Particle Swarm Optimization Approach for Optimum Design of PID Controller in AVR System. *IEEE Trans. Energy Convers.* **2004**, *19*, 384–391. [CrossRef]
18. Mohanty, P.K.; Sahu, B.K.; Panda, S. Tuning and Assessment of Proportional–Integral–Derivative Controller for an Automatic Voltage Regulator System Employing Local Unimodal Sampling Algorithm. *Electr. Power Components Syst.* **2014**, *42*, 959–969. [CrossRef]
19. Zhou, Y.; Zhang, J.; Yang, X.; Ling, Y. Optimization of PID Controller Based on Water Wave Optimization for an Automatic Voltage Regulator System. *Inf. Technol. Control* **2019**, *48*, 160–171. [CrossRef]
20. Alghamdi, S.; Sindi, H.F.; Rawa, M.; Alhussainy, A.A.; Calasan, M.; Micev, M.; Ali, Z.M.; Aleem, S.H.E.A. Optimal PID Controllers for AVR Systems Using Hybrid Simulated Annealing and Gorilla Troops Optimization. *Fractal Fract.* **2022**, *6*, 682. [CrossRef]
21. Mosaad, A.M.; Attia, M.A.; Abdelaziz, A.Y. Whale optimization algorithm to tune PID and PIDA controllers on AVR system. *Ain Shams Eng. J.* **2019**, *10*, 755–767. [CrossRef]
22. Hesham, O.M.; Attia, M.A.; Mekhamer, S.F. Enhancement of AVR system performance by using hybrid harmony search and dwarf mongoose optimization algorithms. *Sci. Rep.* **2024**, *14*, 27177. [CrossRef]
23. Ozgenc, B.; Ayas, M.S.; Altas, I.H. Performance improvement of an AVR system by symbiotic organism search algorithm-based PID-F controller. *Neural Comput. Appl.* **2022**, *34*, 7899–7908. [CrossRef]
24. Sahib, M.A. A novel optimal PID plus second order derivative controller for AVR system. *Eng. Sci. Technol. Int. J.* **2015**, *18*, 194–206. [CrossRef]
25. Shouran, M.; Anayi, F.; Packianather, M.; Habil, M. Load frequency control based on the bees algorithm for the Great Britain power system. *Designs* **2021**, *5*, 50. [CrossRef]
26. Tumari, M.Z.M.; Ahmad, M.A.; Suid, M.H.; Hao, M.R. An Improved Marine Predators Algorithm-Tuned Fractional-Order PID Controller for Automatic Voltage Regulator System. *Fractal Fract.* **2023**, *7*, 561. [CrossRef]
27. Mok, R.; Ahmad, M.A. Fast and optimal tuning of fractional order PID controller for AVR system based on memorizable-smoothed functional algorithm. *Eng. Sci. Technol. Int. J.* **2022**, *35*, 101264. [CrossRef]
28. Sahin, A.K.; Cavdar, B.; Ayas, M.S. An adaptive fractional controller design for automatic voltage regulator system: Sigmoid-based fractional-order PID controller. *Neural Comput. Appl.* **2024**, *36*, 14409–14431. [CrossRef]

29. Micev, M.; Ćalasan, M.; Oliva, D. Fractional order PID controller design for an AVR system using Chaotic Yellow Saddle Goatfish Algorithm. *Mathematics* **2020**, *8*, 1182. [CrossRef]
30. Ekinci, S.; Snášel, V.; Rizk-Allah, R.M.; Izci, D.; Salman, M.; Youssef, A.A.F. Optimizing AVR system performance via a novel cascaded RPIDD2-FOPI controller and QWGBO approach. *PLoS ONE* **2024**, *19*, e0299009. [CrossRef]
31. Omar, O.A.M.; Marei, M.I.; Attia, M.A. Comparative Study of AVR Control Systems Considering a Novel Optimized PID-Based Model Reference Fractional Adaptive Controller. *Energies* **2023**, *16*, 830. [CrossRef]
32. Tabak, A. Modified and Improved TID Controller for Automatic Voltage Regulator Systems. *Fractal Fract.* **2024**, *8*, 654. [CrossRef]
33. Türksoy, Ö.; Türksoy, A. A fast and robust sliding mode controller for automatic voltage regulators in electrical power systems. *Eng. Sci. Technol. Int. J.* **2024**, *53*, 101697. [CrossRef]
34. Furat, S.B.M.; Hameş, Y. A Novel Fast Optimized Sliding Mode Controller for Automatic Voltage Regulator System. Available Online: https://ssrn.com/abstract=4261636 (accessed on 15 November 2024).
35. Mazibuko, N.; Akindeji, K.T.; Moloi, K.; Sharma, G. Improved Model Predictive controller (MPC) for an Automatic Voltage Regulator (AVR). In Proceedings of the 2024 32nd Southern African Universities Power Engineering Conference (SAUPEC), Stellenbosch, South Africa, 24–25 January 2024; IEEE: New York, NY, USA, 2024; pp. 1–6. [CrossRef]
36. Sharma, V.; Kumar, V.; Naresh, R.; Kumar, V. MPA Optimized Model Predictive Controller for Optimal Control of an AVR System. In *International Conference on Frontiers of Intelligent Computing: Theory and Applications*; Springer Nature: Singapore, 2023; pp. 61–70. [CrossRef]
37. Schwenzer, M.; Ay, M.; Bergs, T.; Abel, D. Review on model predictive control: An engineering perspective. *Int. J. Adv. Manuf. Technol.* **2021**, *117*, 1327–1349. [CrossRef]
38. Das, S.; Pan, I. On the Mixed H_2/H_∞ Loop-Shaping Tradeoffs in Fractional-Order Control of the AVR System. *IEEE Trans. Ind. Informatics* **2014**, *10*, 1982–1991. [CrossRef]
39. Modabbernia, M.; Alizadeh, B.; Sahab, A.; Moghaddam, M.M. Robust control of automatic voltage regulator (AVR) with real structured parametric uncertainties based on H_∞ and μ-analysis. *ISA Trans.* **2020**, *100*, 46–62. [CrossRef]
40. Batmani, Y.; Golpîra, H. Automatic voltage regulator design using a modified adaptive optimal approach. *Int. J. Electr. Power Energy Syst.* **2018**, *104*, 349–357. [CrossRef]
41. Mojumder, M.R.H.; Roy, N.K. PID, LQR, and LQG Controllers to Maintain the Stability of an AVR System at Varied Model Parameters. In Proceedings of the 2021 5th International Conference on Electrical Engineering and Information Communication Technology (ICEEICT), Dhaka, Bangladesh, 18–20 November 2021; IEEE: New York, NY, USA, 2021; pp. 1–6. [CrossRef]
42. Jadoua, E.H.; Diwood, N.M.; Azeez, Y.F. The LQR Based On Optimized Tuning PD Controller For AVR System. *Int. J. Appl. Sci.* **2022**, *5*, 8. [CrossRef]
43. Modabbernia, M.; Alizadeh, B.; Sahab, A.; Moghaddam, M.M. Designing the Robust Fuzzy PI and Fuzzy Type-2 PI Controllers by Metaheuristic Optimizing Algorithms for AVR System. *IETE J. Res.* **2020**, *68*, 3540–3554. [CrossRef]
44. Mazibukol, N.; Akindejil, K.T.; Sharma, G. Implementation of a FUZZY logic controller (FLC) for improvement of an Automated Voltage Regulators (AVR) dynamic performance. In Proceedings of the 2022 IEEE PES/IAS PowerAfrica, PowerAfrica 2022, Kigali, Rwanda, 22–26 August 2022; Institute of Electrical and Electronics Engineers Inc.: Piscataway, NJ, USA, 2022. [CrossRef]
45. Al Gizi, A.J.; Mustafa, M.; Al Zaidi, K.M.; Al-Zaidi, M.K. Integrated PLC-fuzzy PID Simulink implemented AVR system. *Int. J. Electr. Power Energy Syst.* **2015**, *69*, 313–326. [CrossRef]
46. Gupta, T.; Sambariya, D.K. Optimal design of fuzzy logic controller for automatic voltage regulator. In Proceedings of the 2017 International Conference on Information, Communication, Instrumentation and Control (ICICIC), Indore, India, 17–19 August 2017; IEEE: New York, NY, USA, 2017; pp. 1–6. [CrossRef]
47. Sivanandhan, A.; Thriveni, G. Optimal design of controller for automatic voltage regulator performance enhancement: A survey. *Electr. Eng.* **2024**, *106*, 3705–3720. [CrossRef]
48. Rao, R.; Savsani, V.; Vakharia, D. Teaching–Learning-Based Optimization: An optimization method for continuous non-linear large scale problems. *Inf. Sci.* **2012**, *183*, 1–15. [CrossRef]
49. Barisal, A. Comparative performance analysis of teaching learning based optimization for automatic load frequency control of multi-source power systems. *Int. J. Electr. Power Energy Syst.* **2015**, *66*, 67–77. [CrossRef]
50. Mohanty, B.; Panda, S.; Hota, P. Controller parameters tuning of differential evolution algorithm and its application to load frequency control of multi-source power system. *Int. J. Electr. Power Energy Syst.* **2014**, *54*, 77–85. [CrossRef]
51. Habib, S.; Abbas, G.; Jumani, T.A.; Bhutto, A.A.; Mirsaeidi, S.; Ahmed, E.M. Improved Whale Optimization Algorithm for Transient Response, Robustness, and Stability Enhancement of an Automatic Voltage Regulator System. *Energies* **2022**, *15*, 5037. [CrossRef]
52. Gozde, H.; Taplamacioglu, M. Comparative performance analysis of artificial bee colony algorithm for automatic voltage regulator (AVR) system. *J. Frankl. Inst.* **2011**, *348*, 1927–1946. [CrossRef]
53. Panda, S.; Sahu, B.; Mohanty, P. Design and performance analysis of PID controller for an automatic voltage regulator system using simplified particle swarm optimization. *J. Frankl. Inst.* **2012**, *349*, 2609–2625. [CrossRef]

54. Aguila-Camacho, N.; Duarte-Mermoud, M.A. Fractional adaptive control for an automatic voltage regulator. *ISA Trans.* **2013**, *52*, 807–815. [CrossRef]

fractal and fractional

MDPI

Article

Applying a Gain Scheduled Fractional Order Proportional Integral and Derivative Controller to a Quadratic Buck Converter

German Ardul Munoz Hernandez [1,*], Jose Fermi Guerrero-Castellanos [1] and Rafael Antonio Acosta-Rodriguez [2]

[1] Electronic Sciences Faculty, Meritorious Autonomous University of Puebla (BUAP), Puebla de Zaragoza 72592, Mexico; fermi.guerrero@correo.buap.mx

[2] Electrical Machines and Drives (EM&D) Group, Department of Electrical and Electronic Engineering, Universidad Nacional de Colombia, Bogota 111321, Colombia; racostar@unal.edu.co or racostar@ieee.org

* Correspondence: gmunoz@ieee.org; Tel.: +52-2222-558-932

Abstract: This work presents a fractional order Proportional Integral and Derivative controller with adaptation characteristics in the control parameters depending on the required output, gain scheduling fractional order PID (GS-FO-PID). The fractional order PID is applied to the voltage control of a DC–DC buck quadratic converter (QBC). The DC–DC buck quadratic converter is designed to operate at 12 V, although in the simulation tests, the output voltage ranges from 5 to 36 V. The performance of the GS-FO-PID is compared with the one from a classic PID. The GS-FO-PID presents better performance when the reference voltage is changed. In the same way, the behavior of the converter with the reference fixed to 12 V output is analyzed with load changes; for this case, the amplitude value of the ripple when the converter is driven by the GS-FO-PID almost has no variation.

Keywords: quadratic buck converter; fractional order PID; simulation; gain scheduling

1. Introduction

Recent advances in the design and construction of DC converters have made great strides in improving power conversion technologies. Included in these advancements are DC ports with a good high voltage, gain ratio, bidirectional conversion capability, and higher overall efficiency. These advances have been very important, especially for applications in renewable sources, storage systems, microgrids, and power electronics [1–4]. In the area of power conversion, there is a wide range of DC–DC converter configurations, given the growing demand for improved efficiency as well as versatility for use in various applications. In this regard, notable advances have been made in the design of transformerless DC–DC converters, which offer tangible benefits in terms of weight, size, and even cost reductions. A notable innovation in this domain is the transformerless quadratic step-down converter, known for its high-voltage gain ratio and continuous input/output current ports [3]. The DC–DC converters are usually placed between a primary power supply and a load. The input includes direct voltage loads capable of powering various digital electronic devices. The system shows nonlinear characteristics due to its variable structure. To address this problem, nonlinear control methods have been used under conditions of time-varying perturbations and load changes [5–7].

Different types of DC–DC converters have been designed, including buck, boost, and buck–boost converters. These converters employ a frequency element for voltage conversion, in conjunction with a diode, plus an inductor and a capacitor. In particular, the buck converter changes a DC input voltage into a lower DC output voltage. This converter is a primary circuit in voltage reducing applications. Theoretically, the voltage conversion ratio (gain) of the buck converter (Vo/Vin) operating in continuous conduction mode (CCM) depends only on the duty cycle (d) of the power switch. The d will vary from 0 to 1, so a wide range of reducing gain is possible [8]. However, it has been found that in practice, the gain of the buck converter is normally limited to at least one-tenth. This is because a very small duty cycle would reduce the driving time of the power switch, and at a very high switching frequency, such a short driving time might not be enough for the switch to function properly. To achieve the large voltage reduction with the extended duty cycle, two or more step-down converters can be connected in series. However, if the cascade approach is followed, the array would require multiple switches. On the other hand, cascaded step-down converter arrays have low efficiency. Proposals for N-stage cascade reducer converters with a single power switch have been presented in different works [9–11]. In these arrays, the voltage reduction ratio is equal to dN. By using a single switch, these converter proposals provide a practical and economical solution for applications seeking to reduce voltage. A quadratic buck converter (QBC) is a subset of the N-stage but single-switch cascade buck converters, with N = 2, which can be considered a double step-down stage.

Although several control techniques have been widely employed in DC–DC converters and their impact on dynamic behavior and change from second order to first order dynamics has been extensively studied, the implications of these techniques are less understood in the realm of cascade converters, which introduce zeros in the transfer function [12].

In recent years, various control techniques have been applied to DC–DC buck converters. For instance, Bonela et al. [13] have introduced a robust non-fragile control approach for these converters that addresses the influence of these uncertainties. This method involves developing a control law that achieves quadratic stabilization of the closed-loop system while also satisfying the $H\infty$ norm constraint for disturbance attenuation. On the topic of robust control, Saavedra et al. [14] have presented a paper with two control design strategies for voltage regulation in a single inductor dual-output DC–DC buck converter system, in which a linear quadratic regulator and a decoupled PI control law were designed to control the power converter system under parametric uncertainties. The controllers are based on a nominal multiple input–multiple output plant model and performance requirements. For their part, Kumari et al. [15] have compared the closed-loop response of a DC–DC buck converter using different optimization strategies such as model predictive control (MPC) and the linear quadratic regulator (LQR). A similar work was published by Raza et al. [16], in which the LQR cost function is converted to the quadratic programming cost function of the MPC. On the other hand, Fetene et al. [17] have proposed a cost-effective observer-based pole placement controller for a PWM voltage-controlled buck converter with a clear step-by-step design approach. The state space averaged model of the DC–DC power converter is used to design pole placement with an integrator compensator and a full state observer. For their part, Nanyan et al. [18] presented an optimal PID controller for the DC–DC buck converter where the sine–cosine algorithm was employed to improve the performance of the converter. Also, novel techniques such as machine learning have been explored [19].

Specifically on the design and control of buck quadratic DC–DC converters, there are recent publications. For example, Trakuldit et al. [8] have published a work where a quadratic buck converter is designed and evaluated. For their part, Lopez-Santos et al. [20] have presented a quadratic buck converter that is applied for battery charging. In that work, a multi-loop controller using three nested loops was used, involving one sliding mode current controller and two conventional PI controllers. On the other hand, Pop et al. [21]

have published a paper where the process of designing a controller for a fourth-order buck converter is presented. In this work, after deriving the control-to-output transfer function based on a matrix state space model, an approximation to a second-degree transfer function is made.

A control option that has not been sufficiently studied is to employ a different classical control for each operational condition. The gain scheduling method in conjunction with the PID offers the advantage of rapid control with the characteristic of adaptability to different operating conditions. Furthermore, if fractional calculation characteristics are employed for the integral and proportional modes of the PID controller, the resulting regulator would increase its performance, given its proven improvement in the control of higher-order systems and systems with nonlinearities, among others. Given the above, this study describes the application of a gain scheduling fractional order PID controller to the regulation of a quadratic buck converter. Then the major contributions of this paper are as follows.

- To present a fractional order Proportional Integral and Derivative controller with adaptation characteristics;
- To apply a fractional order PID to control the voltage of a DC–DC buck quadratic converter (QBC);
- To simulate the system with different output voltages and different loads.

This document is structured as follows: Section 2 provides a comprehensive mathematical model for the dynamic quadratic buck converter (QBC) system. These models serve as the foundation for understanding and controlling the QBC system. Both switched and average models are discussed. Then in Section 3, three types of control are discussed, namely classic PID control, gain scheduling PID, and gain scheduling fractional. This is followed by a discussion in Section 4 about the tuning of the controllers and a comparison of the performance of the controllers, with changes in the reference voltage and load. Finally, in Section 5, some conclusions are drafted.

2. Quadratic Buck Switched Converter Model

It is common that the control strategy starts by obtaining a mathematical model. In this case, an averaged model that requires linearization was obtained. Figure 1 shows the ON and OFF states (Figures 1a and 1b, respectively) that are achieved on the basis of the quadratic buck circuits [22].

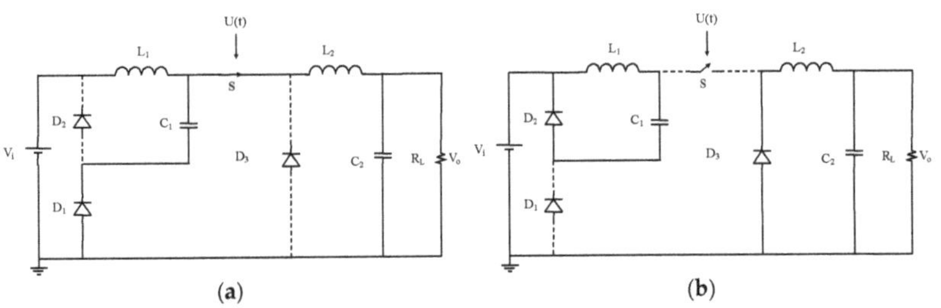

Figure 1. Quadratic buck converter. (**a**) ON state. (**b**) OFF state.

2.1. ON–OFF Equations

The behavior in the ON state, as depicted in Figure 1a, yields the following equations expressed in terms of the state variables:

$$\frac{di_{L_1}}{dt} = \frac{V_g - v_{C_1}}{L_1} \tag{1}$$

$$\frac{di_{L_2}}{dt} = \frac{v_{C_1} - v_{C_2}}{L_2} \tag{2}$$

$$\frac{dV_{C_1}}{dt} = \frac{I_{L_1} - I_{L_2}}{C_1} \tag{3}$$

$$\frac{dV_{C_2}}{dt} = \frac{I_{L_2} - I_R}{C_2} \tag{4}$$

In the OFF state of the quadratic buck converter, as illustrated in Figure 1b, we derive the following equations in the state variables:

$$\frac{di_{L_1}}{dt} = \frac{-v_{C_1}}{L_1} \tag{5}$$

$$\frac{di_{L_2}}{dt} = \frac{-v_{C_2}}{L_2} \tag{6}$$

$$\frac{dv_{C_1}}{dt} = \frac{i_{L_1}}{C_1} \tag{7}$$

$$\frac{dv_{C_2}}{dt} = \frac{i_{L_2} - i_R}{C_2} \tag{8}$$

2.2. Quadratic Buck Switched Model

The equation of the state variable is multiplied by U in the ON state and by (1-U) in the OFF state [8,23].

For i_{L_1}, in this instance, the state variable is derived from the following equation:

$$\frac{di_{L_1}}{dt} = \frac{V_g}{L_1}u - \frac{v_{C_1}}{L_1} \tag{9}$$

For i_{L2}, the state variable is derived from the equation:

$$\frac{di_{L_2}}{dt} = \frac{v_{C_1}}{L_2}u - \frac{v_{C_2}}{L_2} \tag{10}$$

For the variable v_{C1}, we derive:

$$\frac{dV_{C_1}}{dt} = \frac{-i_{L_2}}{C_1}u + \frac{i_{L_2}}{C_1} \tag{11}$$

The state variable v_{C2} is obtained as follows:

$$\frac{dV_{C_2}}{dt} = \frac{i_{L_2}}{C_2}u - \frac{v_{C_2}}{C_2R} \tag{12}$$

2.3. Quadratic Buck Average Model

We substitute the variable u with d, and define the state equations as functions as follows.

For the variable i_{L_1},

$$f_1 = \frac{d_{iL1}}{dt} = \frac{V_g}{L_1}d - \frac{v_{C1}}{L_1} \tag{13}$$

For the variable IiL_2,

$$f_2 = \frac{d_{iL2}}{dt} = \frac{v_{C1}}{L_2}d - \frac{v_{c2}}{L_2} \tag{14}$$

For the variable v_{C1},

$$f_3 = \frac{d_{VC1}}{dt} = \frac{-i_{L2}}{C_1}d + \frac{i_{L1}}{C_1} \tag{15}$$

$$f_4 = \frac{d_{VC2}}{dt} = \frac{i_{L2}}{C_2}d - \frac{v_0}{C_2 * R} \tag{16}$$

Setting the equations provided earlier to zero yields for the function f_1 the following:

$$v_g d = v_{C_1} \tag{17}$$

For the function f_2,

$$f_2 = v_{C_1} d = V_0 \tag{18}$$

For the function f_3, we have:

$$I_{L_2} d = I_{L_1} \tag{19}$$

Finally, for the function f_4,

$$I_{L_2} = \frac{V_0}{R} \tag{20}$$

A transfer function, which characterizes the system's behavior, can now be obtained (37) [22,23].

$$G(s) = \frac{\left(\left(\frac{v_{C1}}{C_1 C_2 L_2}\right)\left(S^2 - \left(\frac{D_{IL1L1}}{V_{C1}}\right)S + \left(\frac{1}{L_1} + \frac{D_{vg}}{R_{C1} V_{C1}}\right)\right)\right)}{S^4 + \left(\frac{1}{RC_2}\right)S^3 + \left(\frac{D^2 L_1 + L_2 - 2}{L_1 L_2}\right)S^2 + \left(\frac{D^2 L_1 + L_2}{C_1 C_2 L_1 L_2 R}\right)S + \left(\frac{1}{C_1 C_2 L_1 L_2 R}\right)} \tag{21}$$

3. Description of the Controllers

Three controllers were compared in this work, namely a classical PID, a gain scheduling PID, and a gain scheduling fractional order PID.

3.1. Classical PID

PID control is a classical type of control widely used in industry, which incorporates an evaluation of the proportion (P), the integral (I), and the derivative (D) of the error signal to obtain the control signal. Different variations of PID that include nonlinear characteristics have been reported for many years [24]. As a form of comparison, a classical PID control (22) was chosen, which was adjusted by the relay method, which will be discussed later.

$$U(s) = K\left(E(s) + \frac{1}{T_i s}E(s) + T_d s E(s)\right) \tag{22}$$

where

U(s) is the control signal;
E(s) is the error of the system;
K is the proportional gain;
Ti is the integral gain;
Td is the derivative gain.

3.2. Gain Scheduling PID

Gain scheduling is a common method used to control nonlinear systems, since it uses linear design methods [25]. Normally, gain scheduled controllers are calculated by conducting an initial linearization of the system to be controlled at a certain scheduling operational point. Subsequently, for every operational point, linear controllers are calculated; each one is acceptable only near to that specific operational point. Then the controller's parameters are calculated at every operational point selected. In the case of the PID controller, the parameters K, Ti, and Td are selected for each of the operating points. The desired voltage level is detected, then the selected parameters are used. Figure 2 shows a schematic diagram of the GS-PID controller. As can be seen, the identification algorithm block receives the reference signal and the output of the system; with these signals, it calculates the point of operation at which the system is located. This information is the input to the gain scheduling block that determines the values of the control parameters that will be applied in the PID controller.

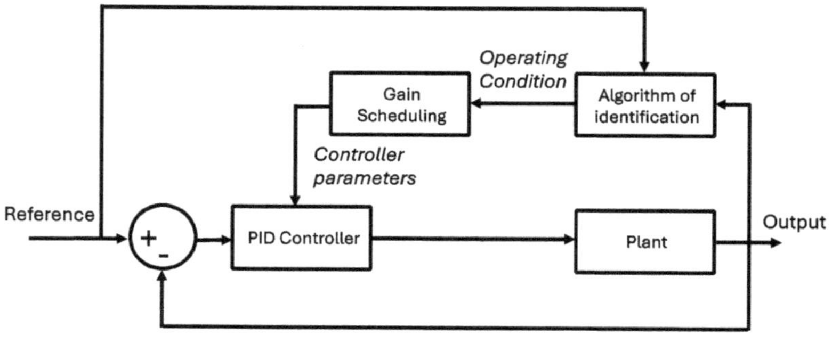

Figure 2. Schematic diagram of the GS-PID controller.

3.3. Fractional Order Controllers

Fractional calculus is a research area that is over 300 years old. There are multiple applications in various areas. Control theory is an important area of application. During the last 30 years, efforts have been made to implement fractional calculus techniques in control theory [26]. In fractional calculus, the integration and differentiation of the fundamental operator of non-integer order ${}_aD_t^\alpha$ are generalized, with a and t as the limits of the operation.

Definition 1. *The continuous fractional integro-differential operator has been defined as:*

$$ {}_aD_t^\alpha = \begin{cases} \frac{d^\alpha}{dt^\alpha} & : \ \alpha > 0, \\ 1 & : \ \alpha = 1 \\ \int_a^t (d\tau)^{-\alpha} & : \ \alpha < 0. \end{cases} \tag{23} $$

Definition 2. *Caputo's definition [27] of a fractional derivative can be written as [28]*

$$ {}_aD_t^\alpha f(t) = \frac{1}{\Gamma(n-\alpha)} \int_a^t \frac{f^{(n)}(\tau)}{(t-\tau)^{\alpha-n+1}} d\tau \tag{24} $$

for $(n - 1 < \alpha < n)$*. It holds an important property: the initial conditions for fractional order differential equations with Caputo's derivative are in the same form as integer-order differential equations.*

Definition 3. *If we consider* $k = \frac{t-a}{h}$*, where a is a real constant, which expresses a limit value, we can write the Grünwald–Letnikov definition as [27]*

$$ {}_aD_t^\alpha f(t) = \lim_{h \to 0} \frac{1}{h^\alpha} \sum_{j=0}^{\left[\frac{t-a}{h}\right]} (-1)^j \binom{\alpha}{j} f(t - jh) \tag{25} $$

where [x] means the integer part of x, and a and t are the bounds of operation for ${}_aD_t^\alpha f(t)$*. This form of definition is very helpful for obtaining a numerical solution of fractional differential equations.*

For zero initial conditions and the lower limit $a = 0$, the Laplace transform of fractional derivatives (Grünwald–Letnikov and Caputo) reduces to [28]:

$$\mathcal{L}\{{}_0D_t^{\alpha} f(t)\} = s^{\alpha} F(s) \tag{26}$$

The fractional differentiation/integration are linear operations:

$${}_0D_t^{\alpha}(af(t) + bg(t)) = a_0D_t^{\alpha} f(t) + b_0D_t^{\alpha} g(t) \tag{27}$$

This work considered Caputo's definition.

3.3.1. Fractional Order PID

In recent years, applications of PID controllers with integral and derivative parts of fractional order have been described. Those studies show an improvement in the performance of the controlled systems [29]. For instance, a generalization of the PID controller was proposed by Podlubny [30,31]; that approach included an integrator of order λ and a differentiator of order δ ($PI^{\lambda}D^{\delta}$ controller), where a better response of this type of controlled was demonstrated in comparison with other classical PID approaches. Frequency domain approaches for fractional order PID controllers have also been studied [32]. Podlubny [31] proposed the fractional order $PI^{\lambda}D^{\delta}$ (FO-PID controller). In his proposal, the integrator (λ) and the differentiator (δ) operators are real. Equation (28) shows the discrete-time transfer function of this controller

$$C(z) = \frac{U(z)}{E(z)} = K_p + \frac{K_i}{(\omega(z^{-1}))^{\lambda}} + K_d\left(\omega\left(z^{-1}\right)\right)^{\delta} \tag{28}$$

where

$\lambda > 0$ is a real integral operator;
$\delta > 0$ is a real differential operator;
K_p is a proportional constant;
K_i is a integration constant;
K_d is a differentiation constant;
$(\omega(z^{-1}))^{\lambda}$ is an equation to produce an integral FO of λ;
$(\omega(z^{-1}))^{\delta}$ is an equation to produce a derivative FO of δ.

In Equation (40), taking $\lambda = 1$ and $\delta = 1$, a classical PID controller is obtained. If $\lambda = 0$ and $K_i = 0$, a PD^{δ} controller is found. In summary, all these types of are particular cases of the fractional order controller, which is more flexible and gives an opportunity to better adjust the dynamical properties of a control system [29].

The performance of a controller has a strong dependence on the parameters selected. For instance, in a PID, its parameters are K_p, K_i, and K_d (38). In the case of a fractional order PID, its parameters also include λ (a real integral operator) and δ (a real differential operator). As in all discrete implementations, the sample time selection is crucial.

3.3.2. Gain Scheduling Fractional Order PID

The case of the gain scheduling fractional order PID is similar to the already explained gain scheduling PID, except in this case, there are two more parameters, λ and δ, that must be obtained. Different algorithms have been published to tune fractional order PID controllers [33–36]. Some methods to tune fractional order PID controllers have been proposed. For instance, Ayres Junior et al. proposed a method that uses pole placement [37]. For their part, Wu et al. presented a tuning method based on probabilistic robustness [38]. For this study, an agreement between rapid response and small overshoot was desired; the technique is based on the one proposed by Awouda and Mamat [39].

3.4. Stability of the Sytem

The stability of systems controlled by fractional regulators has been studied by different authors [40–43]. In the case of this work, Matignon's stability theorem could be used for an analysis of stability [44].

Theorem 3.1 (Matignon's stability theorem) [40,44]. The fractional transfer function $G(s) = \frac{Z(s)}{P(s)}$ is stable if and only if the following condition is satisfied in the s-plane

$$|arg(\sigma_i)| = q\frac{\pi}{2}$$

where $\sigma = s^q$. When s = 0 is a single root of P(s), the system cannot be stable. For $q = 1$, this is the classical theorem of pole location in the complex plane: no pole is in the closed right half-plane of the first Riemann sheet.

4. Analysis of System Response Under the Controllers

4.1. Analysis of the System

The inductance value is determined on the basis of the current ripple pattern within the coil. The equation for inductance when the system is in the ON state is depicted by the gradient of the line originating from 0 to DT_s, while the slope between DT_s and T_s defines the coil's current value in the OFF state. It is convenient to remember that D is the duty cycle and Ts is the working period (inverse of the working frequency, fs). From this,

$$L_1 = \frac{DT_s V_g(1 - v_{C1})}{2\Delta I} \tag{29}$$

and

$$L_2 = \frac{DT_s v_{C1}(1 - D)}{2\Delta I} \tag{30}$$

Capacitors can be calculated using the following equations:

$$C_1 = \frac{-DT_s I_{L1}(1 - I_{L2})}{2\Delta v_{C1}} \tag{31}$$

and

$$C_2 = \frac{DT_s(i_{L2} - i_R)}{2\Delta v_{C2}} \tag{32}$$

Considering the analysis shown in Section 2, the duty cycle is

$$D = \sqrt{\frac{v_0}{V_g}} \tag{33}$$

For the analysis presented in this work, the system was designed with the following parameters:

V_g = 48 V (input voltage);
V_o = 12 V (output voltage);
$f_s = 100 \times 10^3$ (working frequency);
R_L = 5 ohms (load);
L_1 = 0.144 mH;
L_2 = 0.096 mH;
C_1 = 0.0434 mFd;
C_2 = 0.0434 mFd.

The following operating points were chosen: 5 V, 12 V, 24 V, and 38 V. The system was designed to operate at 12 V, so the duty cycle was calculated for that output voltage, although by changing the operating point and the reference voltage, the duty cycle varied.

4.2. Tuning PID Controllers

Using the system's transfer function, the controller was tuned for each operating point. A relay is used to control the system when it is in a steady state; this causes an oscillation in the system. Using the descriptive function, the critical gain of the system can be calculated [45]. On the other hand, the oscillation frequency of the system under the relay is considered the critical frequency. The plant is then fed with a step while it is regulated with only proportional control. The proportional gain value is calculated using the Ziegler and Nichols tables [24]. From the curve of response of the system, considering a second-order plant, W_n and ζ are obtained. To apply this method, the system must have overshooting, which is the case. With the data of the system reduced to the second order, using pole placement, the PID values are adjusted. This method was applied directly to the model circuit of Simulink©.

The tuning algorithm is programmed in a S function. For it to operate, it is necessary for the plant to have a stable behavior under proportional control, so in cascade with the plant, a gain element is placed.

The parameters of the PID obtained for each operation point are shown in Table 1 (GS-PID) and Table 2 (GS-FO-PID).

Table 1. Parameters for the gain scheduling PID.

Voltage	K	K_i	K_d
5 V	0.44437	5347.5	8.715×10^{-6}
12 V	0.44377	5309.2	8.76×10^{-6}
24 V	0.4431	5250	8.8275×10^{-6}
36 V	0.44325	5194.5	8.94×10^{-6}

Table 2. Parameters for the gain scheduling fractional order PID.

Voltage	K	K_i	K_d
5 V	34.82	9.69×10^5	2.495×10^{-4}
12 V	48.88	1.3×10^5	3.877×10^{-4}
24 V	2.769	3.012×10^3	5.549×10^{-4}
36 V	62.36	7.101×10^5	1.195×10^{-4}

Fractional Order Tuning

Awouda and Mamat [39] have described a method to tune PID controllers. Their method involves calculating the gain of the controller (Kc), integral time (Ti), and the derivative time (Td) for PID controlled systems whose process is modeled in first-order lag plus time delay form; in the case of this work, a second-order model, but limited to plants with $\zeta < 1$, has been employed. A Matlab© 2024b program was used to find the optimum value of the PID controller parameters, which can achieve most of the system's requirements, such as reducing the overshoot. The objective function was selected so as to minimize the integral of the time absolute error (IAE) performance index. The procedure can be modified to be applied to fractional order PID controllers. The method can be summarized in six steps as follows [39].

Step 1. The target is to find relations concerning the controller's tuning parameters and the parameters of the process

$$K_c = f_1(KP; L : T) \tag{34}$$

$$T_i = f_2(KP; L : T) \tag{35}$$

$$T_d = f_3(KP; L : T) \tag{36}$$

Functions f_1, f_2, and f_3 must be calculated such that the robustness constraint is satisfied and the load disturbance response is minimized.

Step 2. In both T and L, the unit is time, while for KP, the unit depends on the ratio of output to input. Therefore, dimensionless expressions must be created, using operations between the above factors with appropriate scales, such as

L/T or T_i/L, or T_i/T;

T_d/L or T_d/T;

K_cKP

Step 3. Among the above factors, one is selected that allows the relationships between the adjustment parameters to be found.

For instance,

KcKP = K1 (L/T);

T_i/L = *K2 (L/T);*

T_d/L = *K3 (L/T).*

Step 4. Considering the *L/T* ratio, the optimal values of the adjustment parameters K_c, T_i, and T_d are determined, seeking to minimize the IAE criterion. If the process is of the FOPDT type with a very small, medium, and rather long dead time value, it is convenient to consider values of the dimensionless factor *L/T* between 0.1 and 2.

Step 5. Find the values of K_cKP, T_i/L, and T_d/L agreeing with the values of *L/T*.

Step 6. Solve the equations for *K1*, *K2*, and *K3* using curve fitting techniques.

When considering a FO-PID, it starts with a unit value in the λ parameters and δ, gradually modifying them until the variation in the ITAE is minimal. In the case of the GS-FO-PID, the control parameters are adjusted specifically for each reference value.

The values of λ and δ selected for each operational case are shown in Table 3.

Table 3. The λ and δ parameters for the gain scheduling fractional order PID.

Voltage	λ	δ
5 V	1.25	1.15
12 V	1.25	1.15
24 V	1.14	0.875
36 V	1.7	0.95

4.3. Closed-Loop System Response

4.3.1. Simulation Overview

To evaluate the controllers, a Simulink© platform was developed (Figure 3). The figure is divided into three closed loops, each of them with a different type of control applied to the designed DC–DC quadratic converter. As can be seen, the input signal and reference can be selected as fixed or variable, depending on the type of study desired. The top block shows the GS-PID control, while the fixed PID control is shown in the middle loop. On the other hand, the GS-FO-PID control is shown in the lower loop. The GS-PID is devolved using directly Simulink© 2024b blocks, but the GS-FO-PID has been developed using an S function. Both algorithms use the reference as the factor to change the values of the parameters of the controllers. Figure 3 also shows the block where the analysis of the error integral is performed, in this case, using the IAE criterion.

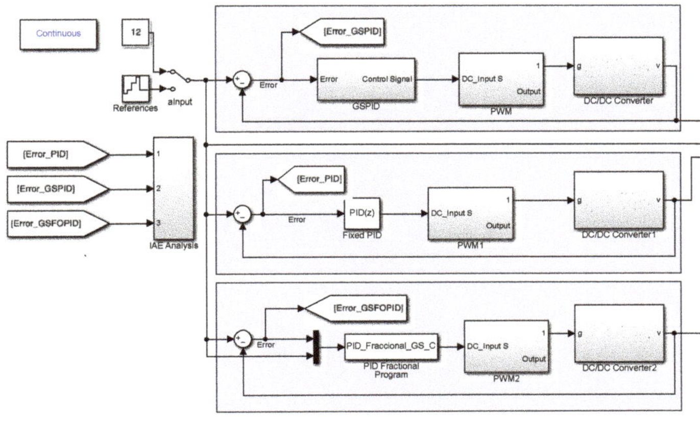

Figure 3. Simulink© models used to evaluate the controllers.

4.3.2. Scenario 1: Reference Voltage Change

To show the performance of the controllers, voltage changes were simulated, starting at 5 V, then going to 12 V, from there to 24 V, and finally 36 V. Changes were made every 0.0025 s; see Figure 4.

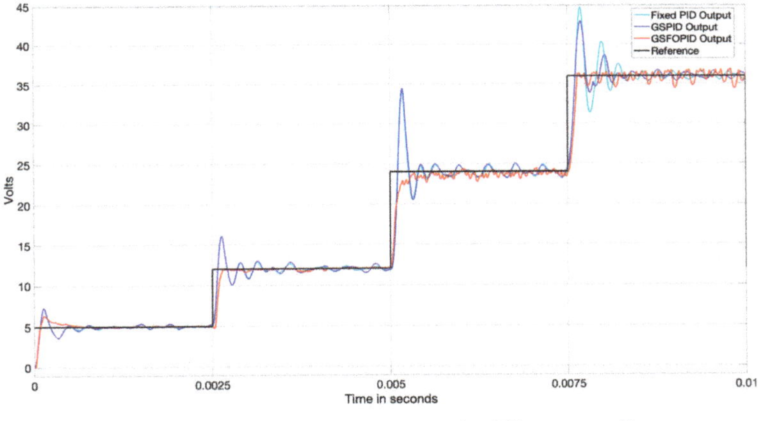

Figure 4. Outputs of the quadratic buck converter under different controllers.

Figure 5 shows in detail the response of the system to changes in the reference voltage. In Figure 5a, the reference voltage is 5 volts. A decrease in overshooting by 40% can be observed when the converter is controlled by GS-FO-PID compared with the response with fixed PID and GS-PID controls, as well as the ripple at steady-state output being greatly reduced under GS-FO-PID control. Figure 5b–d shows the system with voltage references of 12, 24, and 36 volts, respectively. As can be seen, in all these cases, overshooting is practically zero when the system is controlled by GS-FO-PID, while the curve in the state stable is also reduced. However, this increases when the voltage reference is 36 Volts (Figure 5a). However, this amplitude is not higher than the other two controls evaluated. From the above, it can be said that in all cases, an improvement in the substantial reduction in the maximum overshoot is noted when the QBC is regulated by the GS-FO-PID controller. Table 4 summarizes these results. On the other hand, the stabilization time remains practically constant with all controllers. This is because the ripple present at the output—something expected in switched systems—does not allow accurate measurement of this parameter. However, a reduction in the amplitude of the ripple is reduced when the QBC is driven by GS-FO-PID.

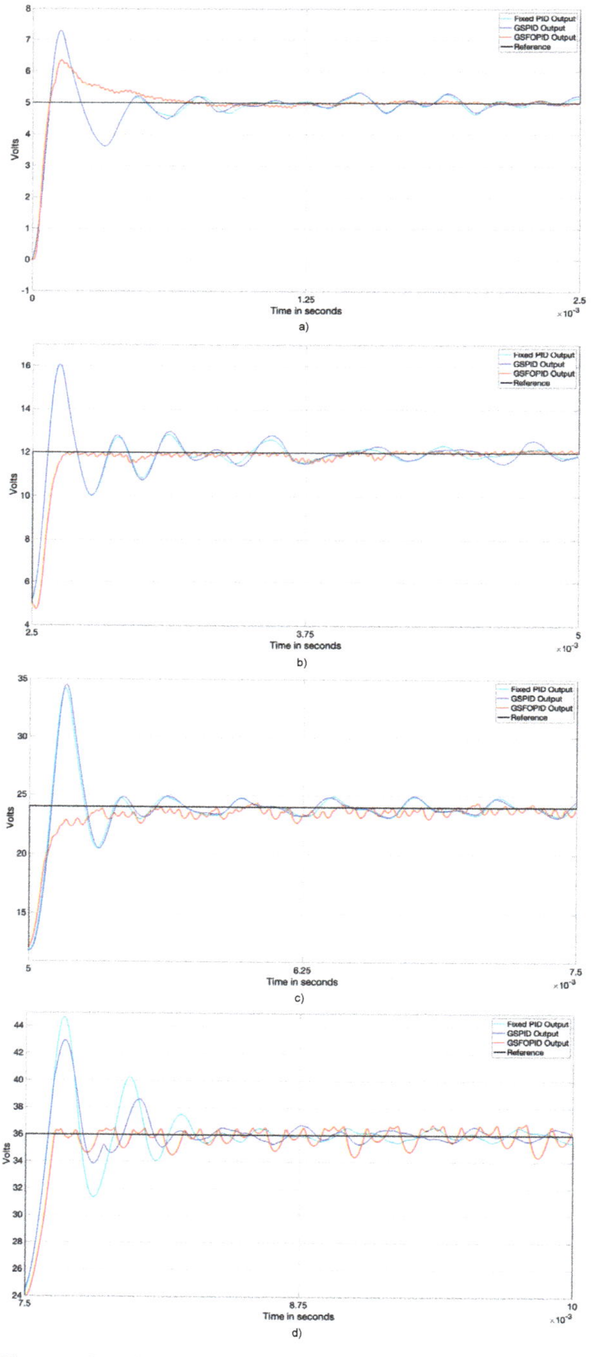

Figure 5. Details of the outputs of the quadratic buck converter under different controllers with voltage change.

Table 4. Overshooting of the quadratic buck converter under the controllers.

Voltage	Fixed PID	GS-PID	GS-FO-PID
5 V	50%	50%	20%
12 V	50%	50%	0%
24 V	65%	68%	0%
36 V	70%	50%	1%

It is expected that the behavior of the quadratic buck converter at 12 V would be the same for a fixed PID as for GS-PID; however, it must be considered that the initial conditions are not the same. By performing a simulation with both circuits starting at zero and uncharged reactive elements, a fairly similar behavior is obtained with both controllers (fixed PID and GS-PID; see Figure 6).

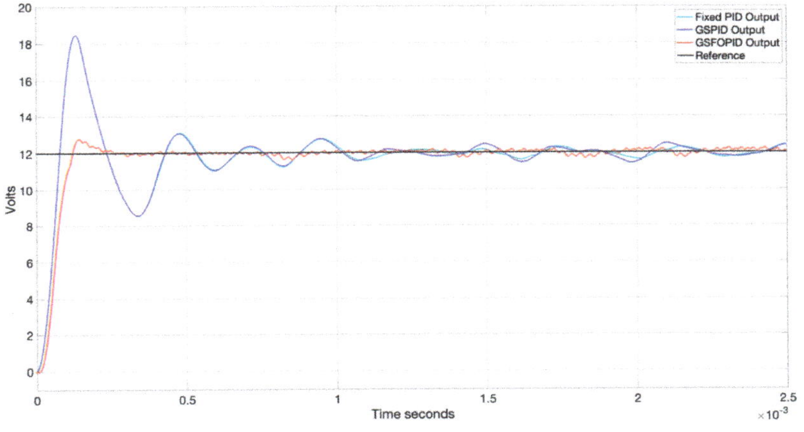

Figure 6. Outputs of the quadratic buck converter under different controllers with voltage change with the initial conditions equal to zero.

In order to measure the improvement in the operation of the quadratic buck converter under the GS-FO-PID controller, an analysis of the error integral was performed. Since an optimal system based on this criterion is a system that has reasonable damping and a satisfactory transient response, an IAE analysis was selected [46]. It was found that in all operational cases, the GS-FO-PID showed a lower value of this index; see Table 5.

Table 5. IAE of the controllers.

Voltage	Fixed PID	GS-PID	GS-FO-PID
5 V	0.8673×10^{-3}	0.8686×10^{-3}	0.5886×10^{-3}
12 V	1.356×10^{-3}	1.448×10^{-3}	0.7999×10^{-3}
24 V	2.895×10^{-3}	2.972×10^{-3}	2.059×10^{-3}
36 V	3.056×10^{-3}	2.484×10^{-3}	2.207×10^{-3}

4.3.3. Scenario 2: Load Change

To verify that the output voltage remains within the regulation limits, a load shift test was performed. Therefore, when the system was operating at 12 V, a sudden change in load was caused at 0.0125 s of simulation, from 5 to 2.5 ohms, which generated an increase in the output ratio in all controllers. However, as can be seen in Figure 7, after an increase in the amplitude of the oscillation, the system returned to a ripple response. In the case of GS-FO-PID, the amplitude value of the ripple almost has no variation. This shows that the system controlled under GS-FO-PID has a better response to changes in load, even in the case evaluated, which was 50%.

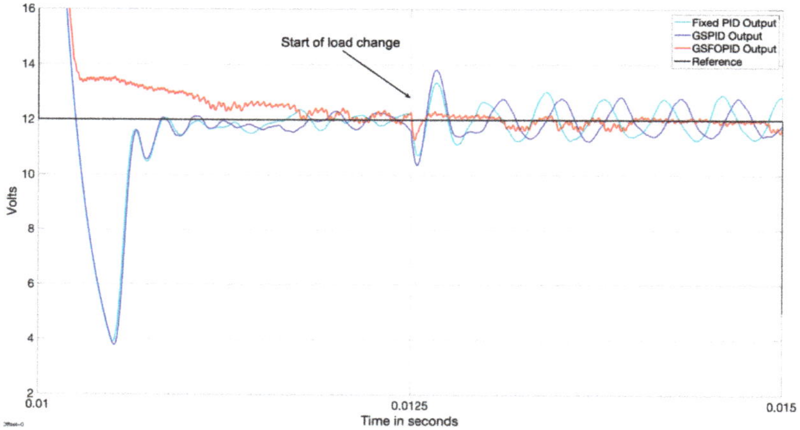

Figure 7. Response of the quadratic buck converter with a load change.

The results of the simulations shown in this section corroborate that the GS-FO-PID controller improves the system's response by reducing overshoots when the system is subject to reference changes. Also, the response to changes in load is improved when the system is controlled with the GS-FO-PID controller compared with the other controls evaluated. This means that the GS-FO-PID improves the dynamic response of the quadratic buck converter to both reference changes and load changes.

5. Conclusions

This paper has shown the effectiveness of a gain scheduling fractional order PID controller driving a quadratic buck converter improving the dynamic response. The simulation results have shown how GS-FO-PID can be applied to a quadratic buck converter to improve its performance, which is possible because GS-FO-PID considers the nonlinear nature of the converter, improving the direct transient responses. In particular, selecting different values of FO-PID parameters depending on the value of the input voltage and the reference allows the GS-FO-PID to produce a good response within each operational case, reducing overshooting. Additionally, it assumes load changes in a good way. As a perspective of this work, further study to determine the operating ranges for each set of GS-FO-PID parameters is necessary. The way in which the parameters for the GS-FO-PID are chosen could be improved by employing autonomous learning strategies, machine learning, and deep learning. Similarly, it is necessary to carry out an exhaustive study on the stability of the regulated system with the controller proposed in this work. The application of this type of control to other power converters, such as the boost and buck–boost, should also be evaluated. In general, these promising results allow and encourage the belief that GS-FO-PID could yield tangible benefits for the operation of DC–DC quadratic converters.

Author Contributions: Conceptualization, G.A.M.H. and R.A.A.-R.; formal analysis, J.F.G.-C.; investigation, G.A.M.H. and R.A.A.-R.; methodology, G.A.M.H.; resources, R.A.A.-R.; validation, J.F.G.-C.; visualization, J.F.G.-C. and R.A.A.-R.; writing—original draft, G.A.M.H.; writing—review and editing,

G.A.M.H., J.F.G.-C. and R.A.A.-R. All authors have read and agreed to the published version of the manuscript.

Funding: This research received no external funding.

Data Availability Statement: No new data were created or analyzed in this study. Data sharing is not applicable to this article.

Conflicts of Interest: The authors declare no conflicts of interest.

References

1. Okati, M.; Eslami, M.; Shahbaz, M.J. A New Transformerless DC/DC Converter with Dual Operating Modes and Continuous Input Current Port. *IET Gener. Transm. Distrib.* **2023**, *17*, 1553–1567. [CrossRef]
2. Okati, M.; Eslami, M.; Shahbazzadeh, M.J. A Non-Isolated DC-DC Converter with Dual Working Modes and Positive Output Voltage. *Electr. Power Compon. Syst.* **2021**, *49*, 1143–1157. [CrossRef]
3. Okati, M.; Eslami, M.M.; Jafari Shahbazzadeh, M.; Shareef, H. A New Transformerless Quadratic Buck-Boost Converter with High Voltage Gain Ratio and Continuous Input/Output Current Port. *IET Power Electron.* **2022**, *15*, 1280–1294. [CrossRef]
4. Damodaran, R.V.; Shareef, H.; Errouissi, R.; Eslami, M. A Common Ground Four Quadrant Buck Converter for DC-AC Conversion. *IEEE Access* **2022**, *10*, 44855–44868. [CrossRef]
5. Pandey, K.; Kumar, M.; Kumari, A.; Kumar, J. Control of Quadratic Buck Converter Using Non-Linear Controller. In Proceedings of the IEEE 17th India Council International Conference (INDICON), New Delhi, India, 11–13 December 2020. [CrossRef]
6. Arteaga-Orozco, M.I. Nonlinear Control of Switched DC/DC Converters: Performance Analysis and Experimental Verification. Ph.D. Thesis, UPC, Institut d'Organització i Control de Sistemes Industrials, Barcelona, Spain, 2007. [CrossRef]
7. Ali, M.S.; Soliman, M.; Hussein, A.M. Robust Controller of Buck Converter Feeding Constant Power Load. *J. Control Autom. Electr. Syst.* **2021**, *32*, 153–164. [CrossRef]
8. Trakuldit, S.; Tattiwong, K.; Bunlaksananusorn, C. Design and evaluation of a Quadratic Buck Converter. In Proceedings of the 8th International Conference on Power and Energy Systems Engineering (CPESE 2021), Fukuoka, Japan, 10–12 September 2021.
9. Carbajal-Gutierrez, E.E.; Morales-Saldana, J.A.; Leyva-Ramos, J. Modeling of a Single-Switch Quadratic Buck Converter. *IEEE Trans. Aerosp. Electron. Syst.* **2005**, *41*, 1451–1457. [CrossRef]
10. Ortiz-Lopez, M.; Leyva-Ramos, J.; Carbajal-Gutierrez, E.E.; Morales-Saldana, J.A. Modelling and analysis of switch-mode cascade converters with a single active switch. *IET Power Electron.* **2008**, *1*, 478–487. [CrossRef]
11. Ayachit, A.; Kazimierczuk, M.K. Steady-state analysis of PWM quadratic buck converter in CCM. In Proceedings of the IEEE 56th International Midwest Symposium on Circuits and Systems (MWSCAS), Columbus, OH, USA, 4–7 August 2013.
12. Mostaan, A.; Gorji, S.A.; Soltani, M.; Ektesabi, M. A novel quadratic buck-boost DC-DC converter without floating gate-driver. In Proceedings of the 2nd Annual Southern Power Electronics Conference (SPEC), Auckland, New Zealand, 5–8 December 2016. [CrossRef]
13. Bonela, A.K.; Sarkar, M.K.; Kumar, K. Robust non-fragile control of DC–DC buck converter. *Electr. Eng.* **2024**, *106*, 983–991. [CrossRef]
14. Montano-Saavedra, A.C.; Barra-Junior, W.; de Medeiros, R.L.P.; Roozembergh-Junior, C.; Sovano-Gomes, A. Experimental Evaluation of the Robust Controllers Applied on a Single Inductor Multiple Output DC-DC Buck Converter to Minimize Cross Regulation Considering Parametric Uncertainties and CPL Power Variations. *Energies* **2024**, *17*, 3359. [CrossRef]
15. Kumari, S.; Rayguru, M.M.; Upadhyaya, S. Analysis of Model Predictive Controller versus Linear Quadratic Regulator for DC-DC Buck Converter Systems. In Proceedings of the IEEE International Conference on Intelligent Computing and Control Systems, Madurai, India, 17–19 May 2023.
16. Raza, M.M.; Rayguru, M.M.; Kaushik, G. Robust Model Predictive Control of a Buck Converter with Constraints. In Proceedings of the International Conference on Computational Intelligence and Sustainable Engineering Solutions (CISES), New Delhi, India, 28–30 April 2023.
17. Fetene, Y.; Ayenew, E.; Feleke, S. Full state observer-based pole placement controller for pulse width modulation switched mode voltage-controlled buck Converter. *Heliyon* **2024**, *10*, e30662. [CrossRef] [PubMed]
18. Nanyan, N.F.; Ahmad, M.A.; Hekimoglu, B. Optimal PID controller for the DC-DC buck converter using the improved sine cosine algorithm. *Results Control. Optim.* **2024**, *14*, 100352. [CrossRef]
19. Bui, V.-H.; Mohammadi, S.; Das, S.; Hussain, A.; Hollweg, G.V.; Su, W. A critical review of safe reinforcement learning strategies in power and energy systems. *Eng. Appl. Artif. Intell.* **2025**, *143*, 110091. [CrossRef]
20. Lopez-Santos, O.; Flores-Bahamonde, F.; Torres-Pinzon, C.A.; Haroun, R.; Valderrama-Blavi, H.; Garriga, J.A.; Martınez-Salamero, L. Multi-loop Sliding-Mode Control for a Battery Charger Using a Quadratic Buck Converter. In Proceedings of the IEEE 8th Southern Power Electronics Conference (SPEC), Florianopolis, Brazil, 4–6 December 2023.

21. Pop, G.M.; Pop-Calimanu, I.M.; Lascu, D. Fourth-Order Quadratic Buck Converter Controller Design. *Sensors* **2024**, *24*, 557. [CrossRef] [PubMed]
22. Acosta-Rodriguez, R.A.; Martinez-Sarmiento, F.H.; Munoz-Hernandez, G.A.; Portilla-Flores, E.A.; Salcedo-Parra, O.J. Validation of Passivity-Based Control and Array PID in High-Power Quadratic Buck Converter Through Rapid Prototyping. *IEEE Access* **2024**, *12*, 58288–58316. [CrossRef]
23. Karaket, K.; Bunlaksananusorn, C. Modeling of a Quadratic Buck Converter. In Proceedings of the 8th Electrical Engineering/Electronics, Computer, Telecommunications and Information Technology (ECTI), Khon Kaen, Thailand, 17–19 May 2011.
24. Ogata, K. *Ingeniería de Control Moderna*, 5th ed.; Pearson Education: Madrid, Spain, 2010. (In Spanish)
25. Orelind, G.; Wozniak, L.; Medanic, J.; Whitemore, T. Optimal PID gain schedule for Hydrogenerators-Design and application. *IEEE Trans. Energy Convers.* **1989**, *4*, 300–307. [CrossRef]
26. Petras, I. Fractional Calculus and Applied Analyisis. *Int. J. Theory Appl.* **2012**, *15*, 282–303.
27. Caputo, M. Linear Models of Dissipation whose Q is almost Frequency Independent -I. *Geophys. J. R. Astron. Soc.* **1967**, *13*, 529–539. [CrossRef]
28. Podlubny, I. *Fractional Differential Equations*; Academic Press: San Diego, CA, USA, 1999.
29. Monje, C.A.; Chen, Y.; Vinagre, B.M.; Xue, D.; Feliu, V. *Fractional Order Systems and Controls: Fundamentals and Applications*; Johnson, Ed.; Springer-Verlag: London, UK, 2010; p. 414.
30. Podlubny, I.; Dorcak, L.; Kostial, I. On Fractional Derivatives, Fractional-Order Dynamic Systems and PIlDu Controllers. In Proceedings of the Conference on Decision and Control, San Diego, CA, USA, 10–12 December 1997.
31. Podlubny, I. Fractional-Order Systems and PILDu-Controllers. *IEEE Trans. Autom. Control* **1999**, *44*, 208–214. [CrossRef]
32. Vinagre, B.M.; Podlubny, I.; Dorcak, L.; Feliu, V. On Fractional PID Controllers: A Frequency Domain Aproach. In Proceedings of the IFAC Digital Control: Past, Present and Future of PID Control, Terrassa, Spain, 5–7 April 2000.
33. Monje, C.A.; Vinagre, B.M.; Feliu, V.; Chen, Y.Q. On Auto-Tuning of Fractional Order PIlDu Controllers. In Proceedings of the 2nd IFAB Workshop on Fractional Differentiation and its Applications, Porto, Portugal, 19–21 July 2006.
34. Monje, C.A.; Vinagre, B.M.; Feliu, V.; Chen, Y.Q. Tuning and auto-tuning of fractional order controllers for industry applications. *Control. Eng. Pract.* **2008**, *16*, 798–812. [CrossRef]
35. Petras, I. Tuning and implementation methods for fractional-order controllers. *Fract. Calc. Appl. Anal.* **2012**, *15*, 282–303. [CrossRef]
36. Milanesi, M.; Padula, F.; Visioli, A. On the implementation of gain scheduling in FOPID controllers. *IFAC Pap. Online* **2024**, *58*, 330–335. [CrossRef]
37. Ayere-Junior, F.A.C.; Bessa, I.; Batista-Pereira, V.M.; Da Silva-Farias, N.J.; de Menezes, A.R.; de Medeiros, R.L.P.; Chavez-Filho, J.E.; Lenzi, M.K.; Da Costa-Junior, C.T. Fractional Order Pole Placement for a buck converter based on commensurable transfer function. *ISA Trans.* **2020**, *107*, 370–384.
38. Wu, Z.; Li, D.; Xue, Y.; He, T.; Zheng, S. Tuning for Fractional Order PID Controller based on Probabilistic Robustness. *IFAC Pap. Online* **2018**, *51*, 675–680. [CrossRef]
39. Awouda, A.E.A.; Mamat, R.B. Refine PID tuning rule using ITAE criteria. In Proceedings of the 2010 2nd International Conference on Computer and Automation Engineering (ICCAE), Singapore, 26–28 February 2010.
40. Chen, Y.Q.; Petras, I.; Xue, D. Fractional Order Control—A Tutorial. In Proceedings of the 2009 American Control Conference, St. Louis, MO, USA, 10–12 June 2009.
41. Choudhary, S.K. Stability and Performance Analysis of Fractional Order Control Systems. *WSEAS Trans. Syst. Control* **2014**, *9*, 438–444.
42. Dewangan, P.D.; Singh, V.P.; Sinha, S.L. Design of FOPID controller for higher order continuous interval system using improved approximation ensuring stability. *SN Appl. Sci.* **2021**, *3*, 493. [CrossRef]
43. Rhouma, A.; Hafsi, S.; Laabidi, K. Stabilizing and Robust Fractional PID Controller Synthesis for Uncertain First-Order plus Time-Delay Systems. *Math. Probl. Eng. Hindawi* **2021**, *2021*, 9940634. [CrossRef]
44. Matignon, D. Generalized Fractional Differential and Difference Equations: Stability Properties and Modelling Issues. In Proceedings of the Mathematical Theory of Networks and Systems Symposium, Padova, Italy, 6–10 July 1998.
45. Astrom, K.J.; Wittenmark, B. *Adaptive Control*; Addison-Wesley: Reading, MA, USA, 1989.
46. Sangeetha, S.; Revathi, B.S.; Balamurugan, K.; Suresh, G. Performance analysis of buck converter with fractional PID controller using hybrid technique. *Robot. Auton. Syst.* **2023**, *169*, 104515. [CrossRef]

MDPI AG
Grosspeteranlage 5
4052 Basel
Switzerland
Tel.: +41 61 683 77 34

Fractal and Fractional Editorial Office
E-mail: fractalfract@mdpi.com
www.mdpi.com/journal/fractalfract

www.ingramcontent.com/pod-product-compliance
Lightning Source LLC
Chambersburg PA
CBHW061815210926
51850CB00006B/707